THE MALARIA PROJECT

THE MALARIA PROJECT

THE U.S. GOVERNMENT'S SECRET MISSION TO FIND A MIRACLE CURE

KAREN M. MASTERSON

New American Library

New American Library
Published by the Penguin Group
Penguin Group (USA) LLC, 375 Hudson Street,
New York, New York 10014

USA | Canada | UK | Ireland | Australia | New Zealand | India | South Africa | China
penguin.com
A Penguin Random House Company

First published by New American Library,
a division of Penguin Group (USA) LLC

First Printing, October 2014

LIBRARY OF CONGRESS CATALOGING-IN-PUBLICATION DATA:
Masterson, Karen, 1964– author.
The malaria project: the U.S. government's secret mission to find a miracle cure/Karen M. Masterson.
p. cm.
Includes bibliographical references and index.
ISBN 978-0-451-46732-4
I. Title.
1. Malaria—history—United States. 2. Antimalarials—history—United States. 3. Government Programs—history—United States. 4. History, 20th century—United States. 5. Human Experimentation—history—United States. 6. Malaria—drug therapy—United States.
WC 750 RC159.A5
616.9'362061—dc23 2014018351

Printed in the United States of America
1 3 5 7 9 10 8 6 4 2

Set in Bulmer MT Std
Designed by Spring Hoteling

For Tom

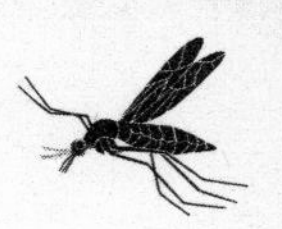

CONTENTS

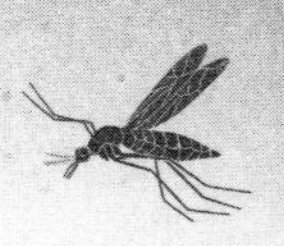

PROLOGUE

A decade ago I knew as much about malaria as I did about professional football—that is to say almost nothing.

This killer disease occupied little of my thinking as I chased down U.S. senators and congressional leaders and wrote daily political stories from Capitol Hill for the *Houston Chronicle*. Little did I know that I would soon cross the globe, go back to school, and spend hours in reading rooms flipping through archived boxes of moldy records. (One box at the National Archives smelled so strongly of ammonia I closed it and returned it to the counter unread.) I did all this to understand why malaria is still around. It is probably *the* most studied disease of all time, and yet it persists, even in the face of the hottest new science in Western medicine. It's both preventable and curable. We've even deciphered its genetic codes. Still, it remains among the top killers of African children—at a rate of two per minute.

By a fluke, malaria crept into my intellectual pursuits. I had quit my job as a national reporter to explore my interests in science and medical writing, both of which I had done before coming to Washington. I accepted a teaching fellowship at Johns Hopkins University, where I studied the history of medicine, and I took a course in mining the National Archives to learn how to find its buried historical treasures.

There, in a hushed reading room of the agency's annex in College Park, Maryland, I learned how to call up records from the building's two million cubic feet of storage space. As an exercise, I searched for archived records on World War II blood plasma replacement studies involving Linus Pauling—two-time

Nobel laureate for chemistry and peace activism. My call slips came back with a box marked with the correct record group number and bearing the letter "P," for Pauling. But the contents were all wrong. Thinking his papers were mixed in under other headings, I thumbed through the entire box, reading through random letters.

One turned my blood cold.

The 1943 letter was to the Massachusetts surgeon general from a physician named George Carden. In it, Carden laid out what sounded like a sinister plan: Federal researchers would use blood transfusions and lab-raised mosquitoes to give malaria to brain-damaged syphilitics and schizophrenics held at Boston Psychopathic Hospital so new drugs could be tested against the resulting infections. No less than the war's outcome was at stake, he wrote, explaining that the military desperately needed a new drug to counter malaria's devastating attacks on U.S. forces in the South Pacific.

I reread the letter a dozen times before my blood warmed to the possibility that I had just stumbled onto a fascinating, if horrifying story. That night I went home and ran Google searches, which, back in 2004, turned up very little. The next day I returned to the archives and met with a specialist in World War II medical research. She helped me navigate the Byzantine reference catalogs used to pinpoint exact call numbers for relevant records. Over the next three days, I retrieved a dozen boxes from the bowels of the archives, each filled with letters, reports, and data sheets on the war's antimalaria program—information that had been classified during and after the war, and, as far as I could tell, hadn't been touched in decades. I also tracked down historian of medicine Leo Slater, who, at the time, was working at the National Institutes of Health. He had an unpublished manuscript on the war's malaria-related work that tracked the involvement of American pharmaceutical companies—a history that has since been published by Rutgers University Press.

With Leo's help, I slowly pieced together a fairly clear picture of what had gone on. The War Department and White House had launched a Manhattan Project–style program to find a cure for malaria, born out of wartime necessity and run by a small army of well-intentioned scientists, many of whom knew precious little about the tricky parasites they studied. All they knew for sure was that U.S. military leaders feared this one disease would force them to surrender to the Japanese—an unacceptable outcome in a war destined to determine the fate of the world.

I hunted for a published history or popular narrative on the subject, and found nothing. The more I searched the more I realized I was in uncharted territory. But to know whether this was something worth pursuing, I needed context and content. When I got to Hopkins that fall, I wrote my schedule to include public health courses that covered the epidemiology and medical history of malaria. I continued my treks to the National Archives—unearthing more and more documents—and took a microbiology course to understand the nature of microbial diseases.

Needing more background, I sought and was awarded a Knight Foundation public health journalism fellowship that funded me to work for three months in the Malaria Branch of the U.S. Centers for Disease Control and Prevention. In CDC's labs just outside of Atlanta, I dissected mosquitoes, studied their salivary glands, and watched under a microscope the sticklike germs that enter the human body when a female anopheles mosquito bites for blood.

My teacher was Bill Collins, a seventy-six-year-old icon in malaria research who, at the time, was the only federally employed malariologist who had been at it long enough to remember what it was like to infect madmen with the disease. He did it in the 1950s and early 1960s for the Public Health Service at the South Carolina State Hospital, where drug experiments begun during World War II were continued.

He still used many of the techniques he learned from his state asylum days. The main difference was that at CDC he infected monkeys instead of humans. He recalled this with remorse. The "good old days," as he called them, allowed scientists to observe malarial parasites in the blood of sick people. He and his colleagues gleaned data that helped the world better understand the microbes' behavior and complex life cycle, ultimately resolving many mysteries that had shrouded the parasites in a type of protective secrecy. The intelligence gave scientists needed insights to develop ideas for a possible vaccine. That he gave a painful and potentially deadly disease to witless syphilitics posed no moral dilemma for Bill. Up until penicillin became available in the 1940s, the madness that came with untreated syphilis could be reversed by malarial fevers—as the fevers ramped up the immune system to kill syphilis spirochetes interfering with brain function. Two decades after penicillin, Bill still used this outmoded malaria treatment—which worked on only a fraction of cases, the rest having no benefit—in South Carolina. He was doing God's work, trying to find a final solution to one of the world's most menacing diseases, until an absence of syphilis patients and ethical concerns shut him down in 1962.

Four decades later, his mission hadn't changed; it just got a lot more difficult. Even though monkeys are our cousins, they are hard to infect with human malaria. So Bill and his colleagues raised parasites to infect primates, and then ran experiments that attempted to extrapolate the extent to which a given drug or vaccine might also work against human malaria.

At CDC's run-down campus on Buford Highway—miles from the glitzy labs that characterize its cutting-edge work on AIDS, Ebola, childhood obesity, and other hot topics—I was trained in the intricacies of manhandling these dangerous microbes. My education started in the insectary, where I helped a Kenyan researcher named Atieli breed colonies of the world's most efficient malaria-carrying mosquitoes. We separated pupae from larvae in a screened-in room set at eighty-four degrees and 80 percent humidity. Just before the pupae molted into mosquitoes, we caught them in netting, put them in small cages with gauze lids, carried them to the primate house, and turned them over to technicians, who pressed the cages against the shaved bellies of monkeys sick with malaria.

After the mosquitoes gorged on infected blood, the insects went to Bill's lab to be placed in refrigerator-size incubators. There the insects stayed while the parasites in their guts sexually reproduced, leaving behind tiny egg sacs. In less than two weeks' time, the sacs burst with offspring that were circulated into the insects' salivary glands. There they awaited an opportunity to escape—which happened whenever the mosquitoes sank their needlelike noses into the flesh of warm-blooded mammals.

I sat next to Bill every morning, watching him handle spherical cardboard cages that held hundreds of highly infectious mosquitoes. They hung inside like razor stubble, bursting into black clouds every time he tried to catch them. This involved pushing a rubber hose through a small opening in the cardboard, with the other end of the hose in his mouth. With a single breath, he'd suck in five or six mosquitoes and then blow them into a small glass jar filled with chloroform—which knocked them out cold. That's when I'd pick up one with tweezers, place it under a magnifying machine, chop off its head, and pull away the gut. Then, under a stronger microscope, he counted the egg sacs to determine the "parasite load" of each batch of mosquitoes. Knowing the load helped Bill determine which were the most likely to induce infections in monkeys primed with experimental vaccines. In this way, CDC ran the first line of tests on potential vaccines developed by the U.S. military and private pharmaceutical companies.

This process was anything but safe—for man or beast.

The monkeys' lives were obviously at risk. They got hit with unnaturally high doses of parasites. If the experimental vaccine or drug being tested was bad or ineffectual, the poor creatures often died. This agitated Bill, but not out of compassion. He was a scientist, which by definition meant he controlled his emotions toward his animals. But an avoidable death angered him. The monkeys were expensive—as much as $5,000 each—and increasingly hard to buy. To Bill, even the most tattered and overused primate could be good for one more experiment. Losing even one to an ill-conceived vaccine preparation concocted by dim-witted drug companies bugged him to no end.

But the process was also dangerous for people. One day in early 2005, while Bill captured mosquitoes infected with a type of human malaria called *Plasmodium vivax*—the same one that crippled World War II forces in the South Pacific—one escaped and bit his lab assistant Doug. He fell sick two weeks later with a high fever and screaming headache. When I asked Doug about it, he shrugged. Sure, *vivax* malaria is horrible. He was seriously ill for more than a week. And he had to report the incident to disease investigators at the CDC. But contracting the disease is a rite of passage worn like a badge of honor, with a history as old as malaria research. Turn-of-the-century photos show professors with cages of mosquitoes strapped to their bare chests, hoping to induce an infection. In this tradition, Doug didn't think of this mishap as a problem. CDC's administrators, however, did. They immediately ordered another layer of screening between Bill and the rest of the world.

Shortly thereafter I joined Bill in the lab. One morning we were dissecting mosquitoes infected with a malarial parasite called *Plasmodium inui*. As Bill sucked the insects into his tube, one flew off over my head. I already knew *P. inui* was a type of monkey malaria capable of jumping species and infecting humans. As I nervously watched the escapee make a U-turn in my direction, I tapped Bill on the shoulder to show him we had a problem. In true Bill Collins form, he looked up from his microscope, spotted the mosquito, and said, "Gosh, ya hate to see that." He then hunched back over his machine, as if enough were said.

I'd been at CDC about a month and tried to assimilate with the culture there, something I've always been good at. But this was different. I couldn't be like the fellow researchers; I couldn't be casual about malaria. I had no interest in living through chills intense enough to rattle my teeth and fevers high enough to damage my organs. I sat paralyzed, scanning the air, knowing that a featherweight bug could at any moment inject devilish microbes into my blood. Next to me, Bill went about his business, clicking slides of mosquito gut under the long arm of his microscope. When I finally locked onto the mosquito's flight pattern and lunged forward to slap the life out of it, I saw the corners of Bill's mouth turn up in a grin—which I took as tacit approval. No matter how skittish my actions looked to seasoned malariologists, I assumed Bill breathed relief: His reporter friend would walk away from his lab without the drama of contracting malaria.

Years later I learned the true meaning of Bill's grin. *P. inui*, I discovered while researching this book, could indeed jump species, but only in rare circumstances, usually requiring an artificial concentration of the microbes injected directly into the bloodstream. Mosquitoes usually are incapable of transmitting a high enough dose of *P. inui* to infect people. The day I read that enlightening journal article I smiled, and remembered the privilege of having spent time sitting next to Bill.

Bill was a round man with bad hips and white hair. His acerbic humor, when he was tickled, would send his voice into a pitch high enough to pass for a girl's. But he also was funny and happy and easy to spend time with. He was extremely generous to anyone who cared enough about his disease to ask questions, and always gave good thought to even the most basic and simplistic inquiries. He wanted everyone to understand his disease, especially the history of it. He hated that malaria in the 1990s got so little respect from global funding outlets, like U.S. AID and the World Health Organization. He remembered malaria's heyday, back when developed nations still had malaria, and, after the war, when U.S. AID and WHO made malaria eradication their top priority. In 1970, when the effort failed and funding dried up, malaria experts were forced to switch fields. There just weren't enough jobs for all the people trained during the eradication era.

Back when Bill was forced to stop his work at the South Carolina State Hospital and move it to CDC, the disease control agency was still flush with cash for malaria research, so this was a good move for him. After 1970, however, malaria jobs there were halved and then halved again. As the agency grew and expanded into stunning new glass buildings with high-tech machines for researching other diseases, malaria moved into the dumpy old buildings. There was no room for the insectary, so it was housed in a trailer on-site.

For anyone researching this disease, Bill was a treasure. He taught me how to induce a malaria infection using methods that had hardly changed since the 1940s. He displayed an attitude toward human experiments that gave me insights into the U.S. Public Health Service's action during World War II. My own anachronistic view of the time was quite critical: The very agency dedicated to the public's good health had reached into state hospitals and penitentiaries to infect the insane and imprisoned with a disease that, at the time, had no cure. The nuances, however, are far more interesting than the headlines. Bill lived those nuances until the early 1960s when he was forced to stop giving malaria to syphilis patients in South Carolina. Bill's personal experiences with and deep knowledge of the war's program—backed by a roomful of his own archived materials—neutralized my biases. And I hope it has allowed me to fairly and accurately retell this seventy-year-ago story, for which there are few remaining witnesses.

FOR a more current view of the disease, I talked my way into a CDC trip to Tanzania with the head of CDC's Division of Parasitic Diseases and one of the Malaria Branch's top investigators. Within days of leaving the United States, we were hours from the nearest pavement, bouncing over deeply rutted dirt roads and creeping slowly toward hard-to-reach villages.

In Tanzania, the perils of trying to impose Western ideals on the African bush were obvious, especially with respect to malaria. Scientists, PhD candidates, and public health graduate students devised elaborate drug-delivery schemes, with the single hope of reaching people in straw shacks elevated above mosquito-infested floodplains, miles from roadways and clinics. On one outing, a Tanzanian researcher and I abandoned our Land Cruiser to bushwhack through tall grasses, looking for three families. We were out where the occasional life was still lost to a hungry lion or rampaging elephant. Waters were low, so the mosquitoes weren't bad. But it was still malaria season, and our survey told us that all three families had young children probably in need of treatment.

Out there, I thought about the Herculean effort we made to reach these families, which included manpower for an entire day, a full tank of gas, wear and tear on an expensive vehicle, and thousands of dollars in survey data designed to instruct us on where to go and whom to look for. Was this the legacy of the World War II malaria program? I wondered. Did the world's first and largest effort to produce a synthetic antimalaria drug lead to this?

The answer, as it turns out, is yes. The World War II program set up a new infrastructure for malaria drug development, creating a colossal shift in the way the world perceived the problem. The United States, for example, went from attacking malaria as a public health challenge that involved lifting people from poverty and getting them off mosquito-infested lands, to seeing it as a problem of drug treatment and delivery. A pill came to replace decades of hard work and coordinated programming, resulting in a paradigm shift that has endured years of skepticism and serious scientific setbacks—including the inevitable resistance that malaria parasites develop against every new drug brought to market.

Out in the bush I stayed in an area with some of the highest malaria transmission rates in the world. An affable and well-known Irish entomologist there—Gerry Killeen—measured exposure rates in surrounding villages as high as two thousand infectious mosquito bites per year, per person. In other words, infectious mosquitoes occupied every part of life. They swarmed the outhouses, guesthouses, meeting rooms, shebeens, restaurants, and even the hospitals and clinics. They darted at your face, attacked your ankles, and stung the thin skin on your fingers and temples. On the walls and drapes of hospital malaria wards, hundreds of anopheles mosquitoes rested after feeding on the sick—soon to flit off and infect others.

While I was there, mosquitoes in my nightmares stung at my eyes until I awoke in a fit. I'd calm down after realizing I was in my small guesthouse, tucked safely under a pristine mosquito net. I figured out soon enough that these subconscious fears were a luxury of Western living. People in the bush survive every day under a blanket of mosquitoes, barely noticing hundreds of daily bites. Far bigger problems haunted their nightmares, like whether they'd have enough water and food for their children. I saw people use bed nets—a frontline defense against malaria—to fish, carry laundry, and cover windows. When I asked individuals why they abused these lifesaving devices, why they would risk causing tears and rips and compromising the nets' usefulness, most told me what they knew a *mzungu*—a white person—wanted to hear: They loved their nets and slept under them every night. But the reality, as explained to me by educated villagers working for Western-funded antimalaria projects, was that the nets were more of a curiosity. Many were never used for their intended purpose.

The attitude toward antimalaria drugs posed an even bigger conundrum. Where I visited, multiple disease loads routinely stole life from families. Parents said saving their children would require an accumulation of wealth sufficient enough to buy their way into the cities—where better housing, water quality, and health care are available. Their only option was to try to earn cash through surplus farming. But too often crops failed from bad weather, or illnesses made family members too sick to bring in the harvest. The parents I spoke to, who survived in mud huts with no hope of employment, were reluctant to spend precious cash on malaria drugs, especially when they knew the disease would return every time the waters ran high. Better to get sick, establish enough resistance to fend off malaria's deadliest symptoms, and save precious cash for bigger problems.

For a better perspective on what I considered an irreconcilable difference between drug delivery programs and the needs as expressed by African people, I interviewed scientists working in the bush. I talked to entomologists, parasitologists, epidemiologists, biostatisticians, and public health experts. While each had theories on how best to fight malaria—combination drug therapy along with insecticide-treated bed nets seemed to be the favorite, with many advocating for a reintroduction of DDT (the banned insecticide made famous by Rachel Carson's *Silent Spring*)—without exception they conceded that malaria is a disease of the poor and can't be beaten without economic development, job creation, higher standards of living, and improved health care delivery.

They said getting people out of malaria-ridden floodplains would save more lives than drugs, bed nets, and DDT combined. But this would require new infrastructure, good governance, better resources, and outside investment—all of which are hard to come by in Africa. Public health advocates, by necessity, focus on what's within their power, especially the development and distribution of new drugs. And perhaps one day they'll even deliver a workable and efficacious vaccine. The goal, as they see it, is not to eradicate the disease, or even treat everyone who gets it. Their job, as defined by international funders, is to save young children. This is a perfectly reasonable design, since the first malaria infections are the most intense. Without having lived long enough to establish a low level of immunity from constant exposure, children under five are the most likely to die. Saving them is the best Western antimalaria programs can do, given the circumstances. But these strategies will never eradicate the disease—they will never pull this scourge from its roots. And as long as mosquitoes are present to pick up the parasites from sick adults, children will remain vulnerable.

THE *Malaria Project* is an intensely human story, one about individuals who dedicated their lives to beating a disease that is older than mankind and likely to survive longer than the human race. It has touched most us one way or another.

My personal story dates back to 1993, after I left my job as a legislative aide on Capitol Hill (first for U.S. Senator John Glenn and then U.S. Representative Tony Hall), and before I went to journalism school. That year I accepted a scholarship to study political science at the University of Cape Town in South Africa. After writing exams, I left the city to backpack through ten countries of southern and eastern Africa in a life-changing trip that reached its pinnacle while I sat with gorillas in the jungles of what was then Zaire.

To fend off malaria, my travel partners and I took a drug called Lariam, and endured side effects that included paranoia and hallucinations. While in Malawi we decided the drug was worse than malaria could ever be, so we stopped taking it—a naive and dangerous decision, as it turned out. I was lucky. Apparently my daily dousing of mosquito repellent protected me. But one friend wasn't so fortunate. She suffered all the classic symptoms—headache, chills, and intense spikes in fever. But her dilated pupils, heavy breathing, and utter listlessness were the most worrisome. A clinician in Nairobi told us we got her treatment just in time; if she had developed cerebral malaria, she could have easily slipped into a coma and died. She was cured by way of an intense round of chloroquine, the "miracle drug" to emerge from the World War II malaria project and deployed soon thereafter in the global eradication effort. Its use, and overuse, triggered resistance to it in many parts of the world, but chloroquine still worked in certain parts of Africa when I was there.

That adventure changed my life and is at least partly responsible for my taking on this book, which put me on so many unexpected paths. This book could have been written from any number of perspectives. I chose to build the narrative from archived records, and draw context from journal articles and academic histories of World War II and of malaria. What is clear from these sources is that this drug program begun during the war created a legacy that explains why our present antimalaria efforts are so heavily weighted toward drug and vaccine development, instead of mosquito abatement, poverty alleviation, and jobs programs—the approaches used to eradicate the disease in the malaria-plagued U.S. South during the first half of the twentieth century. This story is designed to help readers understand the links between past and present, and demonstrate how public health as a discipline needs more than drugs and other magic bullets to succeed—something of which public health professionals are painfully aware.

INTRODUCTION
A Brief History

Many before me have written the natural history of malaria, from prehistoric time to today. So I won't do that here. Suffice it to say that malaria has been around since the dinosaurs and remains entrenched in many parts of the world where the right combination of poverty, mosquito species, and climate persist—which, incidentally, include nearly half the planet. Each year it kills a half million African toddlers; infects up to a half billion people; and costs already impoverished countries billions of dollars in lost workdays and health care.[1] WHO is now shouldering its second major attempt to eradicate it with tools not much different from those used in the first attempt, which began right after World War II when the organization used two breakthrough weapons discovered by wartime researchers: the insecticide DDT and the antimalaria drug chloroquine.

This is the story of how these weapons came to be. To get at it, certain facts about malaria need restating. The fun facts aren't so relevant but are immensely interesting. For example, malaria has inspired fear at least since men started writing—as evidenced by ancient Babylonian tablets that depict a mosquitolike god of pestilence and death, and Hippocrates's twenty-four-hundred-year-old medical teachings, which include a description of malaria's crippling fevers and his hypothesis that they were caused by vapors rising up from polluted waters.[2] Malaria also killed many famous men, including King Tut and Alexander the Great; it scared Attila the Hun out of attacking fever-ridden Rome; it weakened Genghis Khan and probably contributed to his death; it caused great ancient armies to lose wars, annihilated early civilizations, forced medieval farmers to leave crops rotting in fields, and, right up to World War II, presented a force so uncontrollable that military strategists planned to have a second division to replace each one taken down by fever.

The more relevant facts for this story are linked to the different species of parasites that cause the unmistakable symptoms of malaria. Those symptoms are reliable, hideous, and memorable. They start with violent chills, followed by furnace-hot fevers and a burning thirst, then exploding headaches, nausea, stomach pains, and vomiting, and finally sheet-soaking sweats that end with a disquieting delirium. In the kind of malaria that dominates in Africa, the symptoms are worse and often end in death. These include retinal damage, severe anemia, convulsions, coma, and renal failure (nicknamed blackwater fever because the urine turns black right before death). The parasites accumulate so quickly that they block the narrow blood vessels to the brain. The effect is visible as the lack of oxygen to the eyes turns the retina white, and death soon follows.[3]

Malaria is unforgiving and, until the twentieth century, was just about everywhere.

All the different species of parasites that cause malaria belong to the genus *Plasmodia*. They infect many different animals, including primates, birds, bats, and lizards. The four that infect man are called *P.* (for *Plasmodium) falciparum, P. vivax, P. ovale* and *P. malariae*. Another one, a monkey malaria called *P. knowlesi*, may soon join the human four, as it is showing signs of jumping species. Of these human malarias, two cause almost all infections in man: *falciparum* and *vivax*.

Less than thirty minutes after a mosquito spits *Plasmodia* parasites into a person, the microbes can disappear into the liver, where they stay for a week or so, protected from the body's immune system. They feast on cells and multiply profusely. Each stick-shaped microbe becomes thirty thousand more liver-stage germs, gathering strength in numbers, hidden, giving little sign of their presence. Then, when they are ready, they burst from the liver to hunt for hemoglobin—this is called the blood stage. *P. vivax*, the most widespread of the human malarias, cozy up to our red blood cells disguised as friendly molecules and trick a receptor protein called Duffy into letting them inside. Once there, they devour our hemoglobin and expel the iron as waste. Each turns into another thirty or forty. Their bulk ruptures the cells, spilling out new armies of this blood-stage germ, plus thick clouds of iron-based toxic waste that our busy white blood cells (the body's vacuum cleaners) try to suck up. But the parasites keep coming; every other day an exponential number burst out of used-up red cells to march on new ones. They repeat the process again and again until billions of them overwhelm our phagocytes, triggering the immune system's last resort: high fevers.

A small fraction of these marauding microbes change, yet again, into the so-called sexual stage. They are now reproductively distinct females and males, and they must get into a mosquito to procreate and produce offspring. These "shrewdly manipulative parasites" swim to our surface tissue and alter our blood chemistry to actually *attract* mosquitoes; that is, they send a "bite me" signal that increases by 100 percent their odds of being drunk in by nearby mosquitoes.[4] Those taken are the chosen few; they escape their human host for the mosquito's gut, which is the only place they can sexually fuse. Their offspring are later delivered to new human hosts when the mosquito bites for blood. This innovative survival strategy makes them among the most successful germs on the planet.

Falciparum is the most dangerous of the four human malarias, responsible for more than 90 percent of the world's malaria-related deaths. But *vivax* is the more enduring, causing relentless relapses, sometimes for years. *Vivax* does this by launching attacks from the liver in different groups, like regiments. Some stay behind, dormant and protected from the immune system. Just as the host begins to feel better, as his or her red blood cell supply recovers and the malaise subsides, parasites still hidden in the liver awaken and launch another attack. Or maybe they wait a few months, or years—depending on that strain's survival strategy (which is linked to mosquito breeding seasons). Eventually the process starts anew with chills, fevers, burning thirst, head and muscle aches, severe vomiting, and malaise. It happens again and again and again—sometimes over many years—until the infection finally burns out.

MALARIA crippled U.S. soldiers, marines, and sailors during World War II. Nearly half a million of them got sick, with many, many thousands presumed to be over their infections, only to be sent back to the front lines to fall again and again and again from relapses and recrudescence. The early battles in the Pacific boiled down to which side had the means to replace fever-stricken troops. In Bataan, Japan had a clear advantage; in Guadalcanal, after months of uncertainty, the Americans finally gained that upper hand.

Recent genome work on *vivax* and *falciparum* produced pretty good road maps that help explain why they are so different, why they are so menacing during times of war, and why they affect different races of people differently—a feature that both stumped and fascinated malaria researchers during World War II.

The genome work suggests that some two hundred thousand years ago—about the time humans appeared on the scene—*vivax* found its way to tropical Africa to become the greatest killer of early man.[5] It created enormous selection pressure that strongly favored a random mutation in the Duffy receptor—that surface protein *vivax* uses to gain entry into red cells. People who had this changed lock, which *vivax*'s key no longer fit, survived malaria and passed the trait to offspring, who were now also better equipped to survive *vivax*. Eventually a large enough percentage of the population had this mutation that *vivax* dried up. This was a direct genetic hit that made many African people completely immune to *vivax*, a biological advantage that continues to prevent it from infecting a vast majority of Africans and people of African descent.[6]

Vivax had been forced out of Africa. So the parasites adapted. In the great human diaspora, they found hospitable hosts in Caucasians, Persians, and Asians. These races weren't Duffy negative; their red blood cells still had Duffy receptor keyholes that *vivax* easily unlocked. To survive in cooler climates, they "learned" to hide in the human liver for longer and longer stretches, launching their attacks on red blood cells in time for summer mosquitoes to drink them in—somehow sensing that mosquitoes had begun biting.

The exodus of *vivax* might have ended Africa's malaria woes but for the appearance of *falciparum*. Some scientists believe this parasite originated as a bird malaria—and probably a dinosaur malaria—that jumped species millions of years ago to infect African primates. Then, some six to ten thousand years ago, as Africans cut down rain forests to build large settled communities, it jumped to humans. This long lineage makes *falciparum* the most ancient of the human malarias—as far as we know.[7]

What this shows is that *vivax* and *falciparum* adapted well to animal species through our evolutionary processes. This rapid adaptability is seen today as a signature feature of *Plasmodia*, and is the main reason it continues to be one of the most successful disease-causing protozoa on Earth.

Falciparum is so deadly because the parasites can invade most red blood cells, compared to *vivax,* which infects only the young ones (reticulocytes with the Duffy receptor). Untreated *falciparum* can cause severe anemia because it's so destructive to blood. The infected red cells can also clump together. So while *vivax*-infected cells easily pass into the spleen to be destroyed, clumps of *falciparum*-infected cells do not. They remain in the blood, taking up residence in the heart, lungs, liver, brain, kidney, and, in pregnant women, the placenta.[8] They clog capillaries, which prevents blood from getting from tissue to the veins. The brain and cerebral fluid go from pale pink to dirty brown. Blood flow and oxygen to the brain cut off so quickly that a child playing games before lunch can be dead by dinner, with almost no warning.

Survival pressure from *falciparum* favored another random mutation. This one was linked to red-cell protein production. But unlike the direct hit of the Duffy mutation, this was only a partial hit. The mutation shifted the shape of red blood cells, from nice doughnuts to a sickle shape. Children inheriting the trait from both parents were protected from *falciparum*, but their red blood cells were poor carriers of oxygen and they often died from sickle-cell anemia. Those inheriting it from just one parent didn't develop anemia *and* were protected from *falciparum*'s deadliest infections. Those people lived longer and passed the trait to their children.

Today, 10 to 40 percent of sub-Saharan Africans—depending on the region—have this single sickle-cell trait, giving them partial protection. Between this trait and a natural immunity achieved by constant exposure to *falciparum*, most adult Africans are able to survive routine attacks of their local strains. The disease is quite deadly, however, in pregnant women, who lose their acquired immunity during gestation, and young children, who have not achieved it yet. In some areas of Africa, the childhood mortality rate from *falciparum* is as high as 30 percent, while adults with acquired immunity experience it as if it were a routine flu—that over time permanently damages organs, causes anemia and malaise, and contributes to lower life expectancies.

MALARIA'S success has made it one of if not *the* most important diseases of the modern age. Through the centuries it struck seasonally, causing far more illness than death, so it never conjured the hysteria—or headlines in history books—of the other big killer microbes. It was no Black Death, which wiped out half of Europe's population in the 1300s (and may have been a type of plague or a relative of Ebola—the organ-liquefying, hemorrhagic disease of Richard Preston's bestseller *The Hot Zone*). The Black Death, bubonic plague, smallpox, Spanish flu, and other rapid killers were not sustainable over time; they burned through populations then disappeared, sometimes for centuries. By contrast, *Plasmodia* didn't kill as readily, which gave man time to adapt. This won the microbes long-lasting, worldwide success.

Over many tens of thousands of years, these microbes and man formed a "truce" that was more beneficial to the microbes. "What the pathogen wants is a very long-term relationship with the host; it's a smarter approach," explained Wolfgang Leitner, a malaria expert with the U.S. National Institute of Allergy and Infectious Diseases. Malarial parasites create crystallized hemoglobin that is ingested by macrophages—our vacuum cleaner–like white blood cells. This process cleverly paralyzes these cells, instead of killing them. So no signal goes to our immune system to produce more, and we end up with crippled macrophages, less able to fight off the infection.

No one understood any of this until recently. Throughout most of human history, people didn't even know mosquitoes carried malaria; they thought it was carried in foul air rising up from swamps, as Hippocrates had suggested. People felt helpless against the fevers that swept over them every time spring turned to summer. They shuttered windows to block the vapors, not knowing that the still air made it easier for malarial mosquitoes to sneak in late at night for a blood meal.

As the millennia passed, people continued to respond to malaria at a genetic level, selected for conditions that include a blood disorder called thalassemia and deficiency of an enzyme called G6PD. These immunological innovations—caused by random mutations—helped populations survive endemic malaria. Scientists today can often determine whether our ancestors lived in highly malarious regions, and sometimes even pinpoint exact locations, simply by examining our blood.

THE mosquitoes that carry malaria are as relevant to this story as the microbes themselves. This is because fighting the disease in our blood has proved nearly impossible, so many strategies take aim at the vectors, which are a specific type of mosquito called anopheles. Of the more than three thousand mosquito species, about 375 are anopheles. Only about seventy of them carry malaria. And only about forty carry the microbes well. But those forty are just about everywhere. They complete the microbes' life cycle by giving them somewhere to breed.

And breed they do, inside the mosquito's gut. If they never make it to the gut, they die off in our blood, end of story. But to be sucked into a mosquito is to be given the opportunity to procreate and spawn progeny that, with enough luck, find their way back into humans to feed on red cells and make their way back into mosquitoes.

As these microbes complete their life cycle in the mosquito, they produce new, genetically distinct strains that slip into our blood by way of mosquito saliva. In this way, mosquitoes have spawned countless strains of unique malarias. Each settled—like microscopic homesteaders—in its own geographic region of people and mosquitoes.

Falciparum strains, without the ability to hibernate in the liver, needed constant access to mosquitoes to keep from burning out. This geographically confined them to places with year-round mosquito breeding—mainly regions nearest the equator. They were particularly successful in sub-Saharan Africa because the mosquito vectors there lived long and could carry heavy loads of parasites.

MANY biological historians believe that trade, and especially the slave trade, allowed these microbes to spread across the globe. With each shipment of enslaved Africans came trillions of *falciparum* parasites. If they were introduced to a region during the mosquito season, they'd launch brutal though short-lived outbreaks. One season of *falciparum* was deadlier than a dozen seasons of *vivax*. No one knew the cause of the fevers. The anopheline carriers were night feeders with a gentle touch. Gyrating females (male mosquitoes don't bite; they drink nectar) quietly fed while people slept. They injected parasites with their syringe-like noses while extracting the blood they needed for reproduction. Sleeping victims were none the wiser until a little more than a week later, when they'd suddenly feel bone-deep cold, then violent shakes, and finally signature fevers.

Throughout history, humans did the best they could to prevent these cyclical, often deadly fevers. From the Middle Ages to the Industrial Revolution, malaria rode on waves of trade, slavery, and warfare, planting roots across Europe and Asia, intensifying around the Mediterranean basin. A hundred miles of Pontine Marshes plagued Roman life, producing robust outbreaks that killed many popes, cardinals, emperors, explorers, conquerors, and poets—including Dante in 1321 and Lord Byron in 1824. The dreaded fevers eventually made their way into the medical vernacular under many names, including the ague, putrid fever, intermittent fever, the shakes, and *paludisme*—French for marsh fever. But the name that stuck came from the Italian words *mala aria*, meaning "bad air."

Malarial parasites set sail for the New World in the blood of European explorers—mostly *vivax* but some *falciparum* in the people of southern Europe. They created intermittent outbreaks, but no lasting presence. That didn't occur until the Portuguese established sugar plantations in Brazil and began the first large-scale movement of African slaves onto the continent. Africans' full immunity to *vivax* and partial immunity to *falciparum* kept them healthy on mosquito-infested plantations. While native laborers fell with fever, Africans stayed relatively strong and produced huge profits for European overlords. Malaria helped draft what historian James L. A. Webb Jr. calls an "economic and social template" that spread throughout the Americas.[9] The fact that Africans could work harder and longer than indigenous peoples boosted the cross-Atlantic slave trade.

English settlers seeking religious refuge founded Jamestown in the mid–seventeenth century with *vivax* in their blood. The New World mosquitoes picked up the parasites and passed them back and forth between pilgrims and new colonists, establishing a malaria zone up and down the East Coast, from Massachusetts to Florida. When slaves were ripped from their homes in West Africa to work Southern cotton fields, they carried *falciparum*, which, because of the climate, launched only intermittent infections. Still, this added heavy death tolls to seasonal outbreaks, making malaria "the most important killer in the North American colonies," according to Webb.

No one understood why these fevers spread so fast. Everyone simply shuttered the windows in a futile attempt to stop the ague from wafting in with the night air. An ominous dread hung over encampments and settlements whenever the rains ended and temperatures rose.

The microbes traveled deep into America's heartland in the blood of mercenaries, making their way across the U.S. South and up every major waterway, all the way to Canada. Plantation owners and white laborers caught fevers from mosquitoes that bred in the cotton fields, while Africans stayed healthy. The proslavery movement used this as evidence that God had created black people for slave labor.[10] The fact that Africans had genetic protection, evolved over thousands of years, would not be understood for another two centuries.

Native Americans died so rapidly of measles, smallpox, and other European diseases that historians have a hard time measuring the impact of malaria, though they assume it was significant.[11]

IN the eighteenth and nineteenth centuries, improved microscopes allowed clever researchers to see tiny creatures—which they called animacules. This launched the so-called germ theory of disease. Alchemy became chemistry. Sorcery became science. The hunt to find microscopic monsters led to great discoveries by men like John Snow, Joseph Lister, Robert Koch, Louis Pasteur, and many, many more. Stains to color germs made them easier to see, and more discoveries were made. Public health campaigns reduced the reach of terrible killers like cholera, tuberculosis, and anthrax poisoning. The world's first vaccines wiped out smallpox. New methods of pasteurization reduced microbial contamination of milk and wine, and sterilization reduced bacterial infections in hospital operating rooms. Science scored major victories against these enemies, and scientists became superheroes.

Progress against malaria, however, stood still. Discoveries of other parasites had to lead the way so that researchers could comprehend this particular monster. In 1878, Sir Patrick Manson, a prominent Scottish physician studying tropical diseases in China, did just that. He spied tiny worms in the blood of patients who had developed hideous deformities, then saw the same worms in dissected mosquitoes. Clearly, mosquitoes delivered them to man. But scientists at the time still believed that female mosquitoes took a blood meal, released eggs in waterways, and died. Manson assumed his newly discovered worm, which he called filariae, was released into water that people drank. Once inside the bloodstream, the worms grew in size to gum up the lymphatic system and cause severe swelling of the extremities—especially the legs, arms, and scrota. The disease was called filariasis, and these extreme symptoms elephantiasis.

This got Manson thinking that malaria might also be spread by way of dead mosquitoes in drinking water. But no one had spotted a parasite in the blood of malaria-infected people. He encouraged his protégé, Ronald Ross, to look for evidence in support of his malaria hypothesis. Meanwhile, other prominent scientists from France, Italy, Spain, and the United States also attempted to explain malaria, including a claim out of Germany in 1878 that the cause was a newly discovered bacterium, *Bacillus malariae*, located in dirt near swamps. This, of course, was wrong.

Then, as often happens in science, one person stumbled onto an important and defining clue. That was French army doctor Alphonse Laveran, who was sent to Algeria to examine the tissue of soldiers killed by malaria. The men were in North Africa to enforce French rule on the people of Constantine and other nearby cities. Mortality rates as high as 30 percent weakened the army. At that rate, the French feared they would lose hold of the region.

Through two years of devastating death tolls, Laveran performed many autopsies, looking for the killer malaria microbes. He peered down the arm of his microscope, seeing what others had already seen: tiny black specks forming clouds around ruptured red blood cells. He was as stumped as the others, straining to figure out what they were. Then, in 1880, he tried a new approach. He took fresh blood by way of a standard finger prick from a soldier who was sick but still alive. In this blood smear Laveran saw the unexpected. There swam large microbes the size of the red blood cells themselves. Some looked like slugs, while others wriggled with whiplike tails. Not believing his eyes, he pricked another feverish soldier's finger, and another, and another, until he found one with the same gyrating germs.

He had unwittingly captured the sexual phase of *falciparum*. He didn't know that the slug-shaped ones were the females and those with spermlike flagella were the males. He didn't know—nor would anyone figure out for another eighteen years—that these sexually charged parasites had swum to the soldier's surface tissue to be drunk in by biting mosquitoes. But Laveran didn't need to understand their whole story. He identified them as the cause of malaria and threw popular theories into a tailspin.

This didn't happen overnight. Laveran, slow in speech and motion, lumbered through his report, carefully sketching images of his parasites. He couldn't explain how they got into the human bloodstream. Nor could he identify a nucleus, which would have confirmed that they were indeed parasites, not just misshapen blood cells. And he couldn't explain what the organisms meant. The report he submitted to the stodgy elders of the prestigious French Academy of Medicine was disregarded by some as fraudulent or, at best, misguided.

But the drawings and techniques intrigued enough influential people to eventually convince Manson, Ross, and a few others that Laveran had found something important. Another decade and a half passed before Giovanni Battista Grassi, a well-known Italian entomologist and scientific giant on all matters of germ-related medicine, finally accepted the value of Laveran's findings. By 1895 the race was on to figure out what these germs were and how they entered the human body.

An important clue came to light in 1897, when a Canadian enrolled in the first year of Johns Hopkins University's medical school made a discovery. The student, W. G. MacCullum, observed parasites in blood he took from a sick crow fusing into what looked like an egg sac. He presented his paper to the British Association for the Advancement of Science, suggesting that this might be how malaria parasites procreated once being sucked in by mosquitoes. Manson forwarded the paper to Ross in India, who had just found similar parasites in the gut of certain brown mosquitoes.

This helped Ross hone his work. He raised mosquitoes in a lab and fed them on malaria-infected birds, then dissected the insects to see what the microbes did inside the mosquitoes' guts. Hundreds of dissections later he had pieced together the mystery. Extrapolating from bird malaria, he announced the path of human malaria. In 1902 his efforts won him the Nobel Prize in Medicine. As for Laveran's part, he also received a Nobel Prize in Medicine in 1907, but it was for work on several types of disease-causing protozoa, not just malaria.

Other great scientists also contributed, especially Grassi from Italy, whose work was so pivotal he probably should have shared the prize with Ross. But that didn't happen, and Ross's name is the one everyone remembers. He went on to publish a small how-to manual called *Mosquito Brigades*, in which he advocated the spilling of oils and kerosene on swamps and marshes to wipe out mosquito larvae. He helped rid England of the disease. Then he headed south to colonial Africa, where he failed, because he underestimated the durability and sheer volume of mosquitoes there.

In this extreme setting, however, he saw the calculus that *is* malaria; that it's not just a disease but a societal condition based on key factors that include a) the number of infected people; b) the volume of biting mosquitoes; c) the extent to which those mosquitoes have access to infected people during nightly feedings; and d) the extent to which the mosquitoes prefer human blood over, say, cow or pig blood. Today we call this calculus the BCRR—basic case reproduction rate. It measures whether enough unprotected people live among so-called anthropophilic mosquitoes (people biters, not cow or pig biters). Regions measuring 1 or below don't meet the threshold for an outbreak—there aren't enough mosquitoes biting unprotected people (North America, most of Europe, parts of Asia, and a few other regions). Places measuring between 1 and 5 have outbreaks but no permanent malaria (parts of Latin America, Asia and Eastern Europe, parts of the Middle East, and southern Africa). Regions measuring above 5 have endemic, year-round malaria. Parts of India and Southeast Asia, for example, measure between 5 and 10, while some areas of sub-Saharan Africa exceed 1,000—making their malaria the most durable in the world.

By the early 1900s, many well-off regions in northern climates eliminated malaria through mosquito abatement and treatment with quinine—the only known drug for malaria at the time. Socioeconomic changes in cooler areas with shorter mosquito seasons made breaking the infection cycle easier than in southern regions. The upper stretches of the United States and Europe were cleared of endemic malaria, with BCRRs hovering around 1. These regions had occasional outbreaks, when the right combination of mosquito breeding and infected people accidentally reached explosive proportions—especially during war. But for the most part, the days of worrying about malaria were over. That left southern Europe and Asia, the U.S. South, and the tropics. In these regions in the early twentieth century, malaria-carrying mosquitoes had to be tamed. And that is where this story begins.

CHAPTER 1
Lowell T. Coggeshall

Lowell T. Coggeshall arrived in 1901, born to poor farmers in Saratoga, Indiana, where everyone bartered to survive.[1] His parents sold eggs and butter to earn cash for medicine, fabrics, and a few other staples. Everything else they grew. Each season brought the same worries about weather and soil conditions, crop yields and pests, animal husbandry and diseases.[2]

This was to be Lowell's future, but for the fact that he wanted no part of it. Nor did he think he was any good at it.

For starters, he could never get the family's ornery old workhorse, Bird, to pull a plow, so the two exchanged "whacks almost daily." Prone to bouts of recalcitrance, he also did stupid things that earned a thrashing from his stern father, like the time he tossed sunflower seeds into a tree stump near the cornfield instead of planting them in the proper place. His father figured it out when the flowers sprouted out of the stump.

Even summer highlights—the July Fourth parade with decorated vets from the Spanish-American and Civil wars, and the annual twenty-five-mile horse-and-buggy ride to the Jay County Fair—barely held Lowell's interest. And when the community banded together to thresh fields of wheat and oats, and butcher animals for the long winter ahead, he observed like an anthropologist, not a boy learning his trade. He was good-humored and friendly. And he did the work. But he didn't enjoy it.

Nor did he like to study. Had the schoolhouses' three teachers not been his first cousins, he might have skipped school altogether. But as it happened, in

1913—the year before the Panama Canal opened—he moved without fanfare from the first-floor elementary school to the second-floor high school. There he sat day after day, marginally interested in the offerings of rudimentary math, science, and grammar. He had no idea how he would escape this predetermined life; he just knew he would.

HE also had no idea that the Panama Canal had anything to do with his future. He might have followed what the newspapers said: that the French had abandoned the project in 1889 after twenty-two thousand workers died of malaria and yellow fever, and that the U.S. Army took control of it in 1904 and used mosquito-control strategies to finish the work.

But the story was so much bigger than that, and would one day change Lowell's trajectory. For the Americans, at first, suffered the same devastating disease rates as the French, with malaria being by far their biggest problem. Initial campaigns to protect workers were sloppy and inefficient. Screens were improperly installed; loose floorboards and open eaves let in mosquitoes; and antilarvae campaigns were poorly targeted. Army physician William Gorgas had to struggle for the authority and funding to study and fix the problem. So he hired a key person, Dr. Samuel T. Darling—who would soon be a big figure in Lowell's life. Darling trained an army of laborers to build screened-in "isolation cages" for the sick, construct new mosquito-proof buildings, and spray kerosene on breeding grounds of the region's most efficient malaria-carrying mosquitoes.[3] This had the added advantage of also wiping out yellow fever's less robust mosquitoes. Workers stayed healthy long enough to finish the gargantuan engineering feat, delivering the United States an internationally important colonial gem—one that linked two great oceans and substantially shrank the world's shipping lanes.

When the canal officially opened, newspapers splashed Gorgas's face across the front pages. His grandfatherly good looks—thick white hair and mustache, perfect nose and cheekbones, and sharp uniform—made sanitation work sexy. The public saw him as a kind and wise national elder. Men looked up to him and grandmothers swooned over him. His celebrity brought attention and focus to the importance of sanitation work in keeping people healthy. And for that, just weeks after the canal's opening, the army awarded him the coveted position of U.S. Army surgeon general.

From this pulpit he preached his public health religion with his most emphatic sermon on the virtues of mosquito sanitation—how that alone would wipe malaria from the planet in a few short years.

LOWELL showed no sign of being the least bit interested in or inspired by the great Gorgas, but the elders of the heavily endowed Rockefeller Foundation were. They decided to commit $100 million to sanitation projects designed to improve the general health of the world's poorest people, creating the foundation's International Health Commission—which grew to be the largest private-sector investor in health and medical research (on the scale of today's Bill and Melinda Gates Foundation). Rockefeller had already wiped out hookworm in many places by breaking the cycle of infection, which could be done with outhouses and shoes: outhouses to stop infected people from defecating in the streets, and shoes to stop hookworms in the feces from boring into bare feet. Rockefeller partnered with local officials to bring these low-tech, low-cost technologies to poor communities. With the new $100 million infusion, and Gorgas's great success in Panama, the IHC decided in 1913 to take on the U.S. South's entrenched malaria problem.[4]

For there, every summer, severe malaria swept up the eastern seaboard from Jacksonville to Baltimore; reached across Virginia, the Carolinas, Georgia, and the Florida panhandle; almost completely subsumed Alabama, Mississippi, Louisiana, Arkansas, and Tennessee; struck north into southern Kentucky, Missouri, and Illinois, and west into Texas and Oklahoma. It even hit stretches of central New Mexico and California.

Quinine, the only known treatment, had been so overused it stopped working in many regions. And its use often led to side effects, like unbearable ringing in the ears and partial blindness. This was because the antifebrile property in quinine, which came from an exotic tree called cinchona, varied in intensity, depending on a given crop of the trees and the method of preparation. Doctors often misunderstood the concentrations with which they worked. Moreover, the drug worked only against acute attacks and did nothing to prevent or cure infections. This meant people treated with quinine often suffered relapses, as if they had never been treated. And they still carried malaria in their blood, which meant mosquitoes picked up the disease from healthy-looking people and spread it to others. No one fully understood how quinine worked, which created uncertainties that were a nightmare for public health officials to think through, and were a big disappointment for paying customers who took the drug, suffered the side effects, and still weren't cured. But the sick had no other option. A Dutch-run cartel ran sprawling cinchona tree plantations on Java in the East Indies, producing, controlling and fixing prices on 90 percent of supplies. The world was their captive as long as quinine remained the only drug available.

Germans at the Bayer Company tried to change this. In the 1910s, they launched a major effort to synthesize quinine using man-made chemicals. Their dream outcome was to cure malaria with a pill, a "magic bullet" made in their lab, bottled in their factories, and then sold over-the-counter worldwide—like their blockbuster cure-all, Bayer Aspirin. Bayer would supplant the need for slow-growing, hard-to-maintain cinchona trees and put the Dutch out of business.

Bayer advertised that it was working on a miracle cure for malaria, one that was affordable and easier on the body than quinine. This added heft to an assumption that an inevitable drug or vaccine would wipe out or at least reduce the reach of this disease—as was the case for cholera, rabies, tetanus, typhoid fever, and bubonic plague. Edward Jenner's smallpox vaccine by now was more than a hundred years old. Surely a true remedy for malaria would come next.

"Within a decade some biochemical product will be evolved, an antitoxin in the form of a vaccine or a serum, that will effectually dispose of the toxins for malaria," wrote *The New York Times* on July 13, 1913.[5] A month later Paul Ehrlich—the wildly eccentric father of German medicinal chemistry—fed this hope by making an outlandish announcement at the International Medical Congress in London: He said his arsenic-based, magic-bullet pill for syphilis, called Salvarsan, also cured quinine-resistant malaria. Even better, he said, Salvarsan could reverse the resistance. Malaria had met its match, was the message he delivered. The heady participants at the London lecture hall erupted with great applause. "He wore a dusty frock coat," wrote *The New York Times*' London correspondent, "and his voice was shrill rather than powerful. Such was the savant whom thousands . . . greeted with the sort of cheering that men grant a hero."[6]

Ehrlich's assessment of his own drug proved wrong, terribly wrong. Salvarsan had no such powers. It wasn't even good against syphilis—the dreaded corkscrew bacteria that many called the "reward of sin" and that often infected the brain, turning regular people into madmen. But before Salvarsan there was only mercury, which everyone knew was a poison. So finding something better turned Ehrlich into a legend. Before making it, he was just a batty, made-fun-of lab assistant to the great Robert Koch. He arrived at the Koch Institute with a PhD he had earned using dyes to color tissue cells, and broke into microbe hunting by using those dyes to color germs. Then he tested a crazy idea by adding poisonous chemical side chains to germ-seeking dyes and shot them, like "magic bullets," into the sick.[7] Everyone snickered at his tussled appearance and boisterous lab manner. But when he came up with Salvarsan, his reputation was transformed; he became the Koch Institute's genius of chemistry.

His strong opinion that a magic bullet could be made against malaria inspired chemists at Bayer Works' laboratory in Elberfeld, Germany. They started with a drug originally conceived years earlier by Ehrlich, made of a blue dye. He knew the dye stuck to malarial germs because he used it to stain them for microscopy. With that in mind, he added a methyl group as a "side chain," hoping it would kill the germs. Like "magic," he concocted the first chemically derived poison pill against malaria. The concept was ingenious, elegant, and simple—a sign of how cleverly and originally Ehrlich's mind worked when he thought through an attack strategy for germs.

But his drug, called methylene blue, turned urine green, made the whites of eyes blue, and was highly toxic. It couldn't be used by anyone. The concept, however, proved valuable and became an obsession of Bayer's. Chemists tried different versions of this concoction, rotating molecules around the central nucleus. To test each new rotation against infections, they injected canaries with a type of bird malaria, hoping that at least one of their new combinations would work. Hundreds, maybe even thousands of songbirds died in their cages, their little bodies slumped in piles.

THE Rockefeller Foundation's health commission listened to the talk of magic bullets with skepticism. They were all medical doctors and entomologists, not chemists. Their labs were in the swamps, focused on anopheles mosquitoes—those ubiquitous malaria carriers—and the desperately poor people living in shacks, exposed to nightly bites. Many at Rockefeller believed they could end the South's staggering malaria problem by using pyrethrum to kill mosquitoes and kerosene to kill mosquito larvae. This fit with Rockefeller tradition. It made sense to their physicians and insect experts, who were infused with a passion for "public health," which meant serving communities by improving sanitation and curbing disease exposures—the philosophical alternative to "individual health," which focused on magic bullets for the already sick. Rockefeller disciples saw disease prevention as holistic, not to mention cheaper. They were health officers, patrolling for methodologies and opportunities to improve housing, provide safe drinking water, ensure access to healthy foods, put shoes on children, build outhouses to keep feces off the streets, locate wells in appropriate places, and so on. Through this lens they saw an opportunity to prevent malaria by separating man from mosquito.[8]

It helped that the business side of the Rockefeller family—driven by bottom-line decision making—wanted the foundation's public health investments to be results-oriented and cost-effective, while also contributing to better health care delivery systems (like the establishment of clinics with medical staff close to populations too poor or small to have a hospital nearby). Investing huge resources in chemical labs and drug manufacturing made no sense to them when sanitation could *prevent* diseases.[9]

Sanitation by way of killing mosquitoes worked, they believed. This conclusion evolved over several years, beginning with side-by-side studies during horrendous malaria outbreaks in the Mississippi Delta and next door in Arkansas. In the delta they used quinine; in Arkansas they sprayed forests with pyrethrum, spilled kerosene into swamps, did minor ditching, and nailed fine-mesh screens to windows and doors. The studies were deeply flawed. Doctors prescribed extreme doses of quinine to black sharecroppers in the delta, then claimed they were cured, when in reality many had fled to the North and were replaced temporarily by transient black families also heading north, and not yet infected with malaria. Meanwhile, malaria rates in Arkansas were said to have plummeted based on doctors' calls, which were equally unreliable as a measure of infection rates. Then, in 1921, a sanitation officer with the U.S. Public Health Service, M. A. Barber, created an antilarvae cocktail called Paris Green by combining copper arsenite and copper acetate, "mixed with road dust and spread by a hand blower." This worked extremely well against anopheles larvae, measurably devastating mosquito populations and indirectly reducing malaria rates. Rockefeller's scientists thus concluded that prevention worked better than quinine.[10] These studies helped local officials see the value of mosquito abatement—just like in Panama during canal construction.

In the American context this was an either/or proposition, because very little public money was available for antimalaria work. This differed from Italy, where officials bridged mosquito control with treatment plans. The central government there made malaria eradication a national campaign; federal programs paid for mass quinine treatment *and* major land reforms that transformed the Pontine Marshes outside of Rome into farmland. Huge irrigation channels drained water from wetlands and kept it moving, eliminating stagnant pools that had produced the country's abundant and vicious malarial mosquitoes. By the time the Great War broke out in Europe, in early 1914, Italy had reduced its malaria problem—nearly eliminating malaria-related child mortality that had previously reached a staggering 60 percent in some areas.[11]

The United States' central government learned from the Italians, and used their dual-edged strategy around military installations. Lumber companies and the railroads also applied a dual strategy to keep workers well enough to put in a long day. For these efforts, the U.S. Public Health Service partnered with Rockefeller officials. They "sanitized" mosquitoes from the woods and swamps surrounding these industries' work sites, and around federal installations. Demand for screens created a new market for steel. And soon municipalities that could afford the antimosquito work hired Rockefeller-trained experts to run the projects.

So while foundation doctors ran surveys to count the sick—and measure the extent of the problem—Rockefeller recruited young men from university biology labs to be trained in the art of killing mosquitoes. These chemical-spilling brigades grew in size as Rockefeller's programs expanded. Soon the foundation had created a selective vacuum that sucked in the nation's most promising scientists.

Lowell Coggeshall would someday be among them.

LOWELL left high school in 1917 in a graduating class of two, of which he ranked second. While the smarter kid dived into farming, Lowell tried everything but. He worked as a railroad assistant, but got bored. Then he worked as a bank teller, but made too many mistakes. He eventually settled into a winter of trapping muskrats in subzero temperatures just over the state line in Ohio. Barely surviving there, he decided to enlist in the navy. The Great War raged in Europe, and Lowell wanted in on the action.

He made his way to the navy's recruitment center and queued up with other boys like him, trying to be men. At six feet tall he felt himself sturdy enough to serve his country. But he still had a pubescent scrawniness. His skinny neck failed to fill a shirt collar. His clothes hung like a bag on bones. At just sixteen and not much more than a hundred pounds, he was told to go home and gain weight.[12] He'd play no part in this war.

MALARIA drove many outcomes of the Great War, as witnessed by public health officers stationed in Europe. They recorded, for the first time in history, a stunningly rapid spread of malaria in battlefield conditions. Millions of soldiers dug in around local civilians in Macedonia seeded with their own strains of malaria. Then hordes of refugees streamed in from other war-torn countries, bringing in new strains. All were quarantined to camps not far from the fighting, where mosquitoes bred rapidly in the blown-apart landscapes. They rose up to feed on civilians trying to find shelter in barns and under trees, and delivered mouthfuls of malaria to soldiers trapped for months in the trenches. Mosquitoes carried the disease from one army to the next.

Medical corpsmen saw the math: Armies of exposed soldiers mixed with millions of anopheline mosquitoes and thousands of infected villagers equaled major outbreaks. Both Central Power and Allied commanders stopped fighting while whole regiments shook with chills and burned with fever. The British hospitalized 162,512 soldiers for malaria, compared to 23,762 killed, wounded, and missing in action. In one famous cable, a French commanding general reported his entire army too sick to fight—as 80 percent of his 120,000 men had malaria and were hospitalized.[13] Battles commenced when mosquito season ended.

When the Allies finally won, soldiers marched home with the emotional scars of war—and malaria still in their blood. Public health officers predicted the outbreaks to come, which were the worst ever recorded. In Russia, for example, social and political chaos left people homeless and destitute. That, combined with perfect weather for mosquito breeding, sent the basic case reproduction rate—the BCRR—through the roof. The disease easily moved from one insect colony to the next, burning through villages of people, reaching farther and farther across the continent. Within three years, Central Asia hosted the world's worst malaria outbreak on record. It hit six million people, claimed six hundred thousand lives, and spread as far north as Archangel near the Arctic Circle. Anyone who knew anything about malaria watched in stunned fascination. "For the first time, as far as we know, the King of Tropical Diseases set foot within the Arctic Circle," uttered Rockefeller's world-renowned malaria expert, Lewis W. Hackett.[14]

Meanwhile, public health officers in Italy saw their own nightmare unfold. For them, the clock had turned backward. War undid the country's progressive land reforms and quinine-treatment campaigns. Irrigation networks broke down, refilling swamps and marshes. Factories stood still, with countless pots and puddles collecting water for mosquitoes to lay eggs in. Farm animals, an alternative blood source for Italy's species of anopheline mosquitoes, were confiscated for the war and now gone from the hillsides. Millions of mosquitoes rose from the waters and voraciously fed on more than fifty thousand malaria-infected soldiers returning from war. Then the insects delivered malaria to the population at large. This opened an "epidemic highway" that, in 1918, triggered the worst malaria outbreaks in Italian history.[15]

AS nearly a million Europeans and Central Asians died as a consequence of postwar malaria, Lowell found a way out of farming: college. This was how his uncle Don on his mother's side got out of it, and Uncle Don had done well for himself.

Don C. Warren had broken the mold of Saratoga by turning wedding gifts from his parents—$500 and a cow—into tuition at Indiana University's biology department. He later went to Columbia University and became a specialist in a new field called genetics. Don applied his skills to chicken farming, made a bundle mass-producing high-quality chickens and eggs, and traveled to faraway places like California.[16]

The people of Saratoga disliked Uncle Don. His education, elaborate life, travel, and good pay made him stuck-up, if not lazy. Because he rejected the tight world of small-scale farming, he was a role model for no one—except Lowell. Lowell admired how his uncle used education to escape Saratoga's confining standards. So he followed his uncle's steps into Indiana University's biology department.

At age seventeen, Lowell borrowed $100 from his father and left for Bloomington. He brought a small trunk of essentials, including bedding, one suit, and a bag to bring his dirty clothes home in every week. The English Gothic structures organized around IU's stately clock tower felt otherworldly. As did the student body, which was seven times larger than the population of his hometown, and seemed filled with exceptionally smart people. Lowell believed himself to be among the most unsophisticated there, and he might have been. His English professor berated him for "atrocious" grammar and word choice. He couldn't spell. And he couldn't keep up with the rigors of his other courses. Discouraged and feeling outclassed, he played billiards and worked out in the gym. He joined a social fraternity, partied a lot, and followed IU's championship basketball team around the region, by either hitching a ride or jumping the freight train. Once, a conductor dragged him and his buddies into the engine car and forced them to shovel coal. They didn't mind the dirty, backbreaking work as much as getting caught. It meant they couldn't jump the train to get home without risk of jail—which the conductor promised, if they got caught again. So they hitchhiked in freezing sleet.

That first year his dean put him on probation with a threat of expulsion.

If everyone has just one defining moment, this was Lowell's. He could either work harder at academics or go home. That he threw himself into the former is a testament to his fear of the latter. He had to improve his grades or return to the farm and drag a plow. So he nailed himself to his desk, worked around the clock, almost never went home, and took classes through the summer. In doing so, he found he had a knack for science. He used books to bring context and content to his love of nature. The harder he tried, the better he did. He slowly proved to his professors that he could do the work, that he was a budding naturalist with a particular talent for entomology. No more wasting time. He found his calling, and he would take it seriously.

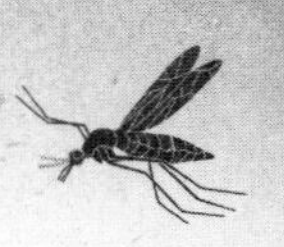

CHAPTER 2
Fever Therapy

The same year Lowell snapped from adolescence into a man of biology, a "brilliant but extremely difficult" psychiatrist by the name of Julius Wagner von Jauregg tried a memorable experiment that would soon transform the study of malaria.[1]

The experiment unfolded across the Atlantic Ocean, in the foothills of the Austrian Alps, while the Great War still raged. This difficult psychiatrist drew blood from a malaria-infected soldier just home from the Balkans and immediately injected the blood into the shoulder blade of another man. Jauregg's stony face and fixed eyes remained unchanged as he pushed the entire contents into the bloodstream of the second man—a thirty-seven-year-old stage actor suffering from the shameful corkscrew bacteria of syphilis.[2] The actor's infection had reached his brain and turned him mad, so his family put him in Jauregg's care. And Jauregg, by filling him with infected blood, hoped beyond all hope that this mental cripple—this babbling idiot prone to obscene acts and violent rants—would soon burn with fever. Then perhaps the madness would end.

The event took place back before Jauregg became the renowned grandfather of psychiatry. Before this tall, statuelike doctor became a Nobel laureate. And before he gave the world an excuse to use many thousands of syphilis patients in malaria experiments. This was back when he was just another man of medicine tied to psychiatry because he had failed to obtain the appointments necessary for internal medicine.[3]

The son of an Austro-Hungarian knight, Jauregg towered over his peers in personality and stature. All fell silent when he entered a room, "as if in the shadow of a titan."[4] His biographer traced this stern, authoritative persona to a once happy childhood gone bad after the untimely death of his mother from tuberculosis. When he was just ten, his two sisters were sent to convents and he and his brother moved with their father to Vienna, where they attended the city's most renowned school for boys, with strict academic standards. Young Julius studied with sons of nobility and earned excellent grades in everything from Latin to the natural sciences, winning him entry into medical school, where he spent five years earning grades with distinction and studying under arrogant and undistinguished professors. He found his place in that sweet spot between intense competence and hard work, with an eye for brilliance in others. He resided where reason took precedence over sentiment, which made him somewhat caustic and short, but never overbearing and rarely wrong.

The famous scientist-turned-writer Paul de Kruif remembered Jauregg as a "piece of granite rock"; his biographer called him "gruff" but "generous," with an "indestructible calmness"; assistants said he was "straight as a candle and as if chiseled in stone," but kind and fatherly, with a "golden humor" and deep humanity.[5] While his students called him a "wooden statue," they attended his lectures in droves, and loved to see his eyes twinkle through that famously dour expression whenever he had cause to tell one of his many off-color jokes, which were always at the expense of Jews or degenerates.[6]

He wore his thick black hair neatly cropped and his handlebar mustache tightly curled and combed. He dressed conservatively in a pressed suit and crossed tie. High chiseled cheekbones framed a long, thin face that, with age, fell southward, creating a jowly, hound-dog effect. His striking appearance made a bold mark on the many medical meetings he attended, even among the famous, at large international conferences, which he called "church fairs." He by far preferred small, productive gatherings of the Vienna Medical Society and other local medical organizations, where he often took charge of a discussion gone astray or gave sage insights during disputes.[7] Everything about him showed off a man in control, but for the outsize mustache and bushy eyebrows, which hung like a shag rug over deep-set eyes.

In the thankless world of psychiatry he saw himself as a hybrid, a gentleman in the trenches—maybe like a pearl amid swine. He could have easily been like other psychiatrists: a man of medicine in title only, unable to practice his trade because scarcely any treatments improved mental illnesses. Psychiatry often meant life as an asylum superintendent, overseeing palliative care—a manager watching over the mentally afflicted until they finally passed away or hanged themselves. In the case of syphilis, death came soon after the bacteria entered the brain. No treatment could slow the progress, which made medical care perfunctory and meaningless.

But Jauregg, in his granite resolve, found this unacceptable. If psychiatry would be his profession, he would find a way to treat mental conditions, which he believed were caused by physiological functions gone awry—not emotional disturbances that could be cured by therapy.

He was among the first in his field to believe that infectious diseases caused, and could cure, different forms of mental illness. At night he went to the Vienna medical library to pore over international periodicals and study biological functions he suspected might be associated with cognitive failings. He conducted autopsies on the deceased to study their spinal fluid and nervous systems, looking for clues that might explain their conditions. He studied and wrote a paper on resuscitating patients who had attempted suicide by hanging, surmising that the convulsions and memory loss that followed probably stemmed from asphyxiation (not from emotional hysterics, as his peers believed). He also observed that the physical shock of being near death and then brought back to life appeared to have a positive effect on a patient's mental state. He described one woman who had been melancholy and paranoid before she was cut down from the rope. After a few convulsions and some minor amnesia she was cured.[8] To him, this was a physiological disturbance correcting another physiological disturbance. His patients weren't crazy; they suffered from biological hiccups that led to mental catastrophes that were probably treatable, if enough study were dedicated to the effort.

For all these reasons, he took good care of his patients, and he even married one—a morphine addict.[9] She was a degenerate by his definition, but still marriageable. This kind of hypocrisy seemed to make him whole. While he believed the insane should be sterilized so as not to procreate, he treated them well. While he told terrible jokes about Jews, he had many Jewish friends, assistants, and students. And while he made fun of the new psychoanalytical theories advanced by his Jewish friend Sigmund Freud, he respected and liked Freud and defended his theories, even as he disagreed with them. This may have been because Freud also showed uncommon respect for the mentally ill—something Jauregg valued above professional partisanship.

Only those who would disrespect patients felt his icy judgment.

One visitor to his clinic, according to his biographer, walked up to him as he towered over his ward wearing the usual white lab coat over a vest and cross-tie, and asked: "Excuse me, sir, where are the madmen?"

With his usual stone face, he pointed to the outside and said: "The madmen are out there," then pointed to his ward and added, "In there are the sick men."[10]

Their ailments humbled his sense of duty and challenged him to find novel treatments—including the use of malaria. He wasn't just testing hunches on the insane because they were easy prey for medical experiments. He hoped to cure insanity. And he believed high fevers could do it.

This hypothesis was broadly shared. He and others had seen outbreaks of typhus, cholera, smallpox, and other fever-causing contagions burn through patients at their respective asylums, always leaving behind death and despair.[11] But they also left some hope by turning one or two lunatics sane, sometimes for short periods, sometimes for good. Jauregg combined his own observations at the Asylum of Lower Austria in Vienna with those made by others at other asylums to write a "landmark" paper in 1887 that advanced the idea of inducing fevers as a possible cure for mental illness.[12]

In 1889, Jauregg tested this observed phenomenon by giving his patients streptococci erysipelas, which caused skin eruptions and high fevers. The results were unclear and he lacked adequate time to flesh out the possibilities. So he encouraged others to take up the cause. He lectured on it and implored colleagues to pitch in—hoping someone would find a reliable way to induce fevers. Then he grew discouraged when he learned that one psychiatrist had used a poisonous ointment on the skin of patients' shaven heads, and left it there for weeks, until red-hot inflammation and oozing puss immunologically forced the onset of fevers. Jauregg openly objected to the method. Others implanted horsehairs under the skin to "provoke abscesses," or applied mustard plasters or Spanish fly, which he called cruel.[13] Eventually he gave up on his fellow psychiatrists and drafted a new study design to try for himself.

But his next attempt didn't set much of an example. He started it in 1890, when a research associate gave him a vial of tuberculin bacteria from Berlin.

Jauregg injected it into brain-damaged syphilis patients—a condition with many names but that was broadly referred to as general paresis, and the sufferers were called paretics. This type of neurosyphilis attacked the central nervous system and usually appeared ten to twenty years after exposure. Symptoms included decreased language and motor abilities, impaired judgment, hallucinations, delusions, violent mood swings, dementia, seizures, obscene behaviors, and muscle weakness that led to a telltale gait. Once these symptoms appeared, patients had two, maybe three years to live. Jauregg chose them as test subjects because of their dreaded condition. Why *not* use fevers to try to save them?

He infected sixty-nine so-called paretics with his vial of German TB and compared them to sixty-nine untreated paretics. Mental health improvements were observable in only those patients who reacted strongly—with particularly high fevers that raged for days. This suggested that intense fevers worked better than regular fevers.[14]

But Jauregg abandoned the project after he learned that patients he had sent home cured of syphilis and dementia later died of TB.[15] Meanwhile, the ethical objections rolled in as "the entire Viennese press printed editorials caustically criticizing his work and holding him to be a potential murderer."[16] It didn't matter that several of his colleagues repeated his experiments and produced the same promising results. If the scientific community saw his data as contaminated—because he mistreated patients—no one would publish his papers or advance his theories.

So he bided his time again. The TB work had narrowed the concept. He'd found his target group. Syphilitics became his cause, and he lectured cautiously about his discovery: "We cannot be reproached for using a procedure which is irrational. We have listened to nature; we have attempted to imitate the method by which nature itself produces cures."[17] He was sure of the concept, but the TB experiments instructed him to move forward gingerly, trying only dead bacillus and staphylococci, neither of which worked well because they failed to produce furnace-hot fevers for days on end.

For that, he theorized that malaria would work beautifully. And the infections could be easily controlled by quinine. His main problem was that Austria had no malaria and he didn't know where to get it. Parasites that cause malaria couldn't be grown or kept alive in petri dishes, so scientists couldn't simply share vials of them.

He eventually moved to the Vienna General Hospital to run its Outpatient

Department for Nervous Diseases—the first stop for patients with signs of late-stage syphilis, and usually a few short months before being committed to an asylum. There he continued testing his other theories, which included the use of iodine to prevent goiters and cretinism—work that led Austrian authorities to add iodine to salt.[18] He started treating battle-traumatized soldiers with electric shock therapy that involved strapping them down and sending a jolt of electricity through their brains—with the hope that it would snap them out of their emotional hysterics.

Jauregg was developing a mixed reputation. On the one hand, he showed flashes of brilliance in thinking through physiological aspects of mental conditions—as evidenced by his iodine work. Even his fever therapy, while dangerous, was innovative and thorough. But the shock treatment was different. Jauregg treated soldiers as he was expected to. It mattered little whether they saw horrifying deaths and bloody dismemberments in grenadelike flashbacks that shattered their sense of safety. They belonged to a war-focused culture that needed men to be strong for the war effort.[19] Jauregg's job was to straighten them out so they could return to the front lines and fight for the Fatherland. This mind-set was Austrian; it was nationalistic; and it put the country before the individual.[20] It also foreshadowed Jauregg's political leanings, which eventually led him to the Nazi party in its early years, before Hitler occupied Austria.[21]

One soldier sued Jauregg's clinic because of his assistant's actions. The suit brought Jauregg's operation and his competence into question—an unbearable burden for a brilliant man who saw himself as cautious and fair-minded. At his trial, his old friend Sigmund Freud came to his defense.[22] Freud testified that all medical psychiatrists had been pressured to act like "machine guns behind the front lines, driving back those who fled." He argued that while Jauregg may have hurt soldiers, the torture was unintended; Jauregg had actually been trying to treat them. And with that, the charges were dropped.

But it left Jauregg shaken. He was no monster. He tried to do no harm.

AMID the controversy, one of his medical colleagues reported that a soldier just admitted for minor nerve damage also shivered with *vivax* malaria.[23] The doctor asked: "Shall I give him quinine?"

To which Jauregg replied, "No!"

Finally, he had a source of malaria.

On June 14, 1917, he drew the soldier's blood and immediately injected it into his demented patient, the former actor.[24] When the actor came down with malaria, Jauregg extracted his blood and injected it into another eight patients also with late-stage syphilis. Then he watched over them as they sweated through days of extreme fevers and shook with bone-cold chills.

One patient died; two worsened and were admitted to the asylum; four regained cognitive function but later relapsed. Only two appeared cured of dementia. One was a thirty-four-year-old man whose cognitive failings had only just begun that month. After eleven attacks of fever, he fully recovered and, at his own request, returned to his army regiment. The other was a thirty-nine-year-old man, also in the early stages of dementia. He suffered through ten attacks of fever before regaining cognitive function. And soon after, he returned to his job as a cleric.[25] As for the actor—the first to receive the transfusion of infected blood—he, too, improved; he gave performances for patients at the asylum and was discharged. But Jauregg later received a letter from a Frankfurt doctor reporting that the patient had relapsed and had to be admitted to an asylum there.

The two patients Jauregg fully cured, however, made history.

SYPHILITICS filled asylums in the late nineteenth and early twentieth centuries, occupying an estimated 10 to 20 percent of beds.[26] The disease first appeared in medical records as an epidemic that burned through Europe beginning in the early 1500s. People back then called it a "pox," because after ulcers appeared on the genitals—today's classic symptom—the sores spread to cover other parts of the body. Symptoms also included nerve damage that caused excruciating pain.[27] People feared it was a new type of leprosy sent by God for man's sexual liberties; others believed Columbus brought it back from the New World. Today, scientists believe syphilis grew from a skin disease called yaws—caused by the same bacterium as syphilis, called *Treponema pallidum*. The mutation from skin disease to venereal disease occurred just as Europe's population had exploded, when there were more houses of prostitution, more sexual freedoms, and more warfare—with mass movement of armies and refugees to spread diseases.[28] Over the next two centuries, severe symptoms of syphilis moderated, became less leprosylike, and therefore less terrifying and prophylactic. It had morphed into an annoyance that people could live with, which advanced its spread.

But then another mutation occurred, one that allowed the bacteria to cross the blood-brain barrier and infect the frontal lobe. Scientists now believe this happened somewhere around 1800, because within the next few decades physicians noticed a tangible increase in dementia among relatively young people. This new type of psychosis caused a somewhat sudden personality change that brought on wild mood swings followed by profanities, inappropriate sexual behavior, and muscle paralysis. Puritans blamed it on masturbation and alcoholism, while scientists tried to figure out the real cause. Meanwhile, those infected brought shame and disgrace on their families and were sent away to lunatic asylums, where they became wards of the state and a financial burden.

Finally, in the early twentieth century, the cause had been identified as untreated syphilis. It had burned through the middle classes at infection rates as high as 20 percent in Europe.[29] So prevalent was syphilis that it spawned its own medical specialty, called syphilology. According to historian Joel T. Braslow, some European asylums reported that 45 percent of their male patients were there because of neurosyphilis. Percentages were much lower in the United States but still ran around 10 percent of male hospital patients.[30]

FINDING something even partially effective against this brain-destroying sexually transmitted disease earned Jauregg accolades and helped repair his damaged reputation.

To continue, he sought permission from a military hospital to take blood from a soldier just home from the fighting and sick with malaria. Jauregg didn't examine the blood before injecting it into four of his syphilis patients. To his horror, the four infections spiraled out of control. Despite large doses of emergency-level intravenous quinine, one patient's red blood cell count dropped precipitously. After thirty-one days of Jauregg trying frantically to control the malaria, the patient died. By that point, Jauregg knew he had accidentally used deadly *falciparum*. Two others also died. Only one patient survived after forty-five days of heavy intravenous quinine. Weeks passed before Jauregg saw the silver lining: his sole survivor had been fully cured of insanity.[31]

By now, Jauregg had grown old. At age sixty his tall, stiff build began to hunch. His stern, stony manner—once considered serious and erudite, forceful and commanding—now came off as grim and difficult. His face appeared longer than ever, drawn down by that thick handlebar mustache and topped by wiry gray shoots coming off his shaggy eyebrows. But those eyes, those sunken, beady eyes, must have lit up at the reality that he had discovered a cure for syphilis. He was right about malaria—it could cure this type of dementia, and maybe other types as well. He may have even bested Salvarsan, the great Paul Ehrlich's magic bullet, which really poisoned more patients than it helped.

So while the *falciparum* fiasco shook him, he continued working with malaria.

CHAPTER 3
Making of a Malaria Warrior

On the other side of the Atlantic, in 1923, Lowell's stellar grades earned him invitations into Phi Beta Kappa and Alpha Xi, a place at the top of his class, and acceptance into IU's doctoral program in zoology. To cover tuition he taught several courses, including undergraduate ornithology. He walked students through the woods, pointing out birds he'd just seen in books that morning while preparing for class. Bird-watching was something Lowell informally did as a kid while avoiding daily chores. Now he had books to put names to species he'd observed his whole life. He took up taxonomy to catalog species for his professors.

Lowell spent his summers at IU's biological station on Winona Lake—about 130 miles north of Bloomington.[1] In exchange for room and board he mopped floors and cleaned windows—and ate fish from the lake, which he cooked outside in an open pit. On weekends he earned money ushering at a nearby tabernacle where firebrand evangelist Billy Sunday preached, and the Redpath Chautauqua ran dances and cultural events. With Prohibition in full force, meetinghouses and Bible thumping gave people something to do, and allowed Lowell to earn cash for school.

He took the job on Winona Lake not for love of limnology—the study of freshwater ecology—but because a limnology professor he liked offered him the position. In return, Lowell worked hard. And he did well. He figured out a clever way to count bluegill eggs in the lake's shallows, feeding important data into his professor's project—counting the lake's bluegill population, the results of which were presented by his professor, Carl Eigenmann (dean of IU's zoology department) at the Ecological Society of America's December 1923 meeting in Cincinnati.[2] When an international limnology expert, Edward A. Birge, needed a graduate student to do similar work the following summer, Lowell's IU professor recommended him.

That's how Lowell ended up working for the somewhat famous Professor Birge—University of Wisconsin's president and a nationally recognized defender of evolution in an ongoing, publicized debate with creationist presidential candidate (and later secretary of state) William Jennings Bryan. Birge's helmet of white hair and gray walrus mustache gave the only clues that he'd reached old age. Otherwise, the tall, athletic water expert remained intellectually and physically nimble. Lowell couldn't believe his good luck when this great man invited him to Wisconsin for the summer of 1923.

In brisk morning fog, Birge and Lowell rowed to targeted sections of Green Lake and dropped anchor. Then the professor attached lead weights to Lowell's waist and ankles, plugged an air hose to Lowell's bulky diver's helmet, and dropped him over the side. As Lowell's wiry limbs disappeared in the frigid water, Birge stayed on top operating the air pump. He'd reel Lowell up after a full five minutes, an impressive duration for anyone, and especially for a kid with no body fat. But Lowell didn't complain. He simply came back to the surface—all bony and determined—warmed his goose bumps in the morning sun, and then plunged back into the dark waters. In the afternoons he helped Professor Birge and another limnology pioneer, Chancey Juday, sort through the specimens. All summer, Lowell lived in a tent on the lake's shore, learning the language and methods of world-class scientific exploration—and, no doubt, tricks for reconciling Darwinism with Billy Sunday's preaching. Most important, he earned a reputation as a hard worker with a good moral compass.

THE fall of 1923 came fast. As Lowell left Wisconsin for a new semester at IU, antimalaria forces gathered in the South. Those forces got an enormous financial boost when the directors of the Rockefeller Foundation decided that the United States' malaria problem weighed too heavily on the country's economic potential; it had to be eradicated. The foundation's secretary-general called malaria "probably the heaviest handicap on the welfare and economic efficiency of the human race." That it stunted the growth of the southeast quadrant of the North

American landmass was unacceptable. So the foundation created a research station in the middle of Georgia.

Heading the effort was Samuel T. Darling, one of the most important brains behind Gorgas's legendary defeat of malaria in the Panama Canal Zone. Darling started out as a medical intern on the canal project in 1904 and within twenty months became its chief pathologist. He managed an onslaught of malaria and yellow fever cases, and quickly realized that malaria was his primary concern. He needed to better understand its mosquito vectors if the United States had any hope of finishing the canal. So Darling hunted them down. He found eleven types of malaria-carrying anopheles within the canal zone, all breeding in different locations and all requiring specific approaches. The standard spraying of insecticides and spilling of kerosene-based larvicides—which presumed mosquitoes all behaved the same—were failing miserably and becoming a major political and logistical problem for Gorgas. His constant naysayers undermined his theory on mosquito eradiation, advocating instead for broader use of quinine, which Darling and Gorgas already knew would doom the project, because quinine only suppressed malaria symptoms and didn't work where huge mosquito populations pumped large loads of parasites into workers already weakened by the disease. This was why quinine failed the French, and why it was failing the Americans. Thousands of fever-stricken workers filled the hospitals. Waves of outbreaks stalled construction. The project fell way behind schedule, and many questioned whether the Americans would succeed.

Darling, by catching mosquitoes, dissecting them, and looking for parasites in their guts, quickly learned that only a few anopheles carried the microbes efficiently, and just one of them was causing the outbreaks: *A. (Anopheles) albimanus.*

More important, he learned that the tactics used to reduce mosquito populations were actually counterproductive for *this* mosquito. Time and money spent to flood swamplands with seawater on the presumption that mosquitoes didn't lay eggs in brackish water applied to the *other* species, whereas brackish water actually attracted the robust and ubiquitous *A. albimanus.* This bland, brown mosquito even liked running streams, meadows, muddy pools, old crab holes, shaded areas, and "the stinking water of sewage"—all inhospitable breeding places for most other mosquito species. *A. albimanus* also outlived all other anopheline species, under all conditions, and could tolerate large loads of parasites. This was the right mosquito for malaria, because it lived long, liked human blood, and bred everywhere. It easily upped the basic case reproduction rate—the BCRR first conceived by Ross—to a level so high that malaria quickly spread to every unprotected person sleeping in and around the work zone.

Darling studied this anopheline intently.[3] He snipped its wings and legs to observe that its "music" actually came from vibrating featherlike hairs around its proboscis. He examined the contents of its gut while the insect was still in larvae form to understand its diet and figure out how to make waters less appetizing to it. He observed that the females' ovaries failed to develop without a blood meal. And he tested different fruit sugars to figure out how to keep lab-raised mosquitoes alive for experiments, noting that raisins and dates enhanced or didn't interfere with digestion, while bananas turned a blood meal into a hard stone that clogged the gut and killed the insect.

Perhaps most profoundly, he observed that quinine-treated patients still produced the sexual stage of malarial microbes—the gametocytes—that were the key to outbreaks. Only at this stage could the microbes be sucked into the guts of biting mosquitoes, where they fused into egg sacs and spawned progeny that slipped into new victims the next time the insects bit for blood. Quinine, by failing to kill off this stage of the parasite, was not the presumed antidote for malaria, as the Europeans believed it to be ever since the 1600s—when Jesuit missionaries observed South American tribes using it to "cure" fevers. Quinine did nothing—nothing—to prevent the spread of malaria. It did, however, reduce the load of parasites in a person's blood, which helped prevent death and could even ward off symptoms altogether.

With this intelligence in hand, he and Gorgas won the necessary political backing, and funding, not only to kill off *A. albimanus*, but also to nail fine-mesh screens to barracks, mess halls, and hospitals, and to quarantine the sick inside screened tents—to stop *their* malaria from getting into mosquitoes.

The strategy worked. Malaria rates dropped to nearly zero.

After the canal project, Darling ran similar studies in Southeast Asia, the South Pacific, and Brazil. Then, in 1920, at the height of his career, his left leg went numb. A month later he was sprawled on a surgical table at Johns Hopkins, his skull sawed open. The risky craniotomy removed a three-ounce tumor from his right parietal lobe. He emerged from surgery with his cognitive skills intact. But he'd lost some motor skills. Thus this meticulous, razor-sharp, internationally recognized man of science was grounded in the United States.

No one was better qualified to run Rockefeller's new Georgia project—

which was the South's first and only malaria research station—than Samuel Darling. And no one would be more influential on Lowell's career path.

DARLING invited Lowell to join the team in early 1924, after he'd written to Professor Birge looking for a graduate student who could be sent into the swamps of Georgia at the height of mosquito season to do the thankless job of catching and dissecting larvae. Birge recommended Lowell.

On a mid-April night, Lowell stared out the window of a Pullman train as it chugged southward toward Leesburg, Georgia. This was his first trip out of the Midwest, an adventure he felt he'd earned by dumb luck—the same dumb luck that landed him on Green Lake with Birge. Here he was about to join an international collection of impressive scientists with a mission to study malaria and beat it. He didn't know much about the disease. But he felt confident he'd catch up quickly.

LOWELL'S train pulled into Leesburg in the early morning, dropping him in this small, dusty town of twenty-eight hundred. The town center was straight out of the nineteenth century, lined with saloons, horse stables, an ironsmith, a hardware store, and an apothecary. He was to report to a run-down antebellum mansion-turned-guesthouse called the Magnolia. Darling housed his staff and lab in this rickety old dog, and even had a few guest rooms for visiting scientists—mostly other malariologists interested in Darling's work.

Wilbur A. Sawyer, a senior scientist just in from Bangkok, wrote his wife: "I have never seen so dirty and neglected a place in any part of the world. All the old bottles and cigarette butts and face powder from the previous generation, or degeneration, are still in the room." The "never-seen" landlady sent her black servant to apologize for the broken shutters and cracked windows. And an indifferent cook served terrible food in a grimy dining room.[4]

The hotel's saving grace lay outside the front door, where a large veranda with elegant pillars caught gentle breezes. All Darling's staff and visiting scientists—from exotic places like Ceylon and China and Malaya—spent the last hours of the day here to relax and talk, turning Darling's weathered porch into an international malaria mecca. The world had only one other malaria field station like it, in Rio de Janeiro.

Bleary-eyed and tired, Lowell began his first day with one of Darling's field assistants on a ride around town. Mules hitched to wagons of baled cotton and watermelons followed black sharecroppers down red clay roads lined with small wooden shacks that passed for people's homes. Dirty, barefoot children ran around with swollen bellies. Nearly 80 percent of Georgia's 2.8 million people lived like this. Half were black, and all were poor. The flat, overgrown landscape—pocked with ugly limestone sinkholes—was only five hundred miles from Indiana, but a century behind in farming and living standards.

Lowell's tour at one point stopped in front of an old oak tree with a broken limb. His guide told him a lynch mob had just hanged a man there. The humidity and heat were heavy enough. Add the assault Jim Crow made on the delicate sensibilities of Yankees like Lowell, and Leesburg felt as far away from Indiana as any place on Earth.

The shock of it all challenged his genetics. For he was a tenth-generation Coggeshall of a lineage begun by John Coggeshall, who had arrived in the Massachusetts Bay Colony in 1636 and, a few years later, founded Newport, Rhode Island.[5] John's sons and grandsons and great-grandsons worked this seaport as daring fishermen and traders. They set slaves free in the 1700s and donated money to help care for the elderly and poor. Some attended Yale; some fought as patriots during the Revolutionary War and as minutemen during the battle for Bunker Hill; some whaled; some farmed.

The one responsible for landing Lowell's clan in Indiana had been authorized during the War of 1812 to seize British cargo ships. During one attempted seizure, the British sank George Coggeshall's ship off the beaches of Virginia. He swam ashore and eventually made his way to Indiana as a homesteader. Other Coggeshalls followed him there. They established a Quaker meetinghouse and began farming. Lowell's grandfather, William Rufus Coggeshall, was a doctor, back when men learned medicine by practicing it, and traveled many miles by horseback to barter services for food and money. According a published family tree, William helped run a local underground railroad for runaway slaves and catered to Union soldiers during the Civil War, when one out of every five needed quinine to treat malarial fevers. Lowell never met this grandfather, because he and Lowell's grandmother had died when his father was still a small boy. William Evart Coggeshall, Lowell's father, went to relatives when his parents died. They used him as a farmhand and kept him out of school. William Evart worked hard, took nothing for granted, and prospered from the land. He joined the Spanish-

American War as a laborer and was quarantined in a cattle car after he and fellow workers contracted malaria. He survived to return to Indiana and marry the daughter of Saratoga's most industrious man: William Riley Warren, who built drills on his farm to pump gas to stoves located in every structure, including the henhouse. When William Evart joined the family, the two men built a hydraulic pump that brought running water into the bathrooms and kitchens. They even joined forces to start a small veal production company that went bankrupt largely because of bad timing. Lowell's dad worked double time on his own farm to repay debts.

While Lowell disagreed with William Evart on many subjects, he admired his integrity. And he agreed with his staunch Republican father on at least one political position: that Jim Crow Democrats of the Confederate South behaved abominably.

Yet here he was, living among them and the harsh effects of their racist ways.

AFTER the disquieting "tour," Lowell returned to his room at the Magnolia, which had a "mosquito bar" nailed to the ceiling and nets that hung to the floor—a result of Darling's efforts to protect his staff from relentless malarial mosquitoes. These insects infested the house after dark through gaps in the windowsills, doors, and floorboards—flitting under, over, and around shabby screens. This aspect of the place had horrified Darling, prompting him to retrofit the beds with these nets.

Lowell met Darling in the late morning of his first day. He'd barely recovered from the oak tree encounter when Darling called him in. Forgoing the niceties of introduction, Darling immediately explained why he was in Georgia: to build on the successes of the Panama Canal control strategies. "Here in south Georgia this fall there will be hundreds of victims who will be unable to harvest their crops. Several persons will die of blackwater fever," he said, referring to the painful and graphic expression of malaria's worst symptoms, in which renal failure—and black urine—precede death. "We know the anopheline mosquito is essential in this transmission but not which species. As a matter of fact, we know very little about the nature of the disease inception and progress. The team will explore every possible aspect. I invited you here because of your experience in freshwater biology, to determine the feeding habits and preferences of the aquatic larval stages—hopefully finding a practical control measure. Are you up to the job?"

Lowell knew neither the language nor the science of malaria. He was tired and culture-shocked. All he could muster by way of a response was an awkward and perhaps unconvincing "Yes, sir."

Lowell felt in the moment that he'd screwed up. He'd failed to impress the man who impressed the world. So he remained silent.

Darling went on. His graying light brown hair, full beard, and kind eyes—once described by a classmate as "the eyes of a poet"—put Lowell at ease.[6] The two ended up talking about Georgia's political atmosphere, racial issues, and squalor. Darling offered an empathetic perspective to help Lowell cope: "We are in the heart of one of the most devastated areas of the Civil War," Darling said. "Although fought over fifty years ago, there are few here who have forgotten its vivid effects, especially from the reconstruction period. Try to understand and be helpful." That most of the state's lynchings happened here in Georgia's cotton belt was beyond the reach of Rockefeller's mandate. The Jim Crow laws and Georgia's active KKK were part of a political disease, not a vector-borne one. The best Lowell could do in helping the poor people of Leesburg was to investigate their malaria nightmare and, hopefully, help find a way to end it.

This compartmental approach to public health, seeing problems in silos instead of as a collection of variables that all played off one another, was the only logical way to proceed. It was expedient, made work in the South doable, and allowed researchers who abhorred Southern laws to function. But it also ignored the root causes of economic diseases like malaria—for only the impoverished lived in such squalor, unprotected from mosquitoes. Many years later, Lowell would come to understand this conundrum. But for the time being, he listened and learned.

The others called Lowell the "limnologist," because he was there to study the waters. Unlike the pristine lakes of the Midwest, these waters were muddy and murky and filled with snakes. He and the other researchers went every day to a work site they called "Swampville."[7] It was really a camp for "Negro" lumber workers who earned dollars a day hacking down exquisite hardwoods for export to the rapidly developing Northern states, all the while making the swamps swampier, the bugs buggier, and the malaria more malarious. The team pulled on waders and stepped into the thick black mud that felt like quicksand, looked like tar, and held on to the ankles like lead weights. The god-awful snakes seemed to swim straight for Lowell. And they were everywhere. No one offered guidance on what to do about the big poisonous water moccasins gliding head-high along the surface water, perhaps because no one knew. They just pulled their waders up and batted at the beasts when they got too close. Even the alligators showed more respect.

One five-foot moccasin sneaked too close to Lowell, so he killed it, earning him immediate street cred with his swamp-savvy colleagues. They extracted the fangs and coiled the beast for display on the "snake shelf" back in the lab.[8]

Creepy nuisances aside, these swamp forests were a focal point of fevers. So Darling wanted to know everything about the mosquitoes that bred there. Lowell worked under team leader Dr. Paul Russell, who a year earlier had helped Darling figure out that even though the region had three anopheline malaria carriers, only one caused a vast majority of outbreaks. This mosquito, with a tongue-twisting name—*A. quadrimaculatus*—bred well in the swamp's still waters filled with life-sustaining plankton. Darling and Russell mapped the mosquito's flight patterns and showed the world that *quadrimaculatus* flew a lot farther than anyone thought—which at the time was presumed to be two hundred yards. They recorded that only the nectar-drinking males stayed that close to breeding waters. The females in need of blood to produce eggs traveled up to a mile and could go even farther, if the winds kicked up.

This was big news in the world of malariology, in part because it contradicted the great William Gorgas. A decade earlier, Gorgas had investigated a malaria outbreak near a new dam on Alabama's Coosa River. The dam had allegedly increased malaria infection rates sixfold, and residents had sued the Alabama Power Company for damages. Gorgas, after roaming around the properties, testified that the dam couldn't be responsible, because malarial mosquitoes could fly only about two hundred yards.[9] Residents were infected by their own mosquitoes, he claimed, which had picked up the disease from infected neighbors.

The jury took thirty minutes to return a not-guilty verdict. The judge absolved the power company of culpability. Gorgas returned to his desk job in Washington. And disappointed residents went home to live with swarms of mosquitoes rising up from the dam's soggy, malarious bogs. They knew Gorgas was wrong.

TEN years later, Darling's team provided the evidence. The Magnolia team scraped data from the mud and the muck, analyzed it, and wrote groundbreaking papers that scientific journals ate up. The papers politely and methodically—without an accusatory tone—corrected the record. They mapped what it meant to be a malarial mosquito in the Deep South. The breakthroughs encouraged others to run similar studies to identify the flying patterns of all malarial mosquitoes, under different conditions, in different regions, until the nation's malaria maps not only showed where the disease occurred, but which mosquitoes carried it and how far they could fly—all reaching at least a mile.

Darling's know-how spread across the South, giving public health officers—at the local, state, and federal levels—a much better understanding of what impounded waters meant to malaria rates, mosquito production, and the public's health in general.

EVERY day, Lowell returned to the Magnolia with larvae and algae from the swamp. He examined them under a microscope he'd borrowed from his lab at Indiana University. (Darling's operation ran on self-sufficiency. Anyone accepting a position agreed to his bring-your-own-microscope policy and, for lead investigators, bring-your-own-Ford-pickup policy.) At night Lowell ate the Magnolia's slop for dinner—which everyone complained about—then sat totally entertained by Darling's stories of autopsies he'd done on malaria-infected corpses in Panama wearing only shorts, a rubber apron, and mosquito boots. The fantastic visual brought history and purpose to the old veranda, usually with blow-by-blow radio broadcasts of baseball games playing in the background.

During the day, Lowell treated the swamps with Paris Green. This emerald poison was used to kill rats and insects in Europe, and by artists to paint verdant countrysides—and may have contributed to Van Gogh's brain damage. Then M. A. Barber earned fame by using it to kill mosquito larvae. But no one understood why or for how long it worked. So Lowell scooped the dust by hand into the swamps, pounds of it at a time. The next day he returned to find all the plankton and half of the larvae dead. Four days later he came back to find no larvae, plus a lot of dead plants. The dust used to paint famous landscapes had burned the life out his corner of Swampville. Though its impact on larvae was quite lethal, when the plankton returned about a week later, so did the larvae.

Lowell observed Paris Green killing the larvae's food first. Then, with their food gone, the larvae ate the green poison and completely died off.[10] His discovery that larvae would become indiscriminate feeders when faced with starvation was groundbreaking and explained why Paris Green was so effective. "They're like people—they eat foods they like best; but when hungry, they'll eat anything available," he said later.

This was a revelation. Most scientists assumed different larvae ate different foods, which they thought explained the range of breeding preferences among anopheles. But the preference had nothing to do with the larvae. The larvae adjusted.

DARLING had his team thinking like mosquitoes, to observe them and learn. Lowell communed with his larvae under a microscope, studied their guts, watched their behaviors, and took guidance from Paul Russell on the minute physical distinctions between different species. Russell had differentiated between the larvae of anopheles by studying the hairs of a thousand specimens, which he called "wrigglers." Then he created a species key that showed research assistants how to identify a *quadrimaculatus* from a *crucians* from a *punctipennis*. The key helped Lowell find and study *quadrimaculatus* so he could add pieces to the giant jigsaw puzzle Darling was putting together on mosquito behavior.

Russell was seven years older than Lowell and lived with his wife in Albany, Georgia—just twelve miles from Leesburg, but a nicer, more livable place, with shops and a theater. Like Lowell, Russell was an unlikely malaria warrior. As a young man, he set out to please his father, a Baptist pastor, by becoming a medical missionary. He went to medical school not to become rich or do research, but to care for the poor. While doing a surgical internship at Bellevue Hospital in New York, he met Rockefeller Foundation people and caught the public health bug. His wife liked the change in plan because it gave them "freedom of mind to grow and be Christians in world service" without being confined "to Baptists only." The Russells were givers who loved diversity. Mrs. Russell enjoyed explaining to others "with narrower views" where the Bible clearly stated that all good people went to heaven, no matter their religion or race.[11] The Russells had progressive blood in their veins. They loathed Jim Crow laws and served black Southerners as friends.[12] This fit in well with the humanity practiced on the Magnolia's porch.

Compared to Darling, Russell was still a rookie. But he had his MD and was on his way to professional greatness—hoping to soon qualify for the Johns Hopkins School of Public Health tropical disease review course (a necessity for Rockefeller's International Health Board).[13] Russell on at least two occasions invited his young "limnologist" friend home "for beans" (Russell's shorthand for dinner), which no doubt gave Lowell a needed break from the Magnolia's horrendous food.[14] Sometimes Darling drove Lowell there, over deeply rutted dirt roads. Then they would return to the Magnolia for a nightcap.

THE steamy nights blended together as the summer progressed with the same daytime and evening routines, except for about two weeks at the end of June and early July, when Darling got stuck in Baltimore. He'd been collecting papers and books at Johns Hopkins's school of hygiene to bring back to Leesburg when a sharp attack of *falciparum* knocked him off his feet. He'd contracted it while sitting on the Magnolia's porch with Russell, ten days earlier.

Russell recorded the night in his diary, which all investigators kept to track their work for Rockefeller, usually written in cryptic notes. Russell scratched out one or two paragraphs for each month, with one- or two-word notations tracking his arrival to and departure from nearby towns to do malaria work. For memorable events he used as many as eight words, like on April 30, 1924, at seven a.m., when he saw a tornado: "en route to Leesburg, saw it pass over." Or when he attended a woman killed by lightning: "joined by local doctor but no hope."[15]

On June 25, the day he learned Darling had malaria, Russell made an uncharacteristically colorful entry: "Darling in Baltimore has malaria." In his shorthand Russell noted how on June 14 they had sat together on the Magnolia's porch for five hours waiting for a midnight train out of Leesburg. He added: "Two supposedly intelligent malariologists, unprotected, in highly malarious area. Should at least have used citronella oil."[16]

Darling fell hard with a fever of 104 degrees Fahrenheit on June 24 and was treated immediately with heavy doses of quinine bisulfate, which caused such horrendous ringing in his ears that he needed to take bromides.[17] He stayed in bed for the next two weeks, unable to get up. When he finally felt well enough to travel, he took a train back to Leesburg and warned everyone to tuck themselves under their bed nets.

AS happened to Russell and to all of Darling's disciples, Lowell's perception was changing. He had arrived to a blighted landscape of shacks and sinkholes. But in a few short months he saw not sinkholes but ponds of larvae. The children no longer appeared as dirty kids, but malaria victims whose bellies swelled not so much from malnutrition but because of their outsize, malaria-swollen spleens. When night fell, it wasn't just bedtime; it was the anopheles' hour. It was time to tuck under bed nets to avoid infectious bites—and to think about the volume of mosquitoes flitting into local dwellings to spread their disease. Lowell had been indoctrinated and would soon be addicted to the pursuits of public health and malariology, a profession *not* for everyone. It involved dirty, smelly, tedious work, living in third-world conditions, and finding patience for investigative research that sometimes meant spending hours peering at the hairs of mosquito larvae.

As August temperatures reached 110 degrees, other researchers got to work. They had "skeeter" catching duty, which involved crawling under houses to count the number of mosquitoes resting in the joists. They stained the insects, then followed them to their breeding grounds, which included roadside drainage ditches, quarry sinkholes, and old abandoned syrup kettles. Then they told Lowell so he could hit those places with Paris Green. Other team members sat bare-chested all night in the middle of a room counting mosquitoes, which was a way of counting anopheles populations. Still others, especially the doctors like Paul Russell, went door-to-door to examine spleens—the body's blood filter known to swell when stressed by malarial parasites—or to prick fingers for blood to look for the parasites, all of which was used to create a "malaria map" of the sick.

Dr. Darling ran a full-service malaria research operation that included many Southern public health officials and medical students, some of whom at first called Lowell a "damn Yank." This core group grew tight and borrowed project trucks on Saturday afternoons to drive to Albany for chocolate soda and a movie, or to visit the Russells for "beans." During these social outings, Lowell kept quiet about his New England ancestry. He took to heart coaching from Darling on how to be more Southern in a place that required it, emphasizing his mother's roots, which were in Virginia and the Carolinas. His efforts paid off. In time, his Southern friends called him "cousin Lowell."

A pragmatic scientist was emerging from the swamps of Georgia. Not only did Lowell learn how to catalog his larvae, he also learned how to suppress his true thoughts in a place where they wouldn't have served him well. In his silence he grew even prouder of his father's forebears.

By the time Lowell packed to go home, Darling had asked him to return the following year, and Lowell accepted. The good doctor had also convinced Lowell that medicine, not entomology, was the correct course of study for someone of his talents and budding interests. He told Lowell that an MD would give him the right to explore insect-borne microbes into the flesh of human hosts. Entomology, by contrast, would stop him at the site of the bite.

The summer ended and Lowell was changed. The Rockefeller Foundation worked its metamorphosis, like it had for so many. The foundation's International Health Board had spawned the country's top public health disciples, among them Drs. Darling and Barber, and others, like Lewis W. Hackett, Fred Soper, Louis Williams, J. A. A. LePrince, W. G. Smillie, Wilbur A. Sawyer, and Thomas Parran. Anyone accepted into this fraternity grew to be giants in their fields. This was already happening to Paul Russell. And it would soon happen to Lowell. A seismic shift shook his ambitions as a Pullman train slowly rocked him back to Indiana.

CHAPTER 4
From Insects to Medicine

Back at Indiana University in Bloomington, Lowell sheepishly approached the dean of the zoology department to drop what he thought would be a bomb: He was abandoning his dissertation in entomology to switch to medicine. His professors barely reacted. "It was very humiliating to be met with almost complete indifference," Lowell recalled years later. The department's main concern was that he return the microscope he had borrowed for the Georgia work. One man, Professor Alfred Kinsey, consoled Lowell's damaged ego. He said the departmental reaction was a mere reflection of the beast. It was standard to switch disciplines within the sciences, so no one was surprised. Kinsey himself was about to switch to reproductive sciences. (Twelve years later he founded IU's Kinsey Institute for Sex Research and went on to write an authoritative and, to some, shocking book on male and female "sexology.")

Lowell decided to stay at IU for his MD because of a girl. He'd been accepted to Harvard Medical School and probably should have gone there. All good Rockefeller investigators went either there or Johns Hopkins, usually both. But he didn't think he could afford tuition. And he was smitten. That romance fizzled. Yet his interest in medicine, especially tropical medicine, grew. He'd already taken most of the courses required of a first-year medical student, so he had time to write up his findings from Georgia. With the help of Dr. Darling, who'd just been elected president of the American Society of Tropical Medicine and Hygiene, Lowell got his manuscript published in the *American Journal of Epidemiology*. This was a striking accomplishment for a graduate student. He was invited to lecture around campus on the terrible afflictions that plagued the South—introduced as "the intrepid young college student who had just returned from the fever-ridden swamps of South Georgia." Then the exclusive Acacia Fraternity invited him to join, which is where he got to know IU's other popular, high-profile figures, including Hoagy Carmichael, who lived in the Kappa Sigma fraternity house next door. Lowell also became a BMOC—big man on campus—befriending other BMOCs, including future journalist Ernie Pyle.

This all went to Lowell's head. No longer did he think of himself as a "country bumpkin." He now belonged to a higher order: a medical student at a prestigious university with well-known friends. This life would not include the hardships his grandfather endured as a doctor in the 1800s—trading medical treatment for a hot meal or warm bed. The germ theory had transformed medicine with "magic bullets" and vaccines and able chemists. Science ruled many professions, including medicine, and Lowell aimed to anchor his science-based medical education in tropical diseases, especially malaria. Darling and Russell, and probably many of the other focused and skilled scientists who traveled internationally and swung through Leesburg to spend time with Darling, made their mark on Lowell.

This was also the year he met his future wife. She took an anatomy course he taught for non-science majors. He'd become a bit cocky and hated to dumb down science for anyone. But the course was part of his assistantship, so he had no choice. As he suffered through questions from his scientifically illiterate students, he also got to know and like them. One, Louise Holland, became his friend and companion whom he took on dates. He still felt it "beneath his dignity" to spend so much time with non-medical students. But he liked Louise and grew to realize she was intensely intelligent in her own right. He probably even fell in love with her, but the pull and ambition of science always came first.

IN May 1925, as Lowell prepared for a second summer with Darling in Georgia, he got terrible news. On tour in the Middle East as a member of the League of Nations' new Malaria Commission—enjoying the kind of exotic travel he promised his young assistants, should they choose tropical medicine—Darling's jeep careened over a cliff in a narrow mountain pass a few miles outside of Beirut. He and two others died instantly. The world of malariology grieved. This great, kind man who had touched so many young scientists was only fifty-two, and there was still so much to learn about malaria. The "President's Review" of the Rockefeller Foundation's 1925 annual report said: "Dr. Darling was an investigator of originality and untiring zeal, an inspiring trainer of men, a notable figure in his chosen field. He has left as his monument research, industry and devotion, which will long animate the men who are continuing his work." This was certainly true for Lowell.

He returned to the South every summer after Darling's death, continuing to build his credentials in tropical medicine, always working on malaria, because that was the one tropical disease the United States still had plenty of.

In the summer of 1926 he tested his hand at "skeeter" killing, working under Louis Williams, a former Rockefeller disciple, now chief of a U.S. Public Health Service project charged with demonstrating antimosquito sanitation work for the U.S. Marines in Quantico, Virginia. Their two-seater de Havilland—which Lowell described as "a tin can with wings"—soared over malaria-infested marshes of the Chopawamsic and Quantico creeks, shooting a stream of Paris Green out the tail, just above the treetops. Lowell hand-cranked the dust over fifteen square miles. A few days later he inspected the ground, collecting dead larvae and dissecting them. The Paris Green particles in their guts proved that the larvae, which marines called "wigglers," had eaten the poison. The thrill of the kill, however, was short-lived. A little more than a week later, the algae were back and so were the larvae—just like in Georgia. Their work demonstrated to marine commanders that spraying every week to ten days could eliminate mosquitoes.

The project made front-page news in the *Washington Post* with headlines that read, "Mosquitoes Killed in Hordes by Poison from Marine Plane," and, "Quantico Camp Makes War on Malaria Pest."[1] The articles praised the green clouds trailing the planes and explained that "leathernecks" used to use oil slicks that killed larvae along with fish, ducks, and other animals.[2] "The science of mosquito slaughter has made great advances since Cuba and Panama were cleaned up," read one editorial. It said the marksmen using this "new medium of murder" killed "skeeters" with "appalling" accuracy. "The number of dead runs up into the millions and the campaign is to be continued every ten days until the last gallinipper disappears."[3]

The news articles never mentioned what went on beyond the Marine Corps area and the resorts, where people lived in shacks unable to afford the spraying—which reportedly cost only about $16 an acre but required airplanes, something the Quantico Marine Corps air base had for its own use, but not for the surrounding population. Eliminating malaria beyond the barracks was not what the Public Health Service was called in to think about. Williams, with his two assistants, one being Lowell, had a defined job to do, which they did well.

When the summer ended, Lowell moved to IU's Indianapolis campus for clinical studies. Louise Holland also relocated there to take a job writing advertisements for a department store. Lowell worked around the clock to cover tuition in a string of odd jobs that included waiting tables at a girls' boardinghouse, selling shoes on Saturdays near the railroad station, keeping attendance records for his medical professors, and preparing petri dishes for biology labs—which decades later he remembered as the time he missed an opportunity for greatness. He and probably most young scientists of the time worked hard to prevent an irritating mold that kept killing off bacteria in the dishes. A clever contemporary in London, Alexander Fleming, also worked with the mold, studied it, and ultimately called it penicillin.

But petri dishes were not Lowell's full-time job, and he didn't think to explore the bothersome mold. He spent most of his time doing rounds as an intern in the emergency room of the city's main hospital, where he mostly administered chloroform, which was the time's only anesthesia. Once, a grocery store clerk showed up with a nasty bite from a large, hairy spider that he said attacked him while he unpacked bananas from Honduras. The surgeon in charge, with no idea how treat the bite of a banana spider, deputized Lowell as the hospital's new tropical disease expert because of his summers in Georgia and told him to take care of it. Lowell also had no idea how to doctor a spider bite, and tried treating it with hot packs. Fortunately, the treatment did no harm, suggesting that the bite probably came from a standard tarantula (the scary but medically insignificant kind that showed up with surprising frequency in Indiana grocery stores) and not a real banana spider—a type of *Phoneutria* that can be deadly.

Lowell finally learned how to make more money in fewer hours by tapping a new state law that required all morticians to study biochemistry, anatomy, and other first-year medical courses—all of which Lowell had taught as part of his assistantships. He and a fellow medical student started a business charging $1 an hour for the instruction, earning them enough money to get through the rest of medical school. By that time, Lowell's sometimes girlfriend, Louise Holland, had moved to San Francisco, then to the Sierras to work as a park ranger.

Maybe Lowell was brokenhearted, but he didn't have much time to think about it.

CHAPTER 5
A Nobel Prize

Julius Wagner von Jauregg had led psychiatry to the light. He created a new paradigm with malaria fever therapy. Psychiatrists were no longer just superintendents of a facility filled with neurosyphilitics who would soon be dead. They had something to offer, which made them more empathic. This transformed institutionalized psychiatric care, says medical historian Joel Braslow.[1] So where neurosyphilis patients were seen by their caregivers as "hopeless," "stupid," "noisy," "untidy," "silly," "vulgar," "vile," "obscene," "lazy," "childish," and generally of loose "moral character" with a sure death sentence, doctors now saw them as victims of a germ and in need of medical attention.[2] Jauregg's treatments led psychiatrists to interact with patients more closely, because malaria therapy required constant monitoring. But more important, doctors now had a treatment that offered hope, and the hope they offered was to the hopeless—a truly satisfying endeavor.

Doctors actually *asked* patients whether they wanted to "take" malaria, as if it were a pill. Many psychiatrists believed that *not* "taking" malaria would lead to certain death.[3] Those in the early stages of syphilitic paralysis—still aware enough to understand their prognosis and the hell they would soon live—"voluntarily sought admission specifically for fever therapy, seeing the asylum as a place of cure rather than an institution of confinement," wrote Braslow. They begged to be treated with malaria.

Granted, they hated the brutal chills and fevers, but many expressed gratitude once it was over. Study after study produced results that matched Jauregg's, with an average of 27 percent full remission, a near equal percentage of incomplete remission, and just under 50 percent showing no improvement, or succumbing to death. These were European studies in the 1920s that encouraged psychiatric hospitals to offer the therapy.

Doctors in Great Britain in 1922 learned of Jauregg's malaria therapy and immediately put it into practice. At the time, roughly twenty thousand people in England and Wales died annually of "general paralysis of the insane"—a type of neurosyphilis.[4] The first British paretics saved by malaria therapy were treated in July at Whittingham Mental Hospital, Lancashire. The treatment quickly spread to another dozen mental hospitals and many private mental homes. Soon the Ministry of Health raised anopheline mosquitoes to infect people for treatment. There, different strains of *vivax* and *falciparum* were imported from India, West Africa, Sardinia, and Italy.[5] Inventory of cases after five years showed that out of nearly sixteen hundred patients treated with malaria, 34 percent died during or soon after; 25 percent were discharged as cured; and the rest remained as they were. By comparison, one hospital tracked 123 patients *not* treated with malaria and all but five died within two years of admission. British psychiatric authorities were so sold on the treatment that the Metropolitan Asylums Board created a special clinic at Horton Mental Hospital specifically for malaria therapy.[6]

Malaria therapy arrived in the United States that same year—1922. The famous psychoanalyst William A. White was the first to use it on his patients at the six-thousand-bed federally run St. Elizabeth's Hospital in Washington, DC, where he was superintendent. He aggressively sought treatments for the insane and jumped on this odd malaria therapy from across the Atlantic. In an interview with a reporter, he said he started it "shortly after I first heard of the malaria method through the Vienna medical press." Then, in the same interview, he went on to incorrectly explain that high fevers killed the syphilis-causing spirochetes.[7]

Jauregg's stony expression must have turned grim at the misleading information coming out of America, for he had done enough autopsies to know that the biological benefit was not derived from the heat of the fevers, but the innate immune response that resulted in fevers.

Nonetheless, White's enthusiasm for the therapy gave it legitimacy in America. In his first study of 103 neurosyphilis patients, sixty-six improved and "only" twelve died. Of a similar-size group that went untreated for the same period, seventy-nine died and only twelve improved. By 1927, St. Elizabeth's had a wait-list for malaria treatment.[8] Other institutions experienced the same kind of demand. They complained that they had a hard time keeping malarial microbes alive to induce infections. Malaria had so dried up around the big cities, especially in the Northern states, that it was hard to find. One medical reporter asserted that "health propaganda against the mosquito" made "good useful malaria cases" hard to find.[9]

So while Lowell Coggeshall and his colleagues at the Rockefeller Foundation racked their brains for ways to wipe out malaria, this new field of psychiatry complained about the disappearance of the disease.

Ever the researcher, Jauregg worked from Vienna to try to find ways to ship malaria-infected blood from laboratory to laboratory. He lamented that public health successes would make the disease in "more enlightened countries" harder and harder to find.[10] But keeping parasites alive in blood proved difficult.

Meanwhile, Dr. White at St. Elizabeth's created an insectary to keep alive a strain of *vivax* he had taken from one of his patients—which he called the St. Elizabeth strain. He cycled it through lab-raised mosquitoes and then had the mosquitoes bite new patients. This rapidly increased the number of patients he could treat and allowed him to collect more data. Consistently, about a third of them would eventually walk home to resume their lives, while roughly half died during the treatment or soon after—some from complications related to malaria, some from the syphilis, some from an underlying problem made deadly by the malaria or syphilis or both. In time, malaria in psychiatric circles became known as the "friendly fever" that saved "doomed derelicts" from their sinful, sexually acquired death sentence.[11]

IN 1924 the Nobel committee considered Julius Wagner-Jauregg for the prize in medicine—which upset the standard because he was a *psychiatrist*.[12] No one in his field had ever won the prize in medicine. But here he was, breaking the barrier by medically treating the mentally ill with malaria! This rattled the medical sensibilities of doctors outside the field of psychiatry. To them, malarial microbes endangered the lives of patients with terrible infections that could make a person comatose in a matter of hours, from which few ever recovered. Others experienced painful complications that turned their urine black and destroyed their kidneys, leading to eventual death. Malaria, the scourge of the planet, the enemy of healers dating back to Hippocrates, had no place in medicine as a treatment, they believed.

But by now Jauregg had given malaria to more than a thousand patients in various states of mental deterioration. He kept meticulous records and controlled for many known variables, impressing even his detractors. With solid data, he showed that syphilitics clearly benefited, but that malaria appeared to offer no relief to his other patients—the schizophrenics and manic-depressives and the like. He also differentiated between the different types of malaria, advocating for the use of only "malaria tertiana" (the old name for *vivax*). At medical conferences around Europe, he lectured against the use of "malaria tropica" (the old name for deadly *falciparum*). He gave detailed guidance to other asylum doctors, instructing them all to use infected blood—not mosquitoes—for their therapy. He warned that malaria transmitted by mosquitoes created full-blown infections that were harder to control with quinine and could lead to relentless relapses.[13]

This last precaution suggests that Jauregg understood more about malaria than many malariologists of the time. Without studying the malarial microbes per se, he surmised correctly that they were complex creatures that, if given their full course of infection, could remain alive, even in the face of quinine treatment. By using only infected blood, he never introduced the sporozoites—the stage that mosquitoes carried in their saliva, which after entering the human body shot straight for the liver, where each turned into another thirty thousand strong that later marched on red cells in regiments, infection after infection, sometimes for years to come. By only using transfused blood, he never infected his patients' livers—he used the blood stage that had already left someone else's liver and was borrowed just for a onetime fever attack. For this, his patients were better off. And so was the population at large, as his patients were not malaria carriers. After their acute attacks burned out, they had no more malaria lingering in their bodies and therefore were unlikely to trigger an outbreak once they left the hospital. Anyone using mosquitoes to transmit the disease could not make the same claim.

He was careful to keep his own well-tested strains going, including one he had taken from a soldier in 1919.[14] He lectured on the value of finding a good, easy-to-control strain to avoid the dangers of tapping an unknown and potentially unpredictable and uncontrollable strain from a naturally infected person—who may have tropical (*falciparum*) or some other form of malaria not as benign as tertiana (*vivax*).

On this advice he was ignored in places like England, where the preferred method for inducing fevers was by mosquito because the insects transmitted only malaria from the donor, not other blood-borne diseases like syphilis. This made a strong case for using mosquitoes, but it also was a sign of things to come. The fever-therapy concept had begun to take on a life of its own.

JAUREGG by now was giving malaria to virtually all neurosyphilitics admitted to his clinic. He separated them into subsets to tease out who benefited most, and produced convincing data that showed patients who suffered from the very early stages of neurological damage—exhibiting only minor slurring or shaky penmanship—could be fully cured 60 percent of the time.[15] From that, he wondered whether malaria might stop syphilis from entering the brain, if caught early enough. As he had only post-brain-invasion cases, he asked other doctors to test his hypothesis. A Dr. Josef Kyrle agreed, reporting widely that the treatment worked—it prevented general paresis!

Jauregg also showed that those in the final stages of general paresis benefited from malarial fevers only in that they could stop the degeneration. Malaria therapy could not reverse damage already done—the telltale gait, delusional megalomania, manic depression, euphoric outbursts, obscene acts, and routine seizures. In all his work, not once was he able to demonstrate that malaria therapy helped mental illness other than that which was caused by syphilis.

Over time, as others tested his therapy, enough experience with it led to standardized approaches. Patients underwent a test called the Wasserman reaction—which could read the presence of syphilis bacteria in the blood—and if it was positive, patients received malaria therapy, sometimes accompanied by a mercury-based medication. If after treatment the Wasserman remained positive, they would undergo a second round of treatments, and so on. About 30 percent of patients could be cleared of syphilis in this way. Another 20 percent benefited but remained infected. And the rest saw no change.

Science journals published hundreds of reports from places using Jauregg's therapy; it spread west and south through Europe and into Russia. It was called a "therapeutic noble deed" with success rates that "stifled most open criticism of the method."[16] The death rate caused from malaria alone—estimated at 10 to 20 percent—bothered a small minority of medical professionals. One was Dr. Bror Gadelius, a Nobel referee and Swedish professor of psychiatry. He felt that injecting a patient with malaria was "criminal" and refused to support Jauregg for the prize.[17] So the 1924 prize in medicine went to Willem Einthoven, for developing the first electrocardiogram. The following year, Gadelius was off the committee, but no prize in medicine was awarded. And in 1926 it went to Dr. Johannes A. G. Fibiger of Copenhagen for his experimental work producing cancer in the mouths and stomachs of rats by feeding them cockroaches infested with nematodes.[18]

The following year, 1927, was Jauregg's. One night in September a Berlin reporter woke him from a sound sleep around one a.m. to tell him he was the first psychiatrist ever to win the prize in medicine (only one other has won it since). His biographer related what happened next: "There was no question of sleep after that and he did what he often did when he could not sleep: he got up and played chess against himself."[19]

At the award ceremony on December 10, Professor W. Wernstedt, dean of the Royal Caroline Institute, told the audience that Jauregg had expelled one disease with another by fighting "evil with evil." In doing so, he delivered an "unusually blessed gift to mankind," one that allowed fathers in their prime to return to work and feed their hungry families. He congratulated the graying, statuesque Jauregg for bringing "wretches" who were doomed "back to life" for a second chance. "It is to such a one, who must be counted as one of the great discoverers and benefactors of mankind, that Alfred Nobel wished his prize to be awarded. . . . I have the great honour to invite you to step before the King, and, accompanied by the heartfelt good wishes of the Institute and the gratitude and admiration of thousands, to receive your prize from the hands of His Majesty."[20]

Three days later Jauregg gave his Nobel lecture, in which he told the story of his discovery, gave credit to others where appropriate, and ended with sage advice. Malaria therapy was malaria, he warned. It would cause fevers up to forty-two degrees Celsius (107.6 degrees Fahrenheit), which were hard on patients already weakened by paralysis, and could lead to death by heart failure. Malaria being malaria, he warned of how easily it escaped into the environment, making this type of therapy a potential source for outbreaks, which placed great responsibility on practitioners to use it safely and cautiously.

. . .

JAUREGG became president of the Austrian Eugenics Society and embraced new ideas coming out of Germany concerning sterilization of the mentally ill (a popular belief put into practice and law in many countries, including the United States). He did a survey of the thirty-four men who passed final examinations with him at school and found they had a total of thirty-four children—"half the total they needed to maintain their number," he complained. He pleaded for policies that would encourage the higher social classes to have large families, including financial incentives.[21] In 1928 he signed a manifesto, along with mayors, two former chancellors, judges, business leaders, and other professors, demanding union with Germany, as part of an Anschluss—reunification or reconnection of Germanic people advocated by the increasingly vocal Nazi party.[22] Historian Otto M. Marx wrote that Jauregg's beliefs—which included a popular perception that masturbation caused mental deterioration—had few consequences in Austria during the 1920s and 1930s, but that a paper he wrote and was published posthumously in 1941 carried many. In it Jauregg proclaimed that "the production of a most favorable genotype was the task of the physician and the legislator as racial hygienist. The elimination of the worst mutations causing congenital diseases had been initiated by the German race protection laws."[23] Marx said historians and psychiatrists must dig deeper into Jauregg's legacy to elucidate "late nineteenth- and early-twentieth-century psychiatry in German-speaking countries. For as Wagner-Jauregg died, psychiatrists who had been the students of his generation perpetrated and abetted the systematic murder of their own patients."[24]

Jauregg could not have fathomed the horrors perpetrated by Nazi eugenics programs. His views on favoring certain genotypes did not involve murdering patients. Instead, he remained focused on treating mental illness and began working on a cure for syphilis, which he believed infected a third of the "world's mentally unsound."[25] This kept him relevant in medical circles, allowing for his methods to spread. It allowed George H. Kirby, director of the New York State Psychiatric Institute, to secure an exemption from a state law that forbade doctors to infect people with harmful diseases. By the time of Jauregg's prize, Kirby had already treated several dozen patients at Manhattan State Hospital on Wards Island. He said the mortality rate among paretics there had dropped from near 100 percent to just 40 percent, with about a third being treated and cured—and about 20 percent or so dying during the malaria treatment.[26]

. . .

MAINSTREAM media called the therapy a major breakthrough—a godsend, an innovation, and a brilliant discovery by a compassionate man—and its acceptance spread to other medical fields. This blew the doors wide-open for malariologists to use it for their own purposes.

The British might have been the first to do so. They publicly established a major malaria therapy program near London at the Horton Mental Hospital.[27] England's Ministry of Health saw Jauregg's therapy "as a means for the study of malaria itself, a matter of great importance in view of the close and constant relations between this country and other parts of the Empire where malaria is endemic."[28] As such, the ministry hired liaisons from the London School of Hygiene & Tropical Medicine to work at Horton as fellows, paid for by the Royal Society.

Germans were also quick to use Jauregg's fever therapy to study malaria at the Kaiser Wilhelm Institute in Munich, at the nerve psychiatric clinic in Frankfurt, and at the state mental hospital in Düsseldorf—all in cooperation with Bayer's chemical factories in Leverkusen and research laboratories in Elberfeld.

The template had thus been established. Jauregg's innovative use of malaria therapy would soon subsume the study of malaria itself.

CHAPTER 6
Divided Loyalties

In 1927, Lowell began his last year of medical school.

He continued teaching courses and working as an anesthesiologist. Pumping chloroform into patients while surgeons operated proved instructive. If the person couldn't afford a hospital stay, surgery was done at home on, say, the kitchen table. On one occasion Lowell gave too much gas to a young victim of a car accident undergoing emergency surgery. Her heart stopped. After several attempts at CPR, Lowell got it going again. But the experience humbled him. Maybe even scared him. He internalized his handling of it and questioned whether he had become too arrogant, that perhaps he had more to learn. In any case, he felt certain he didn't belong in the ER.

Tropical disease studies, by comparison, were less random and more controlled, requiring long, slow investigations that turned into eradication efforts. Working the "tropics" was mostly about prevention and a lot of lab work—not fly-by-the-seat-of-your-pants emergency care. Rockefeller's public health lens made sense to him. Sanitation, improved water quality, mosquito control, and better standards of living—on their own—could dry up many, many disease threats. Without cutting people open, public health officers saved lives—by the hundreds and even thousands. Dr. Darling had completely transformed his vision, and he could see quite clearly through this lens.

When Lowell graduated in 1928, Rockefeller invited him to intern at one of its hospitals on a banana plantation in Honduras owned by the United Fruit Company—home to Indiana's stowaway banana tarantulas. But a severe storm blew the hospital down right before Lowell was to leave, so his first international adventure was canceled. As a consolation, Rockefeller offered him a medical internship at a new student hospital the foundation helped found at the University of Chicago. It wasn't tropical medicine but it was a job, so Lowell took it as a temporary position to hold him over until Rockefeller found him another overseas assignment.

The Billings Hospital, however, he ended up liking—a lot. It was different from those he'd worked at in Indiana, where the doctors' primary focus was on their own moneymaking practices. Billings's new approach raised the hackles of the medical establishment because its innovative policies paid doctors a salary—prohibiting income from private practice—and provided care to the community free of charge. The only catch was that each patient's illness was an open classroom, allowing medical students and doctors to learn from every infection, every organ failure, and every surgery. The program was controversial and, by some standards, had threatened to radicalize the field of medicine. The mavericks on the board were able to forge ahead because they had start-up money from Rockefeller and a healthy endowment from the Sears and Roebuck fortune. The hospital's one full-time doctor practiced Confucianism—a life of continuous self-improvement not for the self but for the good of the whole community.[1] Lowell, who enjoyed new ideas *and* controversy, fit right in.

He worked shoulder-to-shoulder with an influx of Chinese doctors who had arrived in Chicago from a Peking medical school also funded by Rockefeller. They treated pneumonia, removed gallbladders, and provided general health services to university students and the people of Chicago, many from the poorest sections of the city. Studying their live infections provided a great learning lab. It was a unique medical niche. Lowell hunkered down here, and started seeing himself as a Chicagoan—and a medical doctor. The tropical disease intoxication, still lingering from his time with Dr. Darling, started to dissipate.

A few months into his internship, Louise Holland visited her family in Indiana and Lowell asked her to come to Chicago. She'd by then left the Sierras for Alaska, where she flew into Fairbanks with bush pilots after hiking grizzly-bear country. Her independence was unusual—and intriguing. Lowell took her to dinner at The Blackhawk restaurant on North Wabash—favorite spot of the hottest dance orchestra in town: Carleton Coon and Joe Sanders's Kansas City Nighthawks. Perhaps it was Sanders's silky "What a Girl! What a Night!" or just the grand nine-piece jazz ensemble, but something moved Lowell. He proposed on the spot. To his surprise, she accepted. They were married soon after. The plan was set; he'd be a doctor in Chicago and she'd raise their children.

Louise went back to Alaska to wrap up her life there while Lowell spent the summer in the South on another malaria adventure. By now it was an annual event. The disease struck Southern states every summer, and he was one among a small group of scientists able to help—a dependable team member in Rockefeller's small stable of malaria experts.

DURING the summer of 1929, Lowell took a lead role in an antimalaria project, this time in South Carolina. Back he went to the squalor, heat, and humidity, back to the ravenous and relentless mosquitoes, and the thorny social and economic policies that made malaria such a big success. He left the North behind, and all the excesses of the late 1920s that were sustained in a fragile bubble of prosperity. Luxury apartments grew bigger and pearl necklaces hung longer. Flowing, drop-waist dresses of silk fell to T-bar shoes with buckles or bows. Women clacked two-inch heels along Chicago's paved streets toward fabulous clubs echoing with live jazz ensembles that played late into the night. They wore bobbed hair, lots of makeup, and cloche hats that fit snug to the skull. Men used hair tonic to form marcel waves and wore baggy trousers under double-breasted vests. The flapper generation of the roaring twenties obsessed over glamour and glitz. Their eyes stayed fixed on themselves, turned away from the ugliness of the South, where malaria and Jim Crow laws continued to rob millions of people—black and white—of life and liberty. These problems newspapers, for the most part, ignored. They were left to men like Lowell: hired guns paid to create small public-health release valves on the pressure cooker created by a national neglect of the South's general health and well-being.

Lowell's assignment that summer was to work for the South Carolina State Board of Health mapping malaria cases in an area soon to be underwater. The federal government and state authorities had approved a $22 million dam to be built by the Lexington Water Power Company, using clay and earth. It was to be the world's largest hydroelectric dam, an engineering spectacle that caught global attention. The state needed to measure the baseline for potential malaria outbreaks in the event residents, after the waters were impounded, tried to blame their fevers on the dam. This was the standard for legal protection, as established by William Gorgas and the Coosa River dam fourteen years earlier. But the new science showing that mosquitoes flew a mile or more for blood meant power companies had to understand just how entrenched the disease was in the population at large. That was where Lowell came in.

By the time he arrived, some two thousand black workers labored for low wages under the supervision of a few dozen white foremen and engineers. It wasn't yet malaria season, so Lowell and another doctor took care of the workers' general health needs, which piled up between Friday and Sunday nights, when many gambled, drank heavily, and got into fights. On one occasion a young black man named Moses came to Lowell with at least thirty pieces of buckshot blown into his backside. Lowell dug out the pellets with a surgical knife. But with only six extracted, Moses begged Lowell to stop; he needed a break from the agony. A few days later Lowell saw Moses and shouted, "Hey, you were supposed to come back. Does it hurt where you were shot?" To which Moses replied, "Only those ones that you dug out," and off he ran, never to return. On another occasion a family summoned Lowell to their farm, where he found a boy lying on a mule cart, suffering from a red-hot appendix. He quickly boiled sheets, covered an old table in their corncrib, and performed an emergency appendectomy.

When malaria season finally started, he and a fellow doctor drove a Model T up and down rutted dirt roads along the Saluda River, trying to convince impoverished and uneducated families, who knew their homes would soon be submerged by a dam they opposed, that it was in their best interest to consent to a blood smear. He just needed to prick a finger, squeeze a drop of blood onto a glass slide, and thank them for their time. The slides were to be sent back to Rockefeller's lab in nearby Columbia for examination. Once the range of chronic malaria was drawn out on a map, public health officers would come back with quinine to treat the infections.

Not surprisingly, Lowell didn't fully succeed in making his malaria map. He got cold stares as he approached the area's standard two-room shacks, elevated three feet above the spongy soil. If he wasn't chased away by hound dogs he was shouted down by angry residents. It didn't matter that the dam promised to bring electricity—their homes, churches, graveyards, and everything they knew were about to be underwater. No one trusted the power company and, by extension, Lowell. He secured only a few hundred blood slides from the basin's twenty thousand residents, not enough to make a complete survey, but enough to show that malaria was just about everywhere.

When the dam went up it created Lake Murray. A woodland valley of more than fifty thousand acres became a forty-one-by-fourteen-mile flood basin, with five hundred miles of fingering shoreline—perfect for mosquito breeding. This magnificent engineering feat displaced more than five thousand people. Landowners, mainly Dutch families who'd lived there since the 1700s, were bought out. They had little choice, as an act of Congress had created the policies and funding to build the dam. Black squatters were simply told to leave. When the locks closed, a wall 220 feet high and 1.5 miles long plugged up the Saluda River, and the valley filled up like a bathtub. Stone homes stood submerged into the twenty-first century, while shacks broke apart. The dead not counted among the 193 relocated graveyards remained at the lake's bottom. Three Dutch churches, six schools, and numerous suspension bridges disappeared under the crush of dark, debris-filled water.

This project fed into a system of dams that generated five billion kilowatt-hours annually, lighting homes and industries sprawled across 120,000 square miles of the South, involving a geometric escalation of mosquito breeding anywhere dam builders ignored the Public Health Service's antimosquito guidelines—clearing debris from the basin, drainage ditches to prevent pooling, stocking of larvae-eating fish, raising and lowering of water levels timed to kill larvae.

Lowell reported decades later that his involvement in this project, along with that of other Public Health Service officers and the Rockefeller Foundation, actually helped prevent another Coosa River. The malaria rates following the closing of the locks did not escalate out of control; maybe he and his colleagues had actually done something right. And he was correct; malaria rates were ever so slowly going down where people had electricity. No one can really say whether it was public health interventions or just having power that made the bigger difference. With power, people could run fans and air-conditioning, and industries could relocate to the South's cheaper labor force and better weather, which brought jobs.

THE following year, however, Black Tuesday hit. The world economy crashed, triggering economic panic and paranoia. Millionaires, after losing everything, took their own lives instead of adjusting to the new hardships. Over the next few years, a devastating drought turned America's breadbasket into a dust bowl. The jobless migrated to the cities. Blacks fled the South. People starved. And malaria crept back into places that hadn't seen it for years. Somewhat defeated and deflated, Rockefeller pulled out of the South. Its scientists felt that no manner of mosquito control could tackle malaria where people were too poor to screen their own homes, or contribute to a tax base large enough to sustain expensive control measures. The whole effort—which was seriously underfunded from the start, given the size of the problem—amounted to a "one-armed man" trying to "empty the Great Lakes with a spoon."[2]

LOWELL returned to Chicago certain he was through with the disease. He had a family to support and a career in medicine to advance. He took on some of Billings Hospital's administrative work, which he enjoyed. Being a manager of interns and physicians was fun. And he was good at it. He was good at fundraising and he was good at thinking through the challenges of running an experimental hospital that served the indigent. He helped manage a horrendous flu outbreak, financial troubles, and a relentless crush of poverty-related ailments from a city population suffering from severe joblessness.

CHAPTER 7
Germany and Magic Bullets

The 1920s were hard on Germany. Austerity measures to pay off war reparations plunged the middle classes into poverty. Inflation soared and the cost of living grew to twenty times what it had been before the Great War.[1] Housewives took up prostitution to feed their children. In Gelsenkirchen, angry women demanded food rations, and in Hattingen, miners' wives picked up rakes and picks against strikebreakers.[2] The government declared a state of emergency and banned political rallies, so people filled beer halls to meet and complain.

In this atmosphere, little-known conservative radical Adolf Hitler built a coalition of prominent Germans to overthrow the Weimar government in his now famous Beer Hall Putsch, which failed. At his treason trial in early 1924, Hitler greeted friends with assurance: "*Ach! Wir werden schon siegen.*" ("We shall win all right"). He also willingly confessed: "I intended to overthrow the government, but did it for the good of the Fatherland."[3] In the next five weeks he fomented a bewitching message of hope and hatred that artfully blamed everything going wrong in Germany on the Treaty of Versailles, the corrupt Weimar government, and Jewish bankers—advancing a blend of extremism previously confined to small audiences in Bavaria. The trial exposed a hurting population to his hate-filled, oratorical genius.

The international press called him "frothy" and filled with "empty phrases from the vocabulary of Nationalism, Anti-Semitism, and Fascism." The *Manchester Guardian* wrote: "It would be difficult to find a more glaring instance of the intoxicating power of phrases on the perfervid, histrionic demagogue than the career of Adolf Hitler."[4]

To close his defense, Hitler spoke of fantasies: "The army that we are building grows from day to day, from hour to hour. Right at this moment I have the proud hope that once the hour strikes these wild troops will merge into battalions, battalions into regiments, regiments into divisions. . . . You might just as well find us guilty a thousand times, but the goddess of the eternal court of history will smile and tear up the motions of the state's attorney and the judgment of this court: for she finds us not guilty."[5]

The state convicted him of treason and locked him away for nine months, during which he penned *Mein Kampf.* On December 20, 1924, he walked away from the fortress in Landsberg a free man. *The Manchester Guardian* called him a "charlatan,"[6] and *The New York Times* declared he had been "tamed" and reported the following: "It is believed he will retire to private life and return to Austria, the country of his birth."[7]

DURING Hitler's captivity, elections were called twice in Germany. In the spring, voters failed to produce a coalition government, which forced the president to dissolve the Reichstag (the German parliament). This inspired more than twenty-five parties to run in the fall elections, each to win one seat per sixty thousand votes earned (the total number of seats dependent on turnout). Executive committees of each party would then choose representatives to fill the seats they won. Newspapers called the more absurd parties "freaks." One, for example, was comprised of a husband-and-wife team who claimed to be anarchists—running for government. Another party was led by a man who claimed to be a prophet and made up his own religion to prove it.[8]

Most voters chose between the six major parties, which all broke down according to their views of the Versailles Treaty. The Socialists wanted to abide by it, enter the League of Nations, and support disarmament; the Democrats held similar foreign policy views and goals; the People's Party, while detesting the Versailles Treaty, understood the need to abide by it; the National Party promised to work to undo Versailles legally; and the National Socialist Freedom Movement of Greater Germany vindictively opposed Versailles, blamed its very existence on Jews and Marxists, promised to undo it by force, if necessary, and was "anti-everything" that had to do with the postwar German government.[9]

This was Hitler's party, also called the German Fascisti, or Nazis, and labeled the "racialists" for what American newspapers called a "futile hatred of Jews" and foreigners.[10] They were the extreme right, counterbalancing the extreme left of the Communist Party, which also opposed everything in Germany and was closely associated with communists in Moscow.[11]

Hitler expected his party to win seats after his rhetorical successes both during and after his treason trial. But it didn't. The National Socialists actually lost seats.

THE uptick in fervor over Hitler's actions that year probably didn't register much with executives of the dyestuff, pharmaceutical, and industrial chemical companies of Germany. They were forming a massive conglomerate made up of all companies along the Rhine—BASF, Hoechst, Agfa, Weiler Ter Meer, Griesheim, and the gem of them all, Bayer. They created economies of scale to help recover from the brutal terms of the Versailles Treaty. Many executives believed they *had* been stabbed in the back at Versailles and probably felt that this extremist Hitler got at least that part right.

Versailles negotiators treated these companies as instruments of Germany's aggression—the makers of munitions, weaponry, and poisonous gases. None of their assets, patents, and trademarks seized in other countries during hostilities were returned as part of the peace treaty. Corporate executives, who pleaded with treaty negotiators to return their intellectual property rights, felt they had little choice but to band together.

Folding these companies into the same management structure made sense. They had grown out of the same scientific discovery made in the mid-1800s by an apprentice chemist in London, who had been trying to make synthetic quinine from coal tar but instead created chemical messes of bright red and purple. These accidental dyes, as it turned out, adhered to silk far longer than any of the vegetable or animal dyes of the day. From this innovation grew a fabulously valuable industry, anchored in the principals of the periodic table. Chemically derived dyes thus drove the international trade in fabrics and garments. Huge market opportunities materialized at the turn of the century—in perfect timing with a newly formed German state. By then the Franco-Prussian War had ended and four kingdoms fused into a new German Empire (sometimes referred to as the Second Reich). Commerce, ingenuity, industry, and investment flourished along the Rhine. German nationalism and audacious entrepreneurial fervor fueled new start-up industries. And quick-acting management over highly skilled chemists allowed key German dye companies to crush or buy up competitors.

Then came the Great War and the Treaty of Versailles.

And from that came the idea for the conglomerate. It was loosely named the German Dyes Trust, formerly called I.G. Farbenindustrie, and nicknamed I.G. Farben—short for Interessen Gemeinschaft Farbenindustrie Aktiengesellschaft, or, in English, "Community of Interests of Dye-Making Companies, Inc." Fabric dyes were only one of their core moneymakers. Other products included pharmaceuticals, nitrogen-based fertilizers and explosives, and coal- and oil-based by-products.

A Versailles-related hangover for I.G. Farben involved a "quack remedy" company in the United States called Sterling Products. It owned patent and trademark rights to the Bayer Company's products, as well as Bayer's American production and distribution facilities, including a state-of-the-art factory in Rensselaer, New York. All had been seized by the U.S. Alien Property Custodian during the war and sold to Sterling in a 1919 auction for $3.5 million. Sterling held the rights to sell acetylsalicylic acid under the internationally recognized brand name Bayer Aspirin, as well as more than sixty other bestselling Bayer drugs—worth many millions in international market share.

Sterling pharmacists, however, lacked the know-how to make Bayer's products and were unable to read patents written in arcane chemical terms, in German. After three years of hard bargaining, Sterling cofounder William Weiss entered into a legally dodgy agreement in which Bayer and Sterling created a new company to make and market Bayer's products—sharing profits equally. This new business, called Winthrop Chemical Company, had the look and feel of an American business but operated under management of both Sterling and Bayer.

This setup allowed I.G. Farben to circumvent postwar restrictions and reenter lost markets. In return, Bayer's chemists taught the Americans how to make some—not all—intermediate chemicals for aspirin and other medicines and dyes. The remaining components had to be imported from Germany. Winthrop's role was sometimes as minor as packaging a product for sale in the United States. Sterling shared half of Winthrop's profits with the Germans and agreed to stay out of I.G. Farben's markets in South America and elsewhere overseas.

This allowed Sterling to profit off the German patents they had purchased but made Sterling's manufacturing capabilities dependent on I.G. Farben, according to a later Justice Department investigation.[12]

The relationship was as lopsided as the two sides' manufacturing capabilities. Sterling owned the Rensselaer plant, but Bayer had the Leverkusen factory in Germany, which was ten times bigger. "Huge barges came and went from Bayer's riverside wharves, and steam engines shunted between vast sheds and laboratories where dyes and medicines were made and tested, while thousands of workers and scientists and technicians moved purposively about," wrote historian Diarmuid Jeffreys. "Even the company's library was on a jaw-dropping scale, housing tens of thousands of books, journals, and papers on chemical procedures gathered from around the world."[13]

This Sterling-Bayer partnership, and the dependency it created, probably violated the spirit of the sales conditions set by the U.S. Custodian. Even worse, Sterling's agreement to stay out of Germany's overseas markets violated the spirit of U.S. antitrust laws and created what amounted to a pricing cartel. Sterling also kept secret the fact that Winthrop's chemists didn't know how to make some of the products they claimed to be producing—creating a false impression of their technological skills.[14]

Other U.S. companies that purchased German patents and trademarks from the U.S. Custodian, postwar, had similar issues and tried to strike similar deals with I.G. Farben companies, to no avail. DuPont of Wilmington, Delaware, for example, tried and failed. So the company smuggled Bayer's top chemical and dye experts onto a Dutch ship that steamed them across the Atlantic to interpret patents and help produce products—for salaries of $25,000 a year. The new Weimar government and the Bayer Company accused them of stealing a trunk full of documents, betrayal of commercial secrets, and breach of contract—and demanded they be extradited to face charges. Dr. Joseph Flachslander, one of the four to reach Ellis Island with a valid passport for entry, called the charges "absurd." He said his knowledge of Bayer products and dyes was earned by way of his own research and that he brought it all to the United States in his head, not in smuggled documents.[15]

The scandal reverberated through Germany and soon the German government began refusing passports to scientists.

. . .

AMID it all, Bayer's skilled chemists in the hilly riverside hamlet of Elberfeld hammered away at finding a quinine replacement. One of their most revered countrymen, Paul Ehrlich, had passed away in 1915. But his enthusiasm for magic-bullet remedies lived on and inspired Bayer's leaders. No prospect for greatness exceeded that of finding a magic bullet for malaria. It would give the company a new blockbuster product with intellectual properties uncompromised by Versailles.

Ehrlich left behind his old blue drug—methylene blue. It would never replace quinine, because of its toxicity, but it established a method for using dyes against malaria.[16] Bayer heavily exploited the idea shortly after Versailles negotiators gave Great Britain control of Germany's colonial holdings in East Africa, including the country's only cinchona plantations and, therefore, its sole source of quinine.

Company chemists started with what they already understood: that the blue dye molecule in methylene blue stuck to malarial parasites and killed the germs.[17] The chemists used the same amino groups from this blue drug but attached them to a central, quininelike molecule to create what they called aminoquinolines. The positions for the amino groups were numbered on a theoretical ball-and-stick model, and, after rotating them around this central molecule, Bayer's chemists found a stable spot at the eighth position to create a series of what they called 8-aminoquinolines. Then they carefully injected different variations into the wing muscles of canaries that had been artificially infected with a bird malaria called *P. gallinaceum*. Cages fell silent with dead songbirds as the different concoctions poisoned them. Technicians drew blood from survivors to see that small, nonlethal doses allowed the birds to live but that they remained infected by the *gallinaceum* parasites.

So the chemists just kept adding or taking away various atoms. "Add a carbon atom here and toxicity goes up, add a nitrogen there and activity [against malaria] goes up," explained historian Leo Slater.[18] Over the course of a few years, they moved the side chains around hundreds of times. Partner organizations within the I.G. Farben family manufactured different chemical intermediates, making a wide range of component parts quick and cheap to acquire.

The whole enterprise was tricky, as the chemists had to contend with stability. Some molecules bonded better than others; each new compound had a shelf life. Some broke down quickly, especially in tropical heat. Bayer's chemists continued playing around with the structures. Then, in the mid-1920s they struck gold, or so they thought, with a stable compound called plasmochin—which looked a lot like quinine ($Cl_9H_{29}ON_3$ versus $C_{20}H_{24}O_2N_2$, respectively). When they used it on canaries infected with *P. gallinaceum*, the birds actually recovered!

Plasmochin next went to the primate house to be used against monkey malaria, and again it seemed to work. So Bayer, in 1925, sent it to Dr. Franz E. Sioli at the nearby Düsseldorf asylum, who had been using Jauregg's methods to keep a single, safe strain of *vivax* cycling through patients. Sioli had signed an agreement to test Bayer's drugs against his patients' malaria infections whenever the company came up with something worth testing.

Sioli gave plasmochin to sixty of his syphilis patients in different doses, ranging from 0.05 to 0.25 grams. In all sixty, the parasites disappeared from the blood, but those on the higher dose turned blue—not from Ehrlich's blue dye, because it had been removed, but from oxygen deprivation. The higher doses robbed blood cells of oxygen, causing a deadly condition called cyanosis. Those on the lower doses, however, were fine. And Sioli reported that the best results could be achieved on a daily dose of 0.1 gram—taken in five 0.02-gram tablets spread throughout the day.[19]

Sioli's work proved only one thing, however: that this drug worked well against a small number of patients infected with a lab-maintained strain of malaria.

With that feedback, Bayer took the next step. In 1925, plasmochin was tried on small groups of people living in malarious regions of Italy and Spain.[20] Tests were done on children and infants with no untoward effects and appeared to reduce the size of their malaria-inflated spleens. It seemed safe in small to moderate doses, and it saved a few patients suffering from blackwater fever, which by then had been associated with overuse of quinine. This encouraged great enthusiasm for the new drug.

There was more. The drug killed off the mature parasites circulating in the blood—the sexual stage that changes a person's blood chemistry to attract mosquitoes. These parasites disappeared from the blood—all of them—within a week of taking plasmochin.[21] If large studies sustained these results, Bayer would have its magic bullet. Even better, they would have a drug able to sanitize a person's blood of the only stage of these parasites able to escape man and infect mosquitoes. This would eliminate man as a carrier! This, if true, would mean the end of malaria—or at least the human forms of it. So even if Bayer never found a drug to stop transmission from mosquito to man, one that stopped transmission from man to mosquito would break the cycle—malaria would dry up in one season. The excitement must have been palpable, though science journals reporting on the findings showed none of it, as was the standard.

Clearly, this drug inspired hope. But whether it was the magic bullet to end malaria forever remained to be seen.

More tests were needed, so Bayer's collaborators used the drug on several thousand laborers working at a United Fruit Company plantation in Latin America. Severe toxic side effects emerged in people of African descent or anyone with blood deficiencies.

British scientists also tested plasmochin at Horton, where malariologists had won government approval to infect syphilis patients with mosquitoes instead of blood (against Jauregg's strictures). Then they went even further, exposing patients to hundreds of infectious bites in a single day to mimic conditions in intensely malarious places. These patients weren't getting malaria therapy; they were guinea pigs. They were used as surrogate petri dishes, as this London hospital also maintained—in the blood of patients—seven strains of malaria imported from India, West Africa, Sardinia, and Rome. Using them violated Jauregg's standards for care by introducing exotic malaria, including *falciparum*, to patients already weak and sick from syphilis. The research doctors did it because it allowed them to run a range of tests using plasmochin against the very strains of malaria that were interfering with their colonial enterprises. In doing so, they learned that plasmochin and quinine performed differently on these different strains of *vivax* and *falciparum*. Furthermore, when exposed to hundreds of infected mosquitoes, the dose of plasmochin needed to protect against symptoms was too toxic to make it useful.[22]

Plasmochin, as it turned out, was far from a magic bullet.

BUT Bayer, for the first time, successfully moved different atoms and molecules around a base substance to make over twelve thousand different compounds.[23] Chemists for the company now possessed skills unmatched by any other drug company in the world. And the chemists there elevated their work to an art. They moved molecules around in different ways, learning how to make derivatives and new compounds from component parts of other semisuccessful drugs.

In the early 1930s they made a new 8-aminoquinoline using a yellow acridine dye. To stabilize it, they tossed in a chlorine molecule opposite the amino side chain. When they injected this intense yellow liquid into canaries sick with *P. gallinaceum*, it worked—the canaries recovered. So they sent it to Dr. Sioli at the Düsseldorf asylum, where it worked again on syphilitics. They turned yellow, but not from jaundice, from a buildup of the dye in their tissue that went away after the treatment ended. More studies followed showing that this extremely bitter substance could quell symptoms, maybe even as well as quinine.

This was exciting. Medical researchers had another drug to test.

On January 7, 1932, Dr. W. Schulemann from Bayer's Elberfeld laboratories spoke at the British Royal Society of Medicine. He described what he and his team had achieved on the drug front. The discussion led the group into a downward spiral. While comparing notes, they realized that the four life phases of malarial microbes were so distinct that any drug working against one failed against the others. Drug makers were fighting four separate monsters per infection. Plasmochin did little against the blood-destroying stage—which triggered the intense symptoms—while Bayer's new yellow drug appeared to be quite useful against that stage but not against the sexual stage that plasmochin worked against. Nothing they had worked against both. And nothing seemed able to kill the sticklike microbes in mosquito saliva or the tissue stage that would lie dormant, ready to launch relapses.[24]

This conference helped illuminate the complexities of malaria, and suppressed talk of finding a single magic bullet against it.

IN 1930, I.G. Farben completed construction of what was thought to be the largest office building in the world.[25] The sprawling structure smiled down on west Frankfurt with a gleaming marble facade set in a templelike entrance with a portico and large bronze doors. A mammoth spine of a building anchored six wings that jutted into open parkland. Staff used paternoster elevators, stepping on and off the small boxy compartments as they ran nonstop along a conveyor belt moving vertically between floors. This internationally recognized architectural accomplishment became I.G. Farben's symbol of success.

In 1932, the dyes trust showed off its robust demand for products by adding more than five thousand new employees to its 120,000-strong workforce, stationed at more than a hundred plants. They produced all German-made dyes, a vast majority of the country's nitrogen and explosives, nearly half its pharmaceuticals, and a third of its rayon. The trust also diversified to include industrial glues, intermediate chemicals, pesticides, photo supplies, plastics, and a dozen other products.[26]

But then the trust invested many millions of marks in developing synthetic fuels, which failed to produce products able to compete with crude oil. By 1934, business reports openly questioned the trust's value on the stock exchange.[27] Business journalists had picked up on a dire financial problem that grew worse under the Third Reich.

THE Weimar Republic died a slow death throughout the 1920s and early 1930s. Disputes among the many parties led to vicious partisan attacks, including open hostilities that broke into fistfights on the streets and shop floors. As the economy suffered, Hitler reappeared as a changed man who sought legitimacy through elections. He promised security, national unity, a strong military, and a new Germany—restored to its rightful place as a world power. Voters rewarded him with near majorities, which pressured the president to appoint him chancellor. From that post, he pushed through legislation that gave his cabinet full control over the Reichstag and executive branch. And with that, he purged all opposition parties. A year later, he dissolved the presidency and consolidated all power into the chancellorship, making himself führer.

Hitler's brand of nationalism offered the kind of economic protectionism and favoritism that could save I.G. from bankruptcy. But Hitler openly questioned I.G.'s loyalty, pointing out its international ties and, more to the point, its many Jewish executives and chief scientists. Key managers, including Carl Duisberg of Bayer, asked Nazi party members within the I.G. family to convince Hitler of the trust's commitment to the Fatherland and anti-Semitism. I.G. companies donated heavily to the Nazi party and fired their highly skilled Jewish scientists and managers—setting the collective back decades in technical skills. And they openly defended Hitler in the face of news reports that accused him of committing atrocities.[28] By 1933 Hitler approved of I.G. Farben and created internal markets for products the trust could no longer sell abroad.

BAYER'S chemists, meanwhile, continued plugging away at the malaria problem—only now with heavy Nazi pressure to produce products that served the insatiable needs of a rapidly growing military. Hitler made health care a security matter, as healthy soldiers won battles. For them, Bayer produced synthetic morphine called methadone (today's treatments for heroin addiction) and several so-called sulfa drugs that could kill bacteria on contact and cure gangrene and other secondary infections in wounded soldiers—making huge contributions to battlefield medical care.

Bayer also had their new yellow drug atabrine for malaria, which in 1933 made its international debut on the commercial markets. It sold under different names, including mepacrine, erion, and quinacrine. Winthrop Chemical Company manufactured it for the American markets, but Bayer wouldn't allow Winthrop to submit the formula to the American Medical Association for review. Meanwhile, German and British trials showed excellent results with almost no untoward side effects, so the League of Nations' Malaria Commission—which included German malariologist Claus Schilling—endorsed the drug as an alternative to quinine.

Even Rockefeller's public health board—which advocated antimosquito work over drug treatment as a prophylactic for malaria—backed the drug, in part because the foundation's own Dr. Barber (maker of Paris Green) used it on himself to cure an attack of malaria and found it to be superior to quinine. Shortly thereafter he coined the phrase, "Better be yellow than sorry."[29] From 1933 to 1936, the United States purchased more than half a million dollars' worth of atabrine—which represented a fraction of what was spent on quinine, but a sturdy amount nonetheless. The drug showed promise and seemed to have a future.

No one but the executives at Winthrop's parent company, Sterling Products, knew Winthrop imported most of the component parts from Germany. Nor did the U.S. government know that Bayer, for all intents and purposes, owned half of Winthrop, and that Nazi ad agents controlled the company's media strategy. As the world boycotted German products to protest Hitler's bloody political purges and intimidation campaigns, Bayer remained unaffected. Winthrop served as a front company through which Bayer products were sold to American consumers—a setup that would soon attract inquiries from U.S. Justice Department lawyers and the FBI.

BY 1937, problems with atabrine began to surface and legal suits alleged that the drug caused serious side effects. Among them were severe gastritis, hallucinations, and psychosis. The League of Nations' Malaria Commission promptly withdrew its support and warned that because "the effective dose is close to the non-tolerable dose . . . [and] may cause epigastric pains and nervous or mental symptoms . . . atabrine should be used only under medical supervision."[30]

Quinine was not without its problems. Its side effects, by many standards, were bad, including blurred vision and loud ringing in the ears. But it "had centuries of tradition, legend and use behind it," wrote historian Leo Slater. It had been so broadly prescribed—without lasting, life-threatening consequences—that in many countries it was added to water to make tonic and used in salts. With quinine still abundantly available and sufficiently cheap for the middle classes, medical associations refused to cross over in support of atabrine—a sentiment fanned by negative ads propagated by the Dutch-run Quinine Trust. Bayer had little hope of convincing doctors to prescribe it, which meant atabrine had no chance at commercial success. Even worse, serious questions had been raised about its safety, particularly when used for extended periods—as battle would require. Thus, for war preparations, quinine remained the only drug worth stockpiling.

ATABRINE'S shortcomings became Bayer's headache. Hitler wanted company chemists to find something better. And senior managers at I.G.—many of whom pragmatically joined the Nazi party—pressed Bayer chemists to keep working. In the meantime, Bayer commissioned expert German malariologists to study atabrine for military use. Those experts discovered that in low doses the side effects could be reduced and malaria symptoms could be suppressed enough to keep soldiers on their feet and fighting.[31] This meant the Germans now had a secret of sorts. As the world assumed atabrine was too toxic to use, Germans had figured out how to make it work in the event of war.

Malaria rates around the world shadowed each country's economic status. Northern Europe and the Northern American states had rid themselves of the disease in part because the cooler climates made it easier to do, but also because people in these regions did, indeed, live better lives—they built well-constructed homes on lands free of malarial mosquitoes. This dynamic fed a larger bias: that if a place was able to control malaria, its people were stronger and smarter. Superior. Better bred. Educated. The rubric for comparison fit well with wealthier countries' ideals around eugenics and racial purity.

Benito Mussolini felt the North's cold stare of superiority and was ashamed that malaria still plagued his country. Turn-of-the-century mosquito controls had broken down during the Great War, restoring the Pontine Marshes near Rome. These legendary "Marshes of Hell" occupied more than three hundred square miles south and east of the ancient city. Some two thousand severely impoverished malaria-racked residents lived there. The Red Cross estimated that 80 percent of those entering the marshes left infected with malaria because of gigantic and relentless mosquito populations. A summertime trip across the only road, built by Romans in 312 B.C., almost guaranteed a malaria infection. The eerily beautiful landscape was seen by other nations as a symbol of the "deficiencies of the Italian race," according to historian Frank Snowden.[32]

At the First International Malaria Congress in Rome, October 3 through 6, 1925, Mussolini gave the opening remarks, then turned over the podium to experts from countries with colonial holdings. They spent most of the day comparing malaria studies conducted in tropical places where healthy laborers were needed to extract and process natural resources—for the profit of imperial governments and industries. But the final session turned to increasing rates of malaria in Europe, which the British, German, and French blamed on endemic malaria that spread northward from Southern Europe.[33] This must have felt to Mussolini like a sharp jab, as Italians were cultured and erudite, but also infectious and, therefore, looked down upon by leaders with scarcely any homegrown malaria.

To Mussolini, malaria blighted his status as a world leader. So, in 1929, he launched a malaria-eradication campaign that transformed the Pontine Marshes—called the bonifications. This involved draining swamps, prepping reclaimed lands for cultivation, and building screened dwellings for families to move into. Tens of thousands of needed jobs were created, but many laborers died of malaria during the backbreaking work. Future populations prospered, as more than sixty thousand peasants moved into the former marshlands. They received electricity, livestock, grains, and fertilizers, and were loaned money to be paid back once their farms were productive. Mussolini famously rode his red motorcycle in leather cap and goggles through the region's fertile, reclaimed lands, visiting young lovers and lavishing attention on his great success against malaria.

There, the government built roads, hung telephone lines, planted trees, and, most important, established health stations with enough quinine for everyone.

This was the face of the Fascists' so-called rural revolution. The reality was that land reclamation projects started decades earlier and run by socialist peasant associations, mostly in the southern provinces, were stripped of leaders and brutally reprogrammed under the Fascists. Il Duce—as Mussolini liked to be called—took credit for everything ever done to reduce malaria. He reigned victorious.

Few knew that his health ministry also conducted large clinical trials aimed at relieving Italian dependence on Dutch quinine. Some two thousand workers on land reclamation projects in Tuscany and Apulia were intentionally exposed to *falciparum* malaria. Half of them received no treatment so doctors could observe the ravages and record outcomes. The other group received mercury therapy, a treatment well documented as useless against malaria and harmful in its own right. But faithful party doctors assured Il Duce that the treatment worked.[34] As the world inched toward another world war, the Fascists believed they were ready on the malaria front.

CHAPTER 8
Eradication

Lowell Coggeshall—by now Cogg to his friends and Dr. Cogg to his patients—left the Murray Lake dam project to settle back into life at the University of Chicago, where he was swept into a project with Dr. O. H. Robertson, an accomplished and famous doctor who had created the first blood bank during the Great War. Robertson arrived at Chicago to work on a pneumonia vaccine designed to stop pneumococcal infections from turning healthy, soft lungs into rigid, oxygen-trapping casements. These lung infections led to death in nearly half of all cases and were often complicated or facilitated by influenza. In 1936, these two diseases were the leading cause of death in the United States (and today remain among the top-ten killers). Working on such an important disease that touched virtually everyone, not just impoverished Southerners, gave Lowell a needed break from tropical medicine.

Lowell by now also earned a decent living, enough for him and Louise to raise a family. Richard came first. Then Carol. Then Diane. They didn't see their dad much.[1] He worked long hours treating patients, getting involved in the administration of the hospital and the medical school, teaching, and writing papers from the pneumonia studies—at least a dozen of which were published in prominent medical journals. He also developed vaccine-making skills and ran animal studies with pneumococcal germs of differing virulence. And he helped Robertson develop blood serums with antibodies to try to boost the human immune system. He later described this as exciting work because of what he learned and the potential for making a truly remarkable impact on public health, should he and Robertson succeed.

But malaria refused to let him go. In 1935, he was called back to duty, this time for the Tennessee Valley Authority, which had been created in 1933 as part of President Roosevelt's radical New Deal. This "corporation clothed with the power of government but possessed of the flexibility and initiative of a private enterprise" had a far-reaching mandate that included control of the Tennessee River.[2] The TVA became a vehicle to create hydroelectric power—which meant building dams to produce electricity—and restore overfarmed and deforested lands for use in large agricultural ventures. The federal government partnered the Southern states of the Tennessee Valley with private enterprises under a corporate cloak with a goal of ending poverty and malaria.

Like with Hitler and Mussolini, Roosevelt embedded malaria in his social agenda, but under very different circumstances. Roosevelt felt universal health care was a key to securing domestic well-being, not something to be withheld as an advantage granted to a privileged race or loyal partisans. This new securitization paradigm boosted the government's assertion of power in areas of public health. Southern legislators cried foul, calling it an attempt by Roosevelt to supersede states' rights. And they were dead-on. Roosevelt had lost the congressional battle to create socialized medicine as part of the Social Security Act; TVA was just one way of building executive powers to offset that loss. By creating a new security-based prism for matters of public health, coordination of malaria eradication became a federal goal. For that, the U.S. Public Health Service needed a stronger mandate. So he moved it out of the Treasury Department, where it had lingered as an odd-fitting stepchild, and into the Federal Security Agency. This setup later spawned departmental precursors to today's Department of Health and Human Services, Centers for Disease Control and Prevention, and National Institutes of Health.

TVA became the vehicle for tackling malaria, which struck 30 percent of Tennessee Valley residents, making it the number one cause of missed workdays. To get rid of it would be to transform the South from "a belt of sickness, misery, and unnecessary death"—where more than two million annual cases of malaria reduced the region's industrial output by a third, at an estimated annual cost of nearly $40 million—to a place of prosperity. TVA recruited Lowell to convince reluctant state officials to cooperate with the federal program so that places like Swampville in Leesburg could be restored to healthy forest. Lowell used his Southern contacts, which he had accumulated since his days with

Dr. Darling, to convince health departments that eliminating malaria served everyone's best interest.

TVA programs created a new trajectory for malaria. The kind of economic reforms needed to break the cycle of infection finally arrived. The 211,000 men hired to dig thirty-three miles of ditches and clear 544,000 acres of anopheline breeding areas earned a livable wage.[3] Their backbreaking efforts drained more than two million acres of swamplands, destroying prime mosquito habitat, while their earnings paid for better homes with screens and health care services with medicines. Many workers contracted malaria while digging ditches, and a short-lived spike in malaria rates occurred.[4] But they fell again. Electricity and ceiling fans repelled mosquitoes and made life more comfortable, which attracted businesses and entrepreneurs.

Unpopular dams displaced tens of thousands of poor families by drowning many thousands of acres of land and property under heavy black impounded lakes. But where pre-TVA dams often imposed two waves of destruction—dark, still waters that drowned valleys and tore people from their homes and livelihoods, followed by outbreaks of malaria for everyone who remained—dams under TVA's watch went up the right way. They were built according to antimosquito strategies the U.S. Public Health Service had promoted since the 1910s: clearing brush from flood basins, digging ditches to prevent pooling, stocking the right kind of fish to eat larvae, and changing the water level to kill larvae. They weren't optional or easily misapplied, as with the Coosa River dam. They were requirements for companies working with TVA money—which were all of them.

The basic case reproduction rate—first envisioned by Ronald Ross as the formula for predicting malaria outbreaks based on the right combination of mosquitoes and unprotected people—plummeted. The combination of fewer mosquitoes and fewer human carriers (because they sought treatment and lived in screened homes) made the disease unsustainable throughout TVA's territory. There simply weren't enough infected people and mosquitoes to keep the parasites cycling through these two hosts. Malaria retreated into the most impoverished places, where people still lived in shacks, exposed to night-feeding anopheles. National infection rates that had spiked in 1932 following the market crash fell again by 1936. The last reservoirs of the disease would presumably dry up as the economic revolution spread across the South—which is exactly what happened in the following decade.

Historians differ over which of these factors played lead roles versus supporting roles in the elimination of most malaria in the United States. A combination of forces came together to stop malaria from retarding the South's economic potential. Regional and racial inequalities responsible for holding the South to the health and development standards of the nineteenth century were improving. As were the hundreds of Southern counties with no hospital, doctor, or even public health nurse—where pregnancies and births went unattended, and where no one treated the endemic malaria complicated by malnutrition, hookworm, and pellagra. All of these horrific health trends were now reversed.

Malaria wasn't gone from the United States, yet. But it was drying up.

LOWELL left Georgia thinking he was finally finished with malaria.

But then one afternoon in 1936, the director of Rockefeller's International Health Board, the revered Dr. Frederick Russell, offered Lowell a job. Russell was very different from the kind and generous Paul Russell who had worked for Dr. Darling in Leesburg. Frederick Russell ran the Rockefeller health division with a loud voice and bold management. Few turned him down. But for Lowell, leaving Robertson's unfinished vaccine work to go back to the tricky and sometimes unforgiving business of malaria wasn't an obvious next move. So Russell made a convincing offer. This imposing former army brigadier general—credited with creating the first typhoid vaccine and clearing anopheline mosquitoes, and therefore malaria, from Palestine—told Lowell he would have a better mission. Russell wanted Lowell to work on a malaria vaccine.

Much had changed since Lowell left Rockefeller a half decade earlier. Fred Russell's leadership swept in new notions about public health, including the need for a hard-science lab to complement the foundation's field division—where, for decades, public health officers trained community leaders in the medical virtues of sanitation, like the use of toilets and shoes to fight fecal-borne hookworm, and insecticides and larvicides to fight malaria. The health division also built clinics and hospitals to treat and, more important, study people with a broad range of tropical diseases—which served as learning labs in the wild. The New York lab, by contrast, used state-of-the-art controlled environments to conduct hard-core medical research in the same building that housed the world-class Rockefeller Institute, where much of the country's cutting-edge medical technologies were developed and used.

On malaria, Russell had loftier goals than just spilling Paris Green into waterways, or showing local sanitation officers how to install screens, or strong-arming power companies into doing antimosquito work. Russell had convinced the elders on Rockefeller's board that the public health division should work on a malaria vaccine.

This was a bold move. Scientists barely understood the life stages of these odd microbes. Nonetheless, Russell wanted Lowell to find a way to cock the guns of our immune system before mosquitoes spit the microbes into our flesh. Success would surely mean a Nobel Prize—and a rocket ship to the top of Lowell's field. The money, which he always considered, was good enough to make it worth his while. And the assignment intrigued him.

So Lowell packed up his family and moved to Westchester County. He brought with him techniques he learned from Robertson. And within a short time he showed Russell that the presence of malaria did indeed trigger antibodies. What that meant, exactly, would take time to figure out, and ultimately disappoint everyone.

In the 1930s, fewer than a hundred people in the United States understood malarial microbes on a first-name basis. Rockefeller had twenty-six malaria experts on staff, Lowell among them. The Public Health Service had another few dozen. And that was about it. They alone could talk at any length about these microcreatures called *Plasmodia* that grew from the sticklike mosquito stage to the liver stage to the blood stage and finally the sexual stage. And they were forever frustrated at how impossible these germs were to maintain. No one, still, had figured out how to keep them alive in petri dishes. While other microbes—molds, bacteria, helminthes, and viruses—grew nicely in suitable media, malaria remained elusive. They died very soon after being withdrawn from the body in blood, making them very hard to study. Malarial parasites were just stubborn that way: extremely resilient and robust in the wild, but impossible to culture and cultivate in a lab.

So malariology—this intellectually esoteric and isolated subset of medical research—grew more and more dependent on Jauregg's malaria therapy. And everyone doing the work grew totally dependent on psych wards for live infections.

By 1936, at least three American institutions opened their doors to malariologists: Florida State Hospital in Tallahassee, where Rockefeller ran a malaria research station; the South Carolina State Hospital in Columbia, where the U.S.

Public Health Service ran a related station; St. Elizabeth's in DC, where the army in conjunction with Dr. White ran a separate malaria station. These institutions had perfected techniques for keeping malaria germs cycling by moving them back and forth between lab-raised mosquitoes and new patients, generation after generation. Patients were malaria vessels; their veins caged several species of human malaria and two monkey malarias that could survive in people (*P. knowlesi*, the one that today appears to be jumping species to infect humans at an alarming rate, and *P. inui*, the one I describe in the prologue that had escaped from Bill Collins in his lab at CDC). Scientists drew blood to run tests. When the germs died a few hours later, examiners went back to the patients and drew more blood.

These three state hospitals, in effect, were the country's malaria growers.

JAUREGG'S bushy eyebrows and thick mustache had gone gray by the time Lowell learned his techniques. With permission from the chief psychiatrist at Manhattan State Hospital on Wards Island, just over New York's famed Hell Gate Bridge, he opened a research station. Many of the germs he used came from Rockefeller's Tallahassee unit, which was run by Dr. Mark Boyd. He used the standardized St. Elizabeth strain of *vivax* drawn from White's patient at the DC psychiatric hospital many years before—now used around the country to run malaria therapy. It responded well to quinine but still packed a power punch, including extremely high fevers, bone-breaking chills, exploding headaches, nausea, vomiting, sweats, and delirium, again and again and again, every other day until the infection was treated or burned out.

Syphilitics on Wards Island were mixed in with the lifelong mentally ill. They were almost indistinguishable—but for the signature gait. If they weren't suitable for malaria therapy, they simply received the same treatments as the others and displayed the same blur of harmless, monotonous, and often nutty behaviors, pierced by occasional loud rants or violent acts, until death arrived. Lowell carefully chose men and women for therapy—as Jauregg's work instructed—including only those in the earliest stages of neurological degeneration, healthy enough to potentially benefit from fever therapy, and strong enough to survive high fevers.

Lowell proceeded in his usual kind manner with keen focus. He accepted all the presumed rules: Do what you need to do and cause as little harm as possible. The last part, of course, was subjective. Malaria, by definition, caused harm. But it was also the treatment—one evil to kill another.

His operation was small, with fewer than a dozen beds, where he doted on patients through rounds of chills and fevers. Just as they soaked through sheets for the fourth or fifth time, he pricked their fingers for blood and studied the infections under a microscope. He also pressed jars of mosquitoes against their bare flesh to allow the insects to gorge on blood. The process generated new life in two ways. Mosquitoes used the blood meal to produce eggs, which they released in shallow pans filled with nutrient-rich water. Their larvae grew into pupae, and when they took flight as young mosquitoes, they were caught in netting, put in cages, and placed on shelves for future feedings. The microbes also were given new life as mosquitoes sucked in the sexual stage from infected patients to procreate and fill the mosquitoes' saliva with their offspring, ready to be injected into patients during the next biting session.

He had an average of eight infected patients at any given time. No available papers discussed whether the fever therapy changed any of their lives, as that was not Lowell's goal. But reports out of Manhattan State Hospital continued to show that patients undergoing malaria therapy benefited about 30 percent of the time.

Lowell used his lab to apply techniques Robertson had taught him for pneumonia immunizations—injecting live or partially killed microbes as passive immunization. He detected the immune system developing a defense against the initial infection. But the antibodies were highly specific against that type of malaria for that point in time.

First he proved this in rhesus monkeys infected with two standardized strains of monkey malarias: *P. knowlesi* and *P. inui*. To test his theory in humans, he gave these monkey malarias to patients to see whether they generated an immune response that could protect against his lab-raised St. Elizabeth strain. They could not. Defenses he observed were short-lived and specific to each strain—that is, none offered broad immunity against the others. The preparations helped induce antibodies but did not prevent malaria. This portended the main difficulty in developing a workable vaccine—one that today's vaccine makers have spent billions to overcome without complete success.[5]

"I made several original contributions," Lowell modestly told a television interviewer years later. He came away from the experience believing there would never be a vaccine. "One could have malaria all his life, be cured, and six months later could be reinfected, which showed the disease itself doesn't provide immunity," he said.

. . .

ABOUT the time Lowell returned to Rockefeller, German chemists at the Bayer Company made a clever move. Still hammering away at the malaria problem, they reexamined the toxic drug plasmochin from the 1920s—their so-called 8-aminoquinoline that appeared active against the sexual stage of malaria—and moved the amino side chain from the theoretical eighth position to the fourth position. Then they experimented with the other essential pieces, rotating them around to establish stability and minimize toxicity. In a bold move that seemed obvious only after the fact but somehow hard to fathom before, they removed the yellow dye to create a colorless new series of compounds they called 4-aminoquinolines. The first samples from this series went on to kill *P. gallinaceum* in Bayer's canaries—the birds recovered! Clearly Bayer's scientists were on the cusp of finding something truly remarkable.

CHAPTER 9
Claus Schilling

In March 1937 the First International Conference on Fever Therapy met for three days in New York City at Columbia University and the Waldorf Astoria Hotel. Present were psychiatrists and health ministry representatives from twenty-eight countries. Also represented were the U.S. Army and Navy, the U.S. Public Health Service, and the New York City departments of health and hospitals. They shared ideas about this "physical therapy" they called "hyperpyrexia"—or "pyretotherapy"—and bowed to its founding father, Julius Wagner-Jauregg, by making him honorary chairman, even though, at age eighty, he was unable to attend.[1] *The New York Times* called Jauregg's fever therapy "one of the most spectacular discoveries of modern medicine" and noted the value of these dignitaries exchanging ideas on patients who were once no more than the walking dead.[2]

The *Times* also noted, however, that Germany's delegation had to cancel at the last minute for "prevailing circumstances."[3] Their absence was noted and probably expected, as passports for German scientists were hard to get, especially for those not within the Nazi inner circle. Inside Germany, paranoia among intellectuals flourished as arrests without warrants or provocation increased.

Many landed in the Dachau concentration camp, run by the SS (Schutzstaffel, or Protection Squads). Established in 1933 at an old munitions factory adjacent to the Landsberg prison—where Hitler wrote *Mein Kampf*—Dachau was the first in what would soon be thousands of concentration camps (Konzentrationslager, KL or KZ for short). All SS guards received training at Dachau, where they learned how to use terror and brutality to control large prison populations.

In the early years, Dachau held communists, social democrats, trade unionists, political opponents, intellectuals, homosexuals, so-called Gypsies, hardened criminals, so-called asocials, "impure" Germans, and Jews trapped by the Nuremberg Laws (the codification of Hitler's anti-Semitic racial-purity beliefs). Its existence scared the hell out of average Germans—and silenced them, as designed.

By the invasion of Poland, Dachau held more than five thousand prisoners. Most did hard labor in munitions factories. The camp also served as a processing place for Jews who were then transported, in cattle cars, to other "special" camps, as they were built.

Many Germans accepted and even saw the benefit to laws that stripped Jews of homes, businesses, properties, bank accounts, citizenship, and basic rights to human decency. Others, however, were horrified by it. But because new laws criminalized anti-Nazi thoughts and actions, opposition forces went deep underground. Everyone else conformed. This government of ignorant thugs turned decent Germans into quiet conspirators—swept into Hitler's madness.

Within malaria studies, the road to the dark side is traceable.

BEFORE Hitler, malariologists throughout Europe and the United States all took up malaria therapy under dubious ethical circumstances. They all knew they were taking advantage of the mentally ill to serve their own research needs, but most tried to remain within Jauregg's framework for helping patients, who were sure to die from brain damage without the malaria treatment. So these malariologists, understanding that they had to work within the parameters of psychiatry, gingerly did their work. They infected patients, allowing the fevers to burn long enough to work against the syphilis, all the while studying the microbes and testing new drugs. This mutually beneficial setup turned out well for the 30 percent or so of patients cured by the fevers. And tropical disease institutes often covered at least some of the cost of running malaria therapy, which meant thousands of patients received treatment *because* of this dual setup.

When malaria experts came together at international conferences to discuss outcomes and share techniques, they spoke openly about the best way to protect patients. In this way, they were answerable to one another. They spoke the same scientific language, even if their mother tongues were different. They acted like a family, always bickering and questioning one another's findings and practices, while working from the same central assumptions and toward the same goals—to beat malaria. Fever therapy had transformed the field and was declared the "new malariology" at the Third International Malaria Congress in Amsterdam in 1938.[4]

Lowell stood among this new generation of malariologists. He and his colleagues learned from their elders; they watched many proclamations of malaria's demise be proven wrong by the sheer resilience of the disease. They were more careful about their hopes and expectations than their scientific forebears—Ronald Ross, Samuel Darling, Battista Grassi, and the like, all of whom believed the identification of malaria's route for transmission would lead to complete control of the disease. These elders helped illuminate just how complex and tricky malarial germs were. Published research findings became all the more important, because the tiniest of building blocks in the understanding of these microbes was needed if they, as a group, were to ever mastermind a cure against it. They all used malaria therapy and syphilis patients but kept processes in check by comparing notes and ensuring that, to the extent possible, the overall goal in using malaria was to treat patients first.

But as Hitler isolated German scientists from the rest of the world, he also tied them to political and military motives. Heinrich Himmler, the Reichsführer in charge of the SS and Gestapo, and mastermind of the brutal political purges immediately following Hitler's rise to power, made the rules for the health sciences. According to historian Paul Weindling, "Himmler favored human experiments."[5] He supported using prisoners instead of lab animals, especially as war made good lab specimens, like monkeys and rabbits, hard to acquire. Under Himmler's supervision, his SS camp managers set aside buildings for medical work and oversaw huge transport trains loaded with soon-to-be prisoner guinea pigs, shipped from camp to camp, depending on the need.

The Nazi grant-giving process, likewise, favored human-centered experiments, particularly those that sought treatments or new technologies for soldiers. They were deemed "war relevant" and far more likely to receive funding.[6] Under these circumstances, medical researchers felt compelled to design their studies accordingly, and some approached the SS "opportunistically." Claus Schilling, who was not a Nazi party member, was among them.[7]

. . .

SCHILLING had grown old by the time Hitler took control of Germany. Born in 1871, he had become a medical doctor about the time Ronald Ross won the Nobel Prize. By the time William Gorgas and Samuel Darling used Ross's discoveries to wipe malaria from the Panama Canal Zone, Schilling was at the Robert Koch Institute, Germany's most prestigious research lab. He shared the halls with the great Paul Ehrlich. But while Ehrlich shouted at his assistants—always in a cheery, "let's keep at it" sort of way, with scientific papers stuffed in his lab coat pockets and laboratory shelves askew with bright bottles of dyes and "be-penciled and dog-eared chemical journals"—Schilling kept to himself, and was tidier, subdued, and less interested in the potential of his young assistants.[8]

There, at Koch, Schilling was expected to do great things in the world of malariology—make Germany proud. His American contemporaries included Samuel Darling, Lewis Hackett, Louis Williams, and other Rockefeller greats. These men in the 1910s and 1920s explored their scientific art in tropical places to earn reputations. Like Darling, Schilling traveled globally and grew to be his country's chief tropical disease expert—so noticeable and notable that his mere presence at medical conferences merited a mention in medical news journals, even if he presented no paper and said nothing. He gained notoriety by using an immunization model for tackling tropical diseases.

He perfected his skills in Africa while trying to protect cattle from the deadly tsetse fly, which looked like a common housefly but carried a parasite so lethal it prevented colonial interests from penetrating vast swaths of sub-Saharan Africa. In man, the flies spread African sleeping sickness with 80 percent mortality rates, and in livestock it spread nagana with equally high death rates. Both diseases came from the same genus of microbe that, to this day, prevents farmers from utilizing some of Africa's most fertile lands. Schilling was among the first generation of scientists to live in Africa and try to stop this fly from killing so many. He and Robert Koch tried to immunize cattle against it by first running the parasites through lower animals—dogs and rats—which they found would reduce virulence. They then injected the less lethal parasites into cattle to induce immunity.[9] They reported success, but others and even they had a hard time reproducing their results.

While in Africa, Schilling made notable observations on hookworm and the dreaded Guinea worm—a microbe found in dirty water that, once ingested, grew three feet long—with a treatment strategy that centered on grabbing its head as it poked out of an oozing sore and slowly twisting it spaghettilike around a stick, until it was completely out. These and other vile diseases of the tropics caused tremendous losses in productivity of African men, women, and children forced into hard labor or conscripted into fighting armies. Every European country with significant colonial or commercial holdings invested in research to stop these diseases from ruining workers and soldiers. From this one common interest grew a dynamic first-world cohort of investigators funded by their respective governments.

But malaria researchers reigned supreme, because their disease caused the heaviest burden, was the most widespread, and still occupied some parts of the developed world. So perhaps it was natural for Schilling to eventually focus much of his energies on malaria. His work earned him an invitation from Ronald Ross to write a chapter on the problem of malaria in Germany's African "possessions" for a world-famous book, *The Prevention of Malaria*, published in 1910 by E. P. Dutton & Co. in New York.

Schilling worked temporally alongside Samuel Darling—although the motivations behind Rockefeller's work were more altruistic, less about colonialism and more about preventing disease for the sake of preventing disease. Like Darling, Schilling observed and reported on distinctions between species of anopheles. He devised control measures for Tanganyika and Togo, Germany's chief possessions in East and West Africa. But mostly he focused on the parasites. He wanted to find a way to sterilize everyone's blood of malaria, not poison the environment to kill mosquitoes. Within the malariology family, this aligned Schilling more with the Italians than the Americans, and more with the British than the French.

In 1926, a year after Darling careened off the cliff near Beirut, the League of Nations appointed Schilling to its esteemed Malaria Commission.[10] He received financial support from several countries to build up his malaria lab at the Koch Institute, including a grant from the Rockefeller Foundation. He published many reports and wrote a few books, all on attacking malarial parasites either by chemical assault in the blood using new compounds out of the Bayer Company, or by trying to induce immunity. He made many observations that, in the 1920s and early 1930s, significantly contributed to a global effort to understand the disease.

He also did his time in the asylums. In the 1920s, Schilling went to work in

Italy with Professor Alberto Missiroli—who had become Italy's most famous malariologist after the 1925 death of Professor Giovanni Battista Grassi. Schilling set up a lab in the San Niccolò asylum in Siena, where he abandoned all pretense of trying to treat patients with Jauregg's therapy, and even began using nonsyphilitics in his experiments.[11] He felt he needed to break from the norm because his vaccine work required it. He used the blood stage of several different types of *Plasmodia* to see if one could induce immunity against the other stages and the other species of malarial parasites—this included using *falciparum*, which Jauregg begged doctors *not* to use. Schilling, like Lowell Coggeshall, detected the antibodies; he knew he could induce immunity but not in a significant way. One stage did not confer protection against the other stages; one parasite species did not confer protection against the other parasites; and even one strain did not confer protection against the other strains. Stopping the later sexual stage responsible for infecting mosquitoes and spreading malaria to the surrounding community remained as elusive as ever. Whatever defense his patients gained from exposure to his vaccines—blood containing antibodies and/or parasites—had little to no practical value. Most people visiting or living in malarious areas were exposed to multiple strains simultaneously. A vaccine against one would do little to stop an infection.

Nonetheless, Schilling repeated his experiments many times, trying to find a way to induce immunity to give some level of protection. He produced reports that contained valuable information on his techniques, published widely, and made worthwhile contributions.

HISTORIANS damn Schilling for being among the first to depart from Jauregg's fever therapy standards largely because of what he did once he moved to the Dachau concentration camp. To be fair, in the 1930s, this move away from Jauregg's strictures was becoming the norm. It was necessary for anyone interested in keeping up with the science, and turned malaria therapy into an "extremely double-edged affair."[12] Doctors witnessed malaria-induced heart damage, swollen spleens, severe anemia, and jaundiced livers in autopsies of the dead.

The unintended consequence turned an "ample supply of paretic patients . . . into malaria-guinea-pigs," wrote historian Marion Hulverscheidt.[13] It created an obvious and well-documented systematic exploitation of mental patients that went uncontested by the broader medical community. The complicity seemed driven by this increasingly convenient understanding that the work, on the whole, was mutually beneficial. The potential benefit to humanity, should the work result in a magic-bullet cure for malaria, far outweighed any downsides to giving different forms of malaria to dying people.

The more malaria therapy was practiced, the more acceptable it became, and with that came more opportunities to use these patients in all kinds of malaria-related experiments, using all the parasites that cause the disease, so that a whole range of drugs and vaccines could be tested gainst a broad spectrum of infections. Even infected primate blood was used to try to induce immunity in man.

The published reports were written in the esoteric language of malariology. Scientists published hundreds of articles in medical journals that described to the rest of the malaria community new techniques used in infecting patients or trapping parasites or measuring drug activity. Anyone wanting to prove their worth and get credit for time in mental institutions had to publish their data and wait for a response from the community. So Schilling infected Italian mental patients and collected data and published reports, just like everyone else.

BUT as the 1930s rolled by, and Schilling's age wore on his body, his younger colleagues noticed and wondered whether it also wore on his mind. Skinny and frail, with sharp features, graying hair, and a pointed goatee, he looked more like an old bird than an intrepid researcher. When he hit sixty-five years old, he was forced to retire from Koch.

This left him adrift as a consultant in search of places to continue his work. He wrote to the Reich's research committee (or RFR, for Reichsforschungsrat) to request money and got it, which he used to run studies at state hospitals in Wittenau and Herzberge. Growing discouraged with the quality of research he could do on syphilitics, he convinced the head of the RFR, who was a medical professor, to find him medical students to experiment on, and ended up with eight volunteers who gave consent. On them he ran the same experiments (using the blood stage of malaria to try to induce immunity against other stages of the disease) and still got nowhere, so he tried something new. In a report published in June of 1938, entitled "*Die Methoden der experimentellen Chemotherapie*," he explained his methods for infecting the students, which included using saliva that he had painstakingly milked from the noses of infected mosquitoes.[14]

Researchers later questioned the value of his work, noting that all he did was prove that parasite-filled saliva caused infections—something that the knighted Sir Ronald Ross had proven long ago.[15]

Schilling's detractors failed to understand his work. He had made an astute observation: only a few parasites in saliva were needed to transmit malaria, but without the saliva, a very large number was needed. Schilling also noticed that mosquitoes delivered malarial parasites to the skin, not blood vessels, identifying an important part of the infection process. These insights were lost in time and rediscovered only a few years ago as important features of malaria transmission.[16]

Schilling looked deeply at the processes used by parasites to establish long-term infections—how they reappeared to tame antibodies and create a "labile balance" that protected the host from deadly infections while allowing the parasites to feed on hemoglobin and then as a route back into mosquitoes for procreation. He noticed that a person removed from exposure for six months, however, became like a malaria virgin, highly susceptible to the disease's most dangerous symptoms. From this he concluded that people could be vaccinated against severe malaria by giving them controlled injections of parasites, allowing the germs to peak and trigger symptoms, then administering quinine to quiet the fevers. Over time, the patient would become naturally immune. In tracking his findings, he discovered all kinds of caveats. The most significant was that the body's immune response differed for each stage of the parasite. He was able to show, however, that if he injected a person repeatedly over six months with the stick-shaped stage found in saliva, that person when bitten by infectious mosquitoes didn't develop a fever for more than three months (versus two weeks).

With more research, he believed he could *prevent* infections using this method. If the parasites were stopped at this first stage in their life cycle, the rest would never develop; malaria would cease to exist. This line of inquiry was innovative and intriguing.

Schilling applied for more funding to go back to Italy and collaborate with highly regarded experts there. In his letter to the RFR, he wrote: "If experiments conducted in Italy should lead to any applicable results, all German citizens traveling to malaria-infested areas" would benefit. . . . "And should Germany again have colonies, the results of my experiments will be of immediate use. . . ."[17] He had no trouble getting permission to travel, as he was a frequent recipient of Reich financing and a German citizen in good standing with the Nazi party.

. . .

BY 1941, he wrote to the RFR for money.[18] By now, all research requests were run through the German Research Society—the DFG, for Deutsche Forschungsgemeinschaft, which, under Himmler, gave top priority to human experiments. This time Schilling wanted to skip the mental hospitals. He proposed a military-related study, but not one that involved soldiers. Other scientists, including his successor at Koch, Gerhard Rose, had won funding for that—which involved treating *and* studying some thirty-three thousand soldiers felled by malaria while occupying Greece. But military leaders had grown weary of the studies that used troops. So Schilling proposed the use of prisoners, who were accumulating at a rapid rate in concentration camps all over Germany. He claimed to have fully immunized two patients at a state hospital in Volterra by combining live parasites with quinine.[19] "This was cutting-edge stuff; we're going back to these kinds of ideas now," said malaria researcher Wolfgang Leitner.[20]

The RFR asked Rose to review the proposal. He was unequivocal; he told the RFR to deny the funding because he felt Schilling was wedded to "outmoded methods."[21] Schilling showed an unwillingness to update his knowledge through the accomplishments of younger experts. He remained rigid in his scope while younger malariologists, like Rose in Germany and Lowell Coggeshall in America, learned from the previous generations' overly enthusiastic proclamations of malaria's impending demise. Every decade someone in the research community claimed the world was on the verge of eradicating malaria, and no one ever came close. These younger scientists grew into world experts with more respect for what they were up against, knowing that stopping malaria would require a lot of research and a good bit of luck. They curbed their expectations and instead set their sights on basic research that they knew was needed if anyone hoped to one day create a real magic bullet, or series of bullets, against malaria.

Rose submitted his assessment of Schilling to Reichsgesundheitsführer Leonardo Conti—the Reich's health leader. Rose said Schilling's work lacked "war-relevance" and that the German government should not spend resources on it, given other higher priorities in malaria research. "It is undeniable that such a decision may be considered an unnecessary hardship for Prof. Schilling," Rose wrote with regret. Then he asked that he not be asked again "to deliver such a difficult evaluation" of his former teacher and boss.[22]

The committee apparently ignored Rose.

The way Schilling described it three years later at a war crimes hearing, he was at home late one night in the fall of 1941 when a car arrived to take him to the train station. There he met Conti—who would later hang himself instead of face charges at Nuremberg. The two took an all-night train to Heinrich Himmler's office. And in the morning Himmler "ordered" him to set up a lab at Dachau and use the endless supply of bodies there to work out the kinks in his malaria vaccine.

In January 1942, Schilling began his malaria project.

CHAPTER 10
A New Plan

American malariologists also launched a malaria project, but on a different trajectory. It began in April 1939, when Lowell Coggeshall attended his first meeting of a special committee created by the National Research Council to examine the problem of malaria and war.

The Nazis had invaded and occupied Austria and Czechoslovakia. This was just five months from when Germany would send troops, planes, and tanks to slaughter Poles and encircle Warsaw—and Great Britain and France would finally declare war on Germany.

The NRC's job was to anticipate research needs of the War Department, should the United States be dragged into hostilities in Europe. A chief concern on the health care front was vicious endemic malaria in the Pacific, Asian, and Mediterranean regions—all areas mapped out by war planners as likely battle zones. The fevers would attack and disarm occupying troops indiscriminately. A magic-bullet cure would be great. But a good prophylactic would do, as it would allow troops to sit out engagements in a defensive posture and just wait for the enemy to fall back with fever. Then U.S. forces would go on attack in a much stronger position, resulting in overall fewer battle casualties and no malaria-related troop attrition. As far as anyone knew, no government had developed a true prophylactic against malaria, not even Germany.

The Allies at least had Holland and the Dutch East Indies' vast cinchona plantations, which would provide enough quinine to mitigate the intensity of infections. But quinine stopped working when used for too long or in regions with abundant infectious mosquitoes—conditions war created. A drug that could stop infections completely would be as potent a weapon as any in the military's arsenal, and would most certainly shorten the war in favor of the Allies, maybe even by years—which would reduce military and civilian casualties by millions.

But first it had to be developed. The question of how such a drug could be made was brought up at the April 1939 NRC meeting.

For the first time, Lowell found himself on a project with chemists, which was exactly where he wanted to be. His work at Rockefeller proved to him that a malaria vaccine—a preparation with proteins derived from the pathogen—would not come anytime soon, and certainly not in time for war. He steered the NRC from a broad discussion of finding a wide range of chemically derived compounds to treat diseases, which they called chemotherapy, to finding a chemical prophylactic for malaria. This would be a pill troops took daily or weekly to block malaria transmission while they slept under the stars unprotected from thirsty malarial mosquitoes. At the meeting were renowned chemists, including Marston T. Bogert, professor of organic chemistry at Columbia University; Lyndon F. Small, chief chemist at the U.S. Public Health Service; Herbert R. Moody, director of the chemical laboratories at the College of the City of New York and of chemical technology at the NRC; Leonard H. Cretcher of the Mellon Institute of Industrial Research; and Torald H. Sollmann, dean of Western Reserve University School of Medicine.[1]

From this initial meeting grew the NRC-sponsored Committee on Chemotherapy. Lowell and the others invited additional experts—parasitologists, entomologists, hematologists, and medical researchers. For support and funding, they recruited high-end sponsors, including Surgeon General Thomas Parran, who chaired a symposium on malaria in May 1940, and the American Association for the Advancement of Science (AAAS), which was about to run another symposium coming up in December 1940. The group received endorsements and promises of support from the War Department, the U.S. Department of Commerce, the Tennessee Valley Authority, the Pasteur Institute, Harvard, and the American Drug Manufacturers Association, to name only a few.

Lowell undertook the Herculean task of surveying top university labs and commercial pharmaceutical and chemical companies to gauge their willingness to participate.[2] He also recruited other malariologists to bring to the group a deep knowledge of the disease. Others compiled bibliographies, reviews, and digests of all work done thus far to develop antimalaria chemotherapies. The landscape was pretty stark, and almost entirely originating in Germany and centering on Bayer's two synthetic quinines: plasmochin and atabrine.

In August 1940, Lowell and his colleagues wrote a piece for *Science* magazine—the prestigious journal of the AAAS—calling for a national campaign with a media strategy designed to encourage research investments in drug development, specifically for malaria. They argued for a strong national program that put "chemistry in the service of medicine" so that all citizens would benefit. U.S. chemical companies, which were quite small, would grow in size as they filled druggists' shelves with new treatments for everything from the flu to cancer.[3] But anyone watching world events knew the impetus really was the war, and that the United States would soon be in it and in need of battlefield remedies, especially for malaria.

Committee members also canvassed colleagues and spoke and wrote about efforts to bring more experts on board. And they all pressed the War Department, Congress, and the Roosevelt administration to provide needed funds. "Speed is urgently needed," Lowell and his colleague wrote in *Science*, "for the medical profession confidently predicts a malaria peak this year or next."[4] They were, of course, talking about American troops fighting battles across Great Britain's vast colonial empire—most of which lay in tropical malaria zones.

They all, especially Lowell, assumed the Roosevelt administration would open the funding spigot. How could it *not* back the plan? Top administrators were on board. Key leaders in the army and navy medical corps were on board. The NRC was recommending it. And it was visionary, innovative, solution based, and science driven.

But it was also ahead of its time. The big money didn't materialize. And the group remained isolated with only limited resources.

Nonetheless, Lowell and his NRC colleagues forged ahead using Rockefeller's fever therapy lab at the state hospital in Tallahassee; the Public Service's lab at the South Carolina State Hospital in Columbia; and smaller affiliated labs at North Carolina University at Chapel Hill and Emory University in Atlanta. To synthesize new compounds they recruited the chemical labs at Parke-Davis and Company in Detroit; Eli Lilly and Company in Indianapolis; Merck & Co. in Rahway, New Jersey; E. R. Squibb and Sons, in New Brunswick; and, of course, Winthrop Chemical Company in Rensselaer, New York.

. . .

LOWELL forged relationships with these burgeoning commercial companies, which made what they could and imported the rest. From Winthrop, for example, Lowell obtained German-made samples of Bayer's two antimalarials, the odd yellow drug atabrine and its oxygen-choker cousin, plasmochin.

DuPont gave him samples of sulfa-based compounds, which in 1939 had won German pathologist and chemist Dr. Gerhard Domagk a Nobel Prize. Domagk was a university professor and brilliant researcher Bayer had hired in the late 1920s to find drug remedies amid I.G. Farben's huge catalog of already made chemical compounds. In that short time he took off the shelf an old substance from 1908 and turned it into a magic bullet that cured his mice and rabbits of deadly bacterial infections. He gave the formula to Bayer's chief chemist, Heinrich Hörlein, who recognized the sulfa-based compound as something for which Bayer already held a patent that was more than twenty years old—which meant the patent was worthless. What happened next was described by award-winning science journalist Milton Silverman in his book *Magic in a Bottle*. According to Silverman, Hörlein had two company chemists camouflage this simple sulfa compound with hard-to-make molecules from a red tar-coal derivative. He then patented *that* compound, called it Prontosil, and sent it out for clinical testing—which showed a near 100 percent cure rate against a broad spectrum of bacteria, including the sepsis-causing streptococci germs that killed a million and half Americans and Europeans every year. French chemists deciphered Hörlein's "trick." They broke Prontosil into its component parts and tested each against bacteria, quickly seeing that the sulfa compound—a simple mix of aminos, benzene, and sulfonamide (called sulfanilamide)—did all the work. They ran a patent check and discovered that Hörlein himself filed for it in 1909. What a treat for the French. They spoiled Hörlein's deception and robbed Bayer of a potential blockbuster drug. With an expired patent, this amazing antibacterial belonged to the world, license-free![5]

Lowell thought that if sulfa worked so well against really bad bacteria—like the hideous flesh-destroying rods of gangrene and deadly skin- and blood-poisoning spheres of streptococci, staphylococci, and the other coccis—maybe it would work against malaria. Maybe, just maybe, sulfa-based drugs could be made to kill *all* types of malaria, and every stage of this shape-shifting germ—from the sticks delivered in mosquito saliva, to the rings that destroyed red blood cells, to the wrigglers that sexually reproduced in the mosquito's gut. If any of the above panned out, sulfa would make treating and preventing the disease much, much easier. If all of the above panned out, the world would have a magic bullet for malaria—a miracle.

Also intriguing to Lowell was this new antibacterial the British brought to the United States for testing, called penicillin. He was among the first, and may have been *the* first, malariologist to hypothesize that this microbe killer might be broad enough to destroy malaria's *Plasmodia*. The medication was still under development and hard to obtain. Nonetheless, Lowell requested 6.0 grams from Merck & Co. He was told that amount was "unprocurable," but, with the help of A. N. Richards and the White House, Lowell finally got 2.0 grams of it in May 1942. Under heavy secrecy he saw, anecdotally, that it worked well against sporozoites (the stage transmitted by mosquitoes).[6] But further studies would have to wait.

For the time being, he and the others had to learn more about malaria and decided to use their hundreds of state hospital patients for basic research. The infections would help researchers understand why and how the microbes responded to drugs, and why and how they seemingly disappeared so quickly after being delivered by a mosquito—completely undetectable—until they launched their attack on red cells a week or ten days later. What were they doing and why did they do it? Lowell and his colleagues presumed the germs hid in soft tissue—but where? And after an acute attack, where did the parasites disappear to, and why would they suddenly, without warning or provocation, reappear to launch another acute attack? Presumably the initial attack created antibodies against the invaders, yet the body seemed to have no power against the second, third, fourth, and so on, rounds of attack. Why?

Lowell felt that these questions, if answered, might lead the way to a whole new class of treatments, and *away* from the toxic standards—quinine and Germany's quinine look-alikes, the 8-aminoquinolines. The madmen they worked on, however, weren't ideal. The induced infections to treat syphilis produced only normal concentrations of microbes that were difficult to capture in a blood draw. When caught, there were usually only a few swimming around for study. The culprit was the spleen, the body's filter that caught and killed germs as the heart pumped blood through it. Malarial parasites caused the spleen to swell, which gave doctors a simple diagnostic tool for gauging a community's malaria infection rate. For researchers, however, this organ was a hindrance, because it caught and killed a majority of germ specimens.

A clever Indian doctor, B. M. Das Gupta, came up with an answer. He removed the spleen from a monkey, then infected the animal with *P. inui*, creating a rapid buildup of parasites in the animal's blood and intense concentrations of *P. inui* for study. The concentrations were so high that he could even use it to launch malaria in man—something no one had been able to do, because small amounts of these microbes were insufficient to cause human infection.[7]

Das Gupta inspired Lowell. Scientists, including him, had already used *P. knowlesi* to infect man, finding it wasn't as deadly as, say, *falciparum*. The *P. inui* innovation added great value to researchers, because this type of malaria appeared to resemble *vivax*. Scientists now had two close cousins of human malaria, both of which could be used to study the potential for a drug or be used in vaccine preparations—maybe *P. inui* would be the one to induce immunity against them all.

As it turned out, it wasn't and couldn't. Still, here was another animal malaria for screenings—which allowed work to progress without state hospital patients.

THIS shift in focus brought immediate satisfaction. Lowell tried the Germans' sulfa-based compound on rhesus monkeys infected with *P. knowlesi* and *P. inui*, and cured them completely! There was no need to run toxicity tests in lower animals, because this sulfa compound had been rigorously tested as an antibacterial and showed no sign of untoward effects—with the exception of what appeared to be an allergic reaction to sulfa in a small percentage of the population. Otherwise it was completely safe. He tried it on his syphilis patients at Manhattan State Hospital, but their *vivax* infections were unchanged. His chemist friends were moving molecules around to make analogues of sulfanilamide, which he hoped would work better.

For him, sulfa presented the highest potential, in terms of drug development for military use—much better than atabrine or plasmochin, because sulfa clearly was *not* toxic, and it was much easier to make. The other two required medical supervision, which wouldn't work on the battlefield. A safe sulfa drug could be included in medical kits, along with sulfa powders every man would soon carry to stop war wounds from going septic.

But sulfa was just a possibility. Lowell and the others wanted it and other possibilities fully explored. In the big picture, making compounds for the sake of making them would give American chemists needed practice in drug making, and potentially propel them *and* U.S. medicine into the twentieth century. American drug companies were decades behind the British and French, and what felt like light-years behind the crafty chemists at I.G. Farben. In any case, better drugs were desperately needed. Atabrine, plasmochin, and quinine were poisons. They attacked the malaria germs while poisoning the body at levels that were somewhat tolerable. Their value was in their ability to stop infections from turning fatal by reducing parasite counts in the blood. But studies suggested that they were unsafe to take for long periods—which, if true, meant they were no good as a prophylactic. Each was known for triggering severe side effects: atabrine and its intense stomach and intestinal pains, as well as psychosis; plasmochin and its potentially deadly depletion of oxygen in the blood; and quinine and its cardiac disturbances, partial blindness, severe headaches, loud ringing in the ears, confusion, and, when overused, dreaded blackwater fever.

For the United States to find its own new drug would be a godsend. Though funding for a full-blown, federally sponsored malaria project failed to materialize right away, the need was clearly there and the outbreaks were only a matter of months away, depending on the timing of America's entry into the war. This was a medical Manhattan Project that sought to build on basic principles devised by the Germans, but perfected and improved upon by way of American innovation and sheer scientific will. Lowell got support from his University of Chicago friends and former bosses, especially William Hay Taliaferro, dean of the Division of Biological Sciences and the School of Medicine, and editor of the *Journal of Infectious Diseases*.[8] Taliaferro had been elected to the National Academy of Sciences and the American Philosophical Society. He brought great weight to the effort and helped recruit other giants in their fields.

With Lowell, Taliaferro pressed the point that a better lab animal would make things a lot easier—like a mouse, but one that could be infected with malaria.

Lowell already knew from other scientists that dogs, cats, rats, and many other animals could not be infected with malaria (scientists have since found a malaria parasite of tree rats called *P. berghei*, and another one called *P. yoelii* that can infect mice). Monkeys worked well—extremely well—but they were expensive to keep and hard to come by, especially with half the world at war. Even canaries—which Bayer used for preliminary drug screenings—were too expensive at a dollar each. The committee envisioned all the major drug companies feeding experimental compounds into university labs, which would then run large controlled tests on lower species. They needed an animal they could raise for next to nothing and infect with lethal malaria—some small creature that could be shoveled by the thousands into incinerators—until that one drug materialized to save them all.

Studies out of the Pasteur Institute in Paris by Émile Brumpt showed that the bird malaria used in Germany, *P. gallinaceum*, came from a Southeast Asian pheasant—that is, it occurred naturally in the pheasant and could be used experimentally in canaries. The French tried it on chicks to see what would happen, and it killed them all. Perfect! This parasite didn't always kill the canaries, which meant lab technicians had to take blood to see under a microscope whether or not a drug acted against the parasites—which was labor intensive. With chicks, technicians just had to look inside the cages. If the birds were dead, the drug had failed. If they were alive, it had worked. This was ideal. For only pennies each, the Americans would have a cheap, easy-to-work-with animal on which to run countless drug screenings. Lowell made it his mission to import this fantastic chick-killing bird malaria.[9]

The big catch, however, was that *P. gallinaceum* was lethal to adult chickens. The U.S. Bureau of Animal Industry banned its use for fear it would escape malaria labs, by way of mosquitoes, and destroy America's industrial chicken farms—which Lowell's uncle Don helped launch. Lowell wrote letters trying to pry permission out of the bureau, using national security and the impending war as a hook, but got nowhere. He couldn't sleep thinking about the possibilities this parasite offered the malaria project, and he bugged the executives at Rockefeller to help. He was so persistent, the foundation's director of international programs told him, "If you don't quit annoying me, you'll have to go someplace else and work, not here."

Lowell felt hemmed in.

THEN, one day while he strolled through the Bronx Zoo, a favorite pastime, Lowell had an epiphany.[10] What if he could find *P. gallinacium* in the zoo's exotic bird collection? What if there was a bird in New York infected with this malaria? Obtaining permission from the zoo's ornithologist to take blood samples from every Asian bird in captivity, Lowell found his treasure in the blood of a Borneo fireback pheasant. There swam what Lowell thought was *P. gallinaceum*. But soon he realized he'd discovered a never-before-seen type of bird malaria that could infect standard American ducklings. *And* it didn't harm chickens—posing no threat to the U.S. poultry industry. Squibb Institute for Medical Research was the first commercial firm to use it in antimalaria drug screening. It was named *Plasmodium lophurae*, for the taxonomical name of its pheasant host—*Lophura igniti igniti*[11]—but most people called it the Coggeshall strain. Labs across the continent, from Johns Hopkins to the Gorgas Memorial Hospital in the canal zone, soon asked for samples. The Coggeshall strain filled an important gap.

But it had shortcomings. It couldn't infect lab-raised anopheline mosquitoes, so they used the yellow fever mosquito, *Aedes aegypti*, which also carried a type of encephalitis that killed many of Lowell's monkeys. These setbacks he observed and wrote about for prestigious science journals, sharing every detail in this slow march forward in a hunt for a cure. Patient, persistent, and highly competent, he just kept raising ducklings and screening compounds, assigning each a number so they could be cataloged and filed.

One day in 1940 he received a substance from Winthrop. Lowell could tell it was a variation on plasmochin (the oxygen-depriving 8-aminoquinoline). His assistant, John Maier, injected it into ducklings infected with the Coggeshall strain. To their surprise the ducklings lived! Lowell recommended to Winthrop that they do toxicity tests on mice and dogs. If those animals lived, he asked that it be sent back for testing on syphilitics at Manhattan State Hospital, on Wards Island.

MEANWHILE, all-out war hit Europe. In the spring, Germany bombed Belgium and Holland. By June 14, 1940, Nazi tanks rumbled down the Champs-Élysées, taking their revenge for the Treaty of Versailles. French premier Paul Reynaud begged the free world to save his country, while Hitler made a pact with Vichy leader Henri-Philippe Pétain, signing away French independence and her colonial holdings. This put Germany closer to ruling Europe, the Gold Coast of Africa, and the Suez Canal—England's gateway to a vast colonial empire and huge oil reserves.

That same month Italy declared war on France and England. The move sealed an uneasy alliance with Hitler that Italian king Victor Emmanuel III objected to—because he didn't want Hitler waging battles on Italian soil. But Benito Mussolini needed Hitler to oust Great Britain from North Africa, a necessary piece of his plan to rebuild the Roman Empire around the Mediterranean rim. Mussolini neither liked nor trusted Hitler, maybe because the two had much in common. They both led riots and masterfully exploited their country's postwar economic disasters to seize power. Hitler was stronger and forced Mussolini to choose sides—to either fall like France or join him. So Mussolini joined him and sent Italy's ill-equipped troops to march on southern France, East Africa, and Libya—on their way to Egypt.

On the other side of the world, Japan used its control of Manchuria to drive deep into China. Then, in the fall of 1940, Japanese forces set up military bases and naval ports in Indochina, and prepared for attacks on Ceylon, Burma, India, Thailand, Malaya, Singapore, the East Indies, Kuala Lumpur, Manila, and more. The Japanese military's long-term goal of leading an all-Asian economic and military empire was, like Mussolini's ambitions, linked to Hitler's ability to destroy Great Britain. With the Royal Navy gone from the South China Sea and Indian Ocean, all Asian governments would fall in line behind Japan.

As war spread across the globe, Bayer labs couldn't obtain basic supplies, let alone sparrows, canaries and monkeys for screening new experimental compounds. But the company had its new top-secret series, the one it made before Germany invaded Poland; the one company chemists discovered while playing around with the 8-aminoquinolines, trying to make less toxic versions of the partially successful drugs plasmochin and atabrine; the series they made when they experimentally rotated a key side chain to the theoretical fourth position and then cleverly eliminated the yellow dye to produce a series of colorless, tasteless compounds they called 4-aminoquinolines. At first the drugs made from this series appeared too toxic, including one called resochin. But Bayer's chemists kept teasing out different molecules until they added a methyl group to the toxic resochin and arrived at an exciting one they called sontochin. It cured Bayer's canaries. Then it worked on Sioli's syphilitics in Düsseldorf, and on larger groups in Romania.

Bayer had done it again—they had a new drug with new patents. If all went well, this new Bayer concoction would supplant quinine and absorb its worldwide market.

But the world was at war. Bayer made barely enough of the stuff to distribute to Werner Schulemann, who ran preliminary tests on canaries and then tried it on soldiers sick with relapsing malaria, infected during the Greece campaign.[12] Everywhere this drug went it produced great results, and the momentum around it continued to grow.

Making it, however, became impossible.

ABOUT that time, Frederick Russell, who five years earlier had recruited Lowell back to Rockefeller from the University of Chicago, called again. He was now at Harvard and offered Lowell a prestigious professorship there named for Theobald Smith—who first discovered insects as disease carriers in 1889, presaging Ronald Ross's proof that mosquitoes carried malaria. Fred Russell warned that a formal offer would take time but told Lowell to sit tight; his next job would be at Harvard.

While Lowell waited for Harvard's official offer, Thomas Francis—developer of the flu vaccine—offered him a job as chair of a new school of tropical diseases at the University of Michigan, complete with a malaria-therapy lab at a nearby state hospital.

Harvard was expected to pay in the low "five figures" and, with some reluctance, the University of Michigan agreed to meet it. Lowell's new salary would be $10,000 ($160,000 in 2014 dollars).[13] The day after he wired his acceptance, a call came from Harvard offering him the Theobald Smith professorship. But Lowell had accepted Michigan. So, for the second time, he passed up Boston's Ivy League for a solid Midwestern university (the first time being his decision to pass up Harvard's medical school for IU). He moved his family to Ann Arbor that summer and set up a malaria therapy unit at the former Battle Creek Sanitarium less than an hour away. Then he put together what is today one of the country's top schools of infectious diseases. From there he flew into DC for the monthly meetings of the malaria project.

On September 11, 1941, at one of the most important malaria conferences up to that point, there appeared an unmistakable sign of things to come. In attendance was Colonel James Simmons, a decorated Medical Corps physician and tropical disease expert who held many degrees—including a doctorate and medical degree—from several top universities, including Harvard. The conferees all knew him at once by his pressed uniform, large square shoulders, slick combover, and kind but piercing eyes. He was one of the few high-ranking military officers who actually spoke their language of malaria. His presence changed the tenor of the meeting. It meant the War Department was preparing for troop deployments, and indeed, Colonel Simmons said as much. How and where remained undisclosed. But Colonel Simmons said troops would fight in highly malarious areas and that the War Department wanted the committee to speed up its work. He asked questions about mosquito vectors in West Africa and the Far East. And he disclosed that the army had already run preliminary experiments to see if atabrine or quinine could be used dependably to prevent infections—as a true prophylactic. Their experiments, he said, were inconclusive. Then he declared that the War Department's highest priority on malaria was to "get a field prophylactic which can be given without too much administrative difficulty."[14]

Committee members must have been thrilled. According to the minutes, they brought him up to speed in rapid succession—talking over one another, as they were just so damn enthusiastic to be getting real attention from the War Department. They described the setup: Clinical trials at state hospitals were stamped "secret"; promising compounds were stamped "confidential"; and everything else was stamped "restricted." In general, no information could be published without permission from the committee. And they said they expected to have up to 150,000 new chemical compounds or derivatives worth exploring for antimalarial activity.

Several project members, including Lowell, talked of the dire need for basic research, which was not what Simmons wanted to hear. What he wanted to hear was that they had good drugs to work with and that they'd have reports on potential prophylactics shortly. But Taliaferro from the University of Chicago argued that the program would be on shaky ground if it failed to find a *cure*—a drug that could, in fact, eliminate the blood- *and* tissue-stage parasites. Without a 100 percent effective prophylactic—which no one in the room felt plausible, given their limited knowledge of the microbes—troops would still fall from infections. The goal should be finding a radical cure.

By the end of a long and sometimes tedious day of malaria talk, conferees ended up split on this and other smaller issues. But they all agreed on one strategic point: Atabrine would never do; something better had to be found—preferably before U.S. troops shipped out for war.

Colonel Simmons knew something they didn't: that time had run out. He shut down the group's notion that they could move molecules around other molecules until luck produced a magic cure. Such luxuries belonged to peacetime research. Not while Axis troops marched on North Africa, attacked Russia, and occupied Europe. Not while Japan attacked and occupied nearly all of Southeast Asia on its way to claiming China and India, and maybe even Australia. Yes, he agreed that fundamental research was needed. He even announced that the War Department and the White House were prepared to make necessary funds available for a broad research program. But with that money came an expectation that the group would focus efforts on a prophylactic and, first and foremost, figure out how the heck to use this strange yellow drug atabrine.

This, of course, upset the pack. No scientist worth his or her salt took directions well. They preferred to let the science guide them.

But the message Lowell heard—the one that mattered most—was that the malaria project's political stock had just shot up. Lowell was quick to exploit it. Just weeks after the meeting he secured permission to finally import from Mexico City his long-sought-after, chicken-killing bird malaria, *P. gallinaceum*. The victory over the poultry industry's strong lobby was achievable only with the help of President Roosevelt's science advisory team, especially A. N. Richards, who oversaw all medical research linked to war preparations.

At the October 13, 1941, meeting, Lowell announced that he had *P. gallinaceum*, which "created considerable excitement."[15]

The White House help came at a cost, of course. Lowell would not enjoy carte blanche with this dangerous microbe. He was told to keep a lid on it, which he apparently didn't do well, for on October 16, 1941, Lowell received a terse letter from Richards reprimanding him for being too loose with his new chicken-killing germ. "The permit was granted upon my assurance that the precautions described in the report signed by you and Taliaferro would be rigidly carried out," wrote Richards. In that report, they promised to keep tight control of the germ's distribution, limiting it to lead malaria investigators at Johns Hopkins, the University of Michigan, the University of Chicago, the University of Tennessee, and John Maier at Rockefeller, Lowell's colleague there before he left for Michigan. How did G. Robert Coatney at the South Carolina state hospital somehow get his hands on it? Richards learned of the breach and instructed Coatney to kill and burn all chicks and remaining cultures immediately.[16]

"Now, as far as the future is concerned," Richard's letter to Lowell continued, "it seems to me that you as Chairman of the Malaria Conference should inform me officially of the names of the persons to whom the culture may rightly go. . . . I propose also to write to Maier to say that he is to give the culture to no one except on my authorization and that that authorization will be given only on your recommendation. . . . I take it for granted that every effort will be made to impress the workers with the responsibility we all are accepting." The Department of Agriculture had stipulated that all labs using *P. gallinaceum* be located in areas protected from air raids, and have armed guards to keep watch twenty-four hours a day.[17]

Lowell didn't take the reprimand personally. He knew the tight hold on *gallinaceum* would loosen once the U.S. went to war. But for the time being, he would comply with Richards's demands. For Lowell, what mattered most was that he had all the tools he needed to catch up to Germany: ample money, buy-in from the military and the White House, and *P. gallinaceum.*

CHAPTER 11
Africa

In 1940, intelligence officers gaming out war scenarios identified one in which Germany invaded the Western Hemisphere by way of Brazil. Nazi troops would amass in the Vichy French–held seaport of Dakar, Senegal, on the western hump of Africa. They would steam across the Atlantic's narrowest passage—just eighteen hundred miles—to the elbow of Brazil, where they would secure the beachheads and land troops, bombers, and war supplies. Pro-Axis, anti-American German and Italian nationals in South America, and German-run airfields and ports throughout Latin America, would begin an insurgency, as fleets of Nazi fighter planes and battleships cruised northward to bomb the Panama Canal and the U.S. Capitol.[1]

This scenario, presented to Secretary of War Henry L. Stimson by a special joint planning commission, was taken quite seriously and inspired action.

Stimson needed a string of military-ready air bases along the northern crown of South America. But he wanted them built discreetly, so he asked Juan Trippe to do it. As head of Pan American Airways, with dozens of daily flights in and out of Latin America, Trippe could ask for permission as part of an expansion plan for his commercial routes.

Trippe didn't want to do it, though. He already had a monopoly on air service to Latin America; he wanted to expand to Africa and China. This kind of calculated business acumen made him a legend in Washington. He had brazenly stopped attempts to create competition for his airline south of the border, and successfully lobbied against any attempt to impose competition on his federal freight contracts in Europe. The move triggered an antitrust investigation, which gave Stimson an in. If Trippe cooperated, he'd have an opportunity to redeem himself. The deal came with $12 million in loans from the Export-Import Bank that the War Department would secretly repay from an emergency spending account set up by Roosevelt, no strings attached. The money wouldn't even be audited.[2] So Trippe reluctantly accepted.

By March 1941, after a year of carving out runways in dense jungles, Pan Am had a string of military-ready airports from Trinidad to Natal, Brazil. Though the company finished only 38 percent of the work, it was enough for the War Department to establish a defensive posture and prepare for possible battle in the Western Hemisphere.

That same month, German general Erwin Rommel steamed his best infantry and panzer units across the Mediterranean to North Africa. But instead of heading west to Dakar and then over the Atlantic to America, as Stimson had feared, he went to Libya to help Mussolini's shattered army. British forces had captured 130,000 Italian troops on the coastline when Rommel showed up with his Afrika Korps. By May, Rommel had artfully moved his men east, over seas of sand and three-hundred-foot dunes. He refueled panzer tanks by landing cargo planes toting forty-gallon drums of diesel onto the salt flats. And he personally flew low over opposition troops to calculate risks, then pushed commanders forward.[3]

Rommel's men unhinged the Allies' hold on the region, closed off roads, and captured thousands. Panicked, British-led troops fell back in retreat while the Luftwaffe ruled the skies over the Mediterranean, intermittently shutting down supplies moving to beleaguered British forces in Cairo. Cargo ships at different times had to sail around the tip of South Africa, up through the Indian Ocean, and finally to the Red Sea. This was slow, dangerous, and unreliable.

PRESIDENT Roosevelt, still unable to convince Congress to declare war, couldn't stay out of it either. So he used two tools at his disposal: the airfields built by Pan Am Airways, and the newly passed Lend-Lease Act, which allowed the United States to lend the Allies enough war machines and supplies to protect U.S. interests overseas. Some entity, perhaps a commercial airline, would ferry warplanes, fuel, guns, and other supplies from Miami down to Brazil's elbow using the new airfields, then over eighteen hundred miles of open seas to the hump of Africa, and finally across the Sahara to British-led troops in Egypt. It was the reverse of what the airfields were conceived to do, and a brilliant strategy for moving supplies and troops over the safer waters of the South Atlantic. This air route would also branch south into the Belgian Congo to ferry back yellow-cake uranium needed for a top-secret physics project housed under the same funding umbrella as the malaria work.

Pan Am was the obvious choice to run the guns, and this time Juan Trippe wanted the job—it meant breaking into the lucrative air cargo and passenger business over Africa. He braved a flight to London by way of Portugal in June 1941, ostensibly to receive an award from the Royal Aeronautical Society. Once his audience of high-ranking officers assembled, Trippe made his case. Using a wall-size world map in the Royal Air Force's bomb-protected bunker, he traced a ten-thousand-mile air route starting in Miami. Bombs were heard falling on London as his pointer followed a line south to Brazil, then across the Atlantic to West Africa, over the Sahara Desert to Egypt, and beyond, to India, China, and even the Philippines. Trippe explained how Pan Am would build airfields every few hundred miles to service hollowed-out commercial passenger planes ferrying supplies and troops to and from all Allied positions, even if Japan shut down the Pacific. He said his airline had just built multiple airfields in the jungles of South America and could do the same in Africa.

Most important, he emphasized the technical advantages of using high-altitude DC-3s and Clippers flown by pilots trained in celestial navigation—at night without landmarks, over vast stretches of water and sand—to avoid antiair-craft guns. It wouldn't matter which occupying force menaced the lands and seas below; his pilots and planes would simply jump over them. These hazards had downed many RAF aircraft trying to fly supplies across Africa. Trippe said Pan Am could be running guns in sixty days.[4]

At ten o'clock that night, while standing on his hotel balcony watching the flash of German bombs striking far-off targets, a messenger from 10 Downing Street rapped on his door; Trippe was invited to dinner. At midnight he had a second supper with Winston Churchill to explain, again, his plan for running supplies to Egypt. Desperate for a quick solution to England's resupply problem, "weighing strategic expediency against the political risk of inviting this aggres-sive Yank into the African folds of the Empire," as Trippe's biographer put it, Churchill bit.[5]

The route presented a solution to the Middle East bottleneck *and* to

Churchill's abysmal problem in China. The Burma Road had become an unsustainable burden. Monsoons and Japanese bombardments kept shutting it down. And vicious malaria kept incapacitating construction teams charged with keeping it open. Meanwhile, British and American leaders braced for Japanese naval attacks aimed at cutting off Allied supply lines to the Philippines, Singapore, the East Indies, Australia, and New Zealand. Trippe pointed out that Churchill could, with Pan Am's help, resupply them all from the skies.

Trippe landed home a few days later and was met by a marine pilot, who flew him to Washington. The president asked Trippe to work with James Forrestal—the future secretary of defense—on setting up the African air route. By July 1941, Trippe had agreements on two contracts to secretly run guns to Egypt.

THE U.S. Army released its own pilots from active duty to fly for Pan Am Africa. A special school quickly trained more. For management, Trippe recruited graduates from Yale, his alma mater. A nationwide search found an adequate number of aviation experts, meteorologists, radio operators, and technicians. A hefty public relations team controlled the narrative, revealing only that Pan Am would ferry "persons, property and mail" to the British through the lend-lease program.[6] The advertising team stoked pro-Allied sentiments as thousands of men lined up for just three hundred construction jobs in Africa. Many applicants were unemployed rabble-rousers; the crowd became so unruly at Pan Am's headquarters in New York City that police had to patrol the line.[7]

The men hired traveled to Africa by either boat or plane, depending on priority of his job (no women were allowed). Neither mode offered a comfortable ride, but the flights lasted only a few days. The ships, which were cruise liners borrowed into service—including the Grace Line's *Santa Paula* and Alcoa's *Acadia*—were packed to the gills with equipment and men, and sailed for weeks under constant threat of submarine attack.[8] In early September, U-boats sank two of Pan Am's cargo ships, including one that carried several thousand linear feet of window and door screening.[9]

Most of the unskilled crews—needed to pave runways, build barracks, run mess halls, cook meals, and supervise thousands of "natives" employed at $20 a month to do backbreaking work—had barely traveled beyond their home states. "Some wanted mostly to get away from their wives," while others were escaping more serious problems.

"The day the 'Santa Paula' sailed, three sheriffs watched at the pier. A man might walk up the gangway with his hat over his face, another with a duffel bag on his shoulder shielding his face," trying not to get caught, wrote a Pan Am employee who later penned a company history of the Africa work.[10]

In the first few months, women wrote to Trippe directly, demanding child support and court-ordered alimony payments. Claims that could be verified were honored by the legal office and payment was docked from workers' wages. This undoubtedly wore on the morale of the men in these remote reaches of Africa. Construction workers promised $50 a week received only two-thirds pay, leaving them less money for alcohol, prostitutes, and gambling—the anesthesia against maddening heat and humidity, explosive rain showers, relentless biting mosquitoes, giant spiders that dropped at night from the rafters, and deadly cobras that coiled in dark corners of the warehouses and latrines.

The project landed them on a continent filled with dark colonial histories, dangerous animals, deadly diseases, and strange customs. And many couldn't handle it.

PAN Am Africa needed two hubs. The first already existed in Accra, on the Gold Coast. There, the Royal Air Force kept an airfield large enough to land cargo and passenger planes. This was to be the main hub for the whole project. Pan Am's newly acquired DC-3s—passenger planes the War Department leased from other U.S. airlines for the secret mission—hummed in from all-night flights over the Atlantic. Logos were painted over with a "dull sea-gray, and inside them every passenger amenity ripped out"—the seats, sleeping berths, even the floors. First-class accoutrements were left on the runway in Miami.[11] Technicians and engineers packed into the hollowed-out chambers like soldiers, butts on the floor. Knapsacks and the next shoulder over served as pillows as the growling engines rolled the cabin over wild winds and air bumps.

Grizzled and chilled from a cold night over the Atlantic, the men opened the fuselage to Accra's red-clay runway and thick, sweet air. There they stepped backward into the nineteenth century. The British air base ran according to a strict colonial hierarchy built on the backs of African "house boys"—all of whom served their masters in white pants and collared shirts, looking like plantation slaves.

By early October 1941, sixty days after signing contracts, Pan Am pilots began ferrying fighter planes across the Sahara. They reached Allied forces readying to march out of Cairo toward Libya with the mission of shoring up the British Eighth Army as it attempted to liberate thirty thousand troops cornered by Rommel in Tobruk, Libya.

A shortage of cargo planes slowed progress across the Sahara. Pan Am needed its water-landing Clippers to double the effort. But they couldn't land in rough ocean currents around the Gold Coast, or anywhere around the continental hump. Even seaworthy cargo ships had a hard time finding safe harbor on the treacherous seas. To press the Clippers into service, Pan Am needed a large wake-free lake for the planes' bloated bellies to skip across like a stone.

A U.S. Air Corps major suggested a forty-square-mile oblong tidal lagoon, called Fisherman's Lake, which lay hidden in thick jungles just off the Liberian coast—some five hundred miles from Accra. It was far enough from German and Vichy troops occupying nearby Dakar, and isolated enough for secretly unloading personnel and war supplies. This would not be an easy transition. Dense jungles around the lake's shores were enriched by two hundred inches of annual rainfall—five times that of Washington and New York. Its rivers and tributaries destroyed bridges and roads during floods. Overhead, a thick canopy of towering mahogany, ironwood, teak, and walnut trees echoed with the calls of baboons, monkeys, weaver birds, and parrots. This place was paradise for insects and wildlife—not for urban Americans fresh off the streets of New York City. But Pan Am had to make it work to service Clippers, so the shores of Fisherman's Lake had to be tamed.

In early November, as the British Eighth Army fought Rommel's forces in Libya, one of the borrowed cruise liners anchored off the Liberian coast and dropped a gang of twenty street-smart but jungle-stupid construction workers into rowboats. Some two thousand miles from the fierce desert fighting, they quietly bobbed through the rough surf, then along inlets that led to the smooth surface of Fisherman's Lake. They pulled up on its sleepy shores next to a path and disappeared into the thicket until they reached a clearing of mud huts and thatched roofs. A Christian missionary, Father Simmons, had agreed to provide the men with food and shelter until Pan Am supplies arrived.

The plan was for the men to spend the first few weeks setting up camp and building a small airstrip at Robertsport to link the lagoon with Roberts Field (a small airport farther south near the Firestone rubber plantation). Personnel would ferry in on the Clippers and, as quickly as possible, move to Roberts Field, where they'd catch flights to Accra for training and assignment to any of the dozen airfields across the desert—including desolate outposts run by just one radio operator. On their own, the men dumped at Fisherman's Lake drank warm beer and whiskey instead of the locals' murky water. They were often drunk and belligerent by midday, cussing, gambling, and getting into fistfights. The worst of them severely abused the "natives," treating them like mules.

Somewhere around Thanksgiving, Father Simmons visited the Pan Am quarters to discuss the bad behavior, only to be cursed at by a drunken employee and told to leave or be thrown out. The problems reached a pitch when the foreman failed to show up on the job site. A supervisor broke into his room to find him "in bed with a nude negro girl and open bottle of whiskey." In the report on the foreman's dismissal, the supervisor wrote: "[I]t became common knowledge that . . . Mr. Platt had bought this girl from her relatives." Another employee, C. P. Nolan, had severe venereal disease, deemed the African sun too hard on his health, and asked to be sent home just weeks after arrival.[12]

Finally, cinder blocks, plywood, generators, and heavy machinery, taken off the streets of New York specifically for the Pan Am Africa project, arrived on large cargo ships anchored off the coast. Local tribesmen used dugout canoes tied together into makeshift barges to transport thousands of pounds of equipment into the lagoon. The construction gang that had been dropped a month earlier with only cards, alcohol, and money for inexpensive sex—usually purchased from young local girls through their fathers or chiefs—finally got to work. The crews constructed permanent barracks, a mess hall, a kitchen, an infirmary, and outhouses. Pan Am managers smoothed relations with Father Simmons and enforced behavioral standards under threat of docked pay.

OF the supplies sent to the bottom of the Atlantic just two months earlier, most had been replaced. One exception was the fine-mesh screening needed to protect the crews from malaria-carrying mosquitoes in both Accra and Fisherman's Lake. The medical director tried to impose antimalaria measures. But he was up against the same fierce numbers game that ruined Ronald Ross's efforts decades earlier. Pan Am operated with mosquito populations a hundred times that of even the worst areas of the U.S. South; African anopheline mosquitoes carried parasite loads that would kill American mosquitoes; and villagers around Accra and Fisherman's Lake lived with entrenched strains of malaria that had persisted for thousands of years. In these villagers swam vast reservoirs of well-adapted *falciparum* microbes—the world's deadliest kind. Locals' acquired immunity staved off the worst symptoms. They merely appeared lazy and dull-witted whenever their bodies struggled through another attack of malaria. This masked the start of malaria season—because no one appeared *that* sick.

In late November 1941, villagers quietly battled through a round of malaria. Torrents of water that fell all summer and well into the fall filled the rivers, marshlands, lakes, and spongy fields. Construction of the airfields created even more breeding sites. Everywhere, mosquito larvae jerked about, maturing into pupae and molting into flight. Every night the insects spit malaria from the locals into the men of Pan Am, while they slept half naked in the jungle's unrelenting heat. A little over a week later they felt chills, fever, splitting headaches, and nausea. The medical officer in charge gave them quinine, which probably saved their lives—as deadly cerebral malaria develops most readily in malaria virgins who lack access to treatment.

In early December, a crew of pilots, radio technicians, and meteorologists whose Clipper had just skipped down onto Fisherman's Lake found nearly everyone on the ground sick with fever. Work had come to a standstill. George Kraigher, Pan Am's head of operations, met the Clipper just after dawn. His usual commanding pep talk was dulled by malaria. Body on fire, hair a mess, and uniform askew, he offered a brief welcome and then warned the newcomers that they would "all" get malaria.[13]

TRIPPE was never one to wait and see. While planning to dismiss his medical director, he asked for names of possible replacements. Someone told him of a skinny professor of tropical medicine at the University of Michigan named Lowell T. Coggeshall. This doctor could be trusted; he was already part of the national security apparatus through the National Research Council. He knew how to control mosquitoes and infections. So Trippe called Lowell, who was still in his first semester at the University of Michigan, asked him to join this secret mission in the jungles of Africa.

For Lowell's purposes, Pan Am's dilemma posed interesting research possibilities. This would be a great chance to study and treat wild malaria. No more cultivated strains in the blood of the insane. Pan Am offered carte blanche in managing the company's health disaster. Lowell could test atabrine and sulfa drugs, and any other compounds coming from the National Research Council's malaria project.

Of course he would accept.

Meanwhile, Clippers continued to land needed personnel at Fisherman's Lake for a quick stopover before heading east to the more comfortable guesthouses in Accra. But layovers lasted days as ground crews struggled through fevers and other illnesses. Venereal disease, for example, had become as rampant as malaria. Men often acquired them together during evening jaunts to buy sex in local villages. Newly arrived employees were moved out as quickly as possible to avoid exposure—but usually not soon enough.

By the bombing of Pearl Harbor, more than half of Pan Am Africa's seven hundred employees suffered bone-crushing chills, sheet-drenching fevers, unbelievable nausea, aches, and delirium. The mess hall in Accra was empty when a telegram brought word that the Japanese had attacked American soil—and that the United States had entered the war.[14]

According to Pan Am records, no one died of malaria—a near miracle, given that they had *falciparum*. The medical team used rapid quinine treatment, fed intravenously in the worst cases. They could barely keep up as hundreds of men streamed into the small hospital in Liberia and the larger one in Accra. The whole venture ground to a halt. Trippe's five DC-3s, conscripted by the War Department from different airlines, were in various states of disrepair, with different engine types, brakes, tires, wheels, landing gear, and hydraulic systems. Pan Am engineers tried to retrofit mismatched parts and solve complicated engine problems, but couldn't think straight under the fog of malaria. Pilots showing no outward signs of the disease were grounded anyway for fear that a rapid onset while navigating the Sahara's featureless deserts and high-altitude sandstorms would result in a crash—and loss of precious employees and aircraft.

Pan Am managers hired a small army of local Africans to dig miles and miles of drainage ditches in an effort to reduce mosquito populations—to no avail. One Pan Am employee observed: "At this rate they were going to spend 12 million dollars digging ditches clear across Africa, and they were getting no results. Everybody, more or less, had malaria or was sure to get it."[15]

By Christmas Day, malaria brought operations in Accra to a complete standstill. In less than three months—from October 1 to December 26—outbreaks caused an estimated loss of 1,894 days of labor in Accra alone, and showed no sign of letting up. Whether in the hospital or laid out in barracks, nearly every employee had at least one turn with the disease. To the malaria novice this looked like an intractable catastrophe. Trippe and Kraigher, backed against a wall by a microbe, made plans to move their main hub in Accra to the dry, mosquito-free city of Khartoum.

Back in the United States, Lowell packed for Africa and fielded directives from Pan Am managers used to telling contractors what to do. Trippe's top corporate troubleshooter, S. B. Kauffman, told Lowell to order a certain type of lizard from the Philippines. Kauffman couldn't say how the lizards prevented malaria, but he was certain of their efficacy.[16]

Lowell got other advice as well, which he politely ignored. Instead he gave Kauffman a list of needed supplies, in order of importance: enough small-gauge screening to cover every building at every airfield on the route to Egypt; mosquito nets for all beds; long pants, shirts, mosquito boots, and helmet netting for all workers; an endless supply of Paris Green for still waters; crankcases and crankshafts for drainage work; and access to heavy earthmoving equipment.

Lowell's orders were given high priority. The U.S. Air Transport Command, which oversaw Pan Am's work, bumped bullets and bombs to make space for Lowell's screens. Then Lowell left Michigan for Miami and the long flights to Africa.

CHAPTER 12
Bataan

Three hours after the Japanese bombed Pearl Harbor they hit U.S. facilities in the Philippines, eventually driving American and Filipino forces out of Manila and onto the Bataan peninsula. This stump of malaria-ridden jungle and sun-soaked beaches formed the upper mouth of Manila Bay and served as gateway to the basin's calm waters, where American naval vessels moored between missions in the Pacific.

On the peninsula, monkeys, wild pigs, pheasant, and quail lived amid giant banyan trees, elegant mahoganies, heavy vines, thick ferns, and palm groves. Bataan's twenty thousand residents farmed mostly rice and survived year-round bouts of malaria by acquiring a level of immunity that made an attack no more virulent than a standard flu. About 80 percent of them carried malaria in their blood, according to a Rockefeller study done in the early 1930s by Paul Russell. Russell had left Leesburg, Georgia, several months before the tragic death of his friend and boss, Samuel Darling, and set off for projects in Ceylon and Malaya. Just before Christmas 1929 he landed in the Philippines as chief of the Philippine Bureau of Science to make surveys of malaria-infected people, and bring them treatment and mosquito sanitation. He showed that anopheles on the main island of Luzon, around Manila, bred in the foothills not in swamps, which led to effective antimosquito work that substantially reduced infection rates. On the Bataan peninsula, Russell made grueling treks through the hills, along a defensive line plotted by the U.S. Army in a war plan designed to fend off a Japanese attack. But there he recorded such widespread infection rates and heavy mosquito breeding that military officials decided it was too challenging to try to sanitize and they left it alone.[1] By 1934, Rockefeller whisked Russell off to other projects—back to Ceylon, then India, Singapore, and Java, running investigations, doing surveys, and teaching malaria courses. Meanwhile, Bataan remained relatively untouched, even as rapid growth took root across the bay in the capital city of Manila.

The surveys were folded into what the War Department called War Plan Orange-3—code name for the Philippines in the department's color-coded planned defense of U.S. territory. Short of maintaining a large, ready force on Luzon, American planners foresaw no way to hold on to Manila. So they planned to deny Japan's navy access to Manila Bay. To do this they would occupy the Bataan peninsula, which bit down on the bay's opening, and they would maintain heavy artillery on Corregidor, a tadpole-shaped island two miles off the southern tip of Bataan and set in the middle of the bay's open mouth. Troops in these two strategic positions would bombard Japanese ships trying to pass into the bay. Under WPO-3, forces on Corregidor would endure even the angriest attacks in bunkers and bomb shelters built into the rocky hills. On Bataan, forces would simply dig in. WPO-3 had the peninsula stockpiled with food, artillery, munitions, and quinine. Soldiers would supplement food supplies with the natural resources of the mountains, while the terrain itself would allow vastly outnumbered American-Filipino forces to hold the line and "bloody the Japanese until the Pacific Fleet steamed into Manila Bay" from Hawaii.[2]

The plan assumed Japan's first target would be the Philippines.

In 1940, world events placed new urgency on war planning in the Pacific. General Douglas MacArthur—Allied commander of the Philippines—was promised a significant increase in his strength there and, with it, believed his additional forces could hold on to Manila until backups arrived from Hawaii. MacArthur planned for supplies to be stockpiled in other areas, including depots in and around Manila, south of the city in Los Baños, and on the island of Cebu.[3] When the Japanese bombed Pearl Harbor first, then the Philippines, MacArthur's defenses fell apart. He had yet to receive the promised increase in strength. His planes were caught on the ground and destroyed. And U.S. Navy vessels were bombarded and forced to scatter, then they fled for the East Indies. Reinforcements at Pearl Harbor were destroyed or damaged, and army forces in the U.S. had to remain on the West Coast to defend air bases, aircraft plants, and oil supplies from possible attack. Top command had nothing left with which to help MacArthur.

Within two weeks, American and Filipino troops were forced onto the Bataan peninsula, as earlier predicted. Some stockpiles survived the cross fire and made it to Bataan, but many did not. There simply weren't enough trucks and barges to carry everything under emergency conditions. Drugs and medical supplies were abandoned in Manila "because of limited time, extreme congestion of the single road into Bataan, and limited shipping facilities."[4]

On January 24, the War Department issued a press release commending General MacArthur for "brilliant" counterattacks that were a "smashing success." This spin said nothing of the doomed realities on the ground. That same day, the Japanese drove the general's troops down toward Mariveles on the southern tip of the peninsula, and the main U.S. Army hospital in Limay had to be abandoned.[5] MacArthur's men carried medical supplies to two makeshift medical outposts carved into the jungles, named Hospital No. 1 and Hospital No. 2. Allied footing was bad but not entirely lost. Evacuation of the sick and wounded was still possible. The men had their personal supplies of quinine to fend off malaria. Despite being weakened by serious food shortages and dysentery, they fought valiantly to slow the advancing Japanese forces, and they crushed sneak attacks from the beaches.

Many Japanese troops mysteriously withdrew in February with the Allies still holding on to the tip of the peninsula. Imperial military leaders had expected to take Bataan in a matter of weeks, so troops weren't equipped for protracted jungle warfare. U.S. military hospitals in Central Luzon now taken over by the Japanese were packed to capacity with soldiers from Japan's 14th Army—90 percent of whom had malaria and dysentery.[6]

This temporary withdrawal only delayed the inevitable. Food supplies were dwindling; every water buffalo in the region and every horse of the 26th Cavalry had been slaughtered for food. The men resorted to monkey meat and mules, and ate a lot of rice, stretching rations wherever possible. They were losing weight. Their immune systems were down. Worst of all, their quinine had run out. By March 7, malaria struck and men fell by the hundreds.[7]

Transportation officers didn't know what to do. At one point, "it was necessary to render Pool 4 at Mariveles non-operative due to an epidemic of malaria. Men with raging fevers drove vehicles and on several occasions became unconscious behind the steering wheel," wrote eyewitness John Whitman.[8] Shorthanded medical staff, already overwhelmed by battle casualties, cleared large swaths of jungle and threw together bamboo cots on which to lay the sick. At Hospital No. 2, doctors treated the worst-off patients in an A-frame structure of sackcloth walls and corrugated metal roofing. Everyone else was shoulder-to-shoulder under the trees. Hospital No. 1 had even less, with wards made of sheets strung between trees. Doctors tried to segregate the infectious from the injured, but couldn't. Mosquitoes bred in the torn-apart landscape, picked up parasites from the sick, and infected nearly everyone else. "Ninety-five percent of admissions developed malaria by cross infection before they left the ward," Whitman wrote.[9]

Fever-ravaged men of all ranks overwhelmed the two hospitals. Their combined capacity was a little more than two thousand patients, but each was forced to take nearly forty-five hundred. Brigadier General Vicente Lim, the highest-ranking Filipino under MacArthur, commanded his diseased and demoralized troops as his body burned with fever. Fevers also reduced the medical staff, making hospital management difficult. Patients died unattended and unnoticed until they were "found and tagged for removal," wrote Whitman. Supplies were ad hoc. "At one point, cases of Kotex arrived, which were used as abdominal packs for stomach wounds."[10]

The Pan Am air route across Africa already should have extended from Egypt to India, then over the Himalayas into China, and to MacArthur's troops in the Philippines. But it was still incapacitated by malaria.

Meanwhile, the scenes at Bataan's two hospitals were mild compared to the medical catastrophe unfolding on battlefields. A vast majority of the sick stayed at the front lines because motor fuel had run out. Ambulances, jeeps, and supply trucks stopped running. Soldiers collapsed in their foxholes, shivering with bone-rattling chills one hour and furnace-hot fevers the next. About 10 percent lapsed into comas and never woke up.

Meager reserves of quinine pills and liquid for IVs soon ran out. Doctors were left with bitter quinine powder that caused vomiting. The situation became so dire that on March 10 the commanding officer of one of the hospitals wrote: "The General Staff should understand the extreme gravity of the malaria problem and give priority to quinine above that of any other critical item. . . . It is my candid and conservative opinion that if we do not secure a sufficient supply of quinine for our troops from front to rear that all other supplies we may get, with the exception of rations, will be of little or no value."[11] The next day General MacArthur fled Corregidor by boat and then flew to Australia.

The Allies had no way of replacing or resupplying the diseased men of Bataan. Upon landing in Australia, MacArthur learned he had no replacement troops. By the time MacArthur made his famous declaration, "I shall return," another hospital commander reported that the malaria admissions on Bataan had reached a thousand soldiers per day, and that an emergency shipment of quinine had arrived by air from a medical depot on the nearby island of Cebu, but that it offered little relief because malaria patients had stopped responding to it.[12] They were too weak from malnutrition, fever, and exhaustion.

On April 2, refreshed Japanese troops returned to Bataan. They found more than 80 percent of MacArthur's forces laid out in beds, ditches, and clearings. News that MacArthur had just received the Medal of Honor for his defense of Bataan probably didn't register with the delirious and exhausted troops. Most of his men, whose veins and arteries swarmed with blood-destroying microbes, could not lift their heads, let alone fight. They had little choice but to surrender.

Visible from Bataan's beaches was the island of Corregidor, heavily bunkered and not part of the surrender agreement. MacArthur ordered the men there to continue attacking Japanese vessels trying to enter Manila Bay. To stop the bombardments, Japan planned to launch an attack from the now surrendered beaches of Bataan. That meant all seventy-five thousand American and Filipino POWs had to be moved out. They were marched under a searing sun to prison camps more than a hundred miles to the north. Japanese commanders pressured their rank and file to keep the massive migration of weak and thirsty captives moving. Too few soldiers for the job made each life a burden. Faced with this impossible situation, the overtaxed, tired, and war-weary Japanese soldiers employed cruel and severe tactics. As thousands of prisoners fainted, stumbled, and collapsed without the strength to get up, the Japanese finished them off with bayonets, machetes, or clubs. Some were beheaded, many had their throats cut, and others were run over by trucks—until they resembled smashed tomatoes, as one survivor recalled.

News that more than ten thousand dead bodies littered the roads leading out of Bataan mortified the U.S. public. American leaders tried to keep the details in close confidence, but the horrors leaked out. A reporter for the United Press wrote on April 17: "When the end came, 1,000,000 rounds of .30-caliber ammunition was blown up by our troops. . . . There was ammunition aplenty to fight off the enemy. . . . There was courage aplenty, too, to pit against the Japanese in those terrible days before Bataan collapsed. But there was no quinine to fight the deadliest of our enemy there—malaria. It was malaria and dysentery—not Japanese tanks or dive-bombers or bayonets—that told the final story."[13]

On April 23, *The New York Times* scolded the War Department for concealing the truth. "The reader at home might easily have thought that in spite of the constant danger the men on Bataan were leading an excitingly interesting, and at times almost jolly, life. In memory of the dead and in honor of the living, most of them now prisoners, it is well to have some of the bitter truths. . . . If there is blame for the lack of food and medical supplies it can be assigned later. . . . We want no information that will help the enemy, but neither do we want a romantic version of what war means. If war were not horrible—such is the fantastic paradox of our struggle—we should not be engaged in it: we are in it to the end. And if our soldiers can endure what happens in the front line, surely it is their right that we civilians should be fully aware of what they are undergoing."[14]

CHAPTER 13
War on Mosquitoes

In late January 1942, while MacArthur still held Bataan, Lowell said good-bye to his wife, son, and two young daughters and left for Africa.

He first flew to Miami, then boarded one of Pan Am's famous water-landing Clippers. It purred over the Caribbean islands, with stops for fuel and maintenance in Puerto Rico and Trinidad. Once in Natal, at Brazil's elbow, he and the other passengers and crew rested in a comfortable brick staff house with mosaic floors, red tile roof, and spacious guest rooms. Fresh food—eggs, meats, fruits, and vegetables—were grown on the property, which was alive with roaming chickens, ducks, and turkeys, and lined with lush mango trees and spindly coconut palms.[1] Here, the southern hemisphere's balmy summer had arrived.

After dark, Lowell and the others climbed into the hollowed-out cabin of a DC-3, headed for Accra. Packed with canned food and fuel, the plane took off into the black of night to avoid enemy fire. Passengers huddled for warmth as the temperature in the fuselage dropped. The engines snarled and growled "like a dog being held back from a fight," observed one passenger.[2] Everyone rolled with the plane, some nodding to sleep on a neighbor's shoulder.

The "jump to Africa" often meant flying through terrific rainstorms that drummed out the engines, and created giant air pockets that knocked supplies onto the men's laps and heads. Below, in the black churning seas, German U-boats patrolled. On one flight, pilot Dan Fowlie reported spotting seventeen of them. "We had a lot of sport for a while, diving on them, until we encountered a big one, and rather than submerge he surfaced and opened fire on us!" Fowlie made a dash for the coast and landed safely.[3]

On Lowell's flight, dawn appeared on the horizon just as the African coastline came into view. The rising sun sparkled in the rivers and drenched the deep red earth in a peach tone. Ten thousand feet below lay remnants of some thirty castles built over hundreds of years by European occupiers who had robbed the land of ivory, gold, raw materials, and, worst of all, people. Just up the coast from Accra were the notorious slave trading centers, where some two to three million men, women, and children were sold to mercenaries headed for the Americas—with malaria in their blood. The slave trade was long gone, but whites still ruled this land. The British owned the Gold Coast and turned Accra into an anthill of militarized activity. The militias of the Vichy French and their German overseers occupied neighboring countries to the north, east, and west.

Looking down, one could see a vast landing field with adjacent military-style barracks on one side and circles of thatched-roof huts on the other. The temperature jumped as the plane descended. When the fuselage door opened, passengers were hit with a blast of hot, sweet air. Everyone was escorted to the mess hall for food, where black houseboys in white jackets served orange juice, scrambled eggs, cereal, and canned milk, while calling Lowell and every other white person "master." The racial hierarchy here was unapologetic. Whites enjoyed generator-provided electricity, concrete floors, marvelous African carvings, furniture of wicker and ebony wood, and white tablecloths, while Africans, at the end the workday, walked in darkness toward their villages.

By the time Lowell finished his breakfast, a searing sun burned through his clothes, which were issued by Pan Am—khaki shorts and collared shirt plus an elephant hat with wide brim and vents that whistled faintly when caught by a hot breeze.

Outside the compound, Africans were Africans. The Krobo people wore their dark chocolate skin mostly bare but for loincloths tied at the waist by braided fabrics. Some wore long, heavy necklaces of beads and shells, and armbands just above the elbow. Older girls had special virginity beads partially obscuring their bare breasts, and headdresses that fit close to their shaved heads, like cloche hats off the streets of Chicago, but of earthen materials that smelled of dirt and perspiration. Young children ran naked through tight clutches of thatch-roofed huts made of mud and sticks.

War, however, was changing this place. The flat coastline had been militarized. Many men from the villages, instead of serving their "masters" in white jackets, joined the Royal West African Frontier Force, which came with khaki uniforms and bayonets. These men served as sentries and sat for hours high up in the palm trees lining the sandy shore, looking out for German U-boats and ships. Or they dug slit trenches and bomb shelters, which were never used because they filled too quickly with snakes and scorpions.[4] These peaceful farmers were pressed into service through agreements made in Accra between British viscounts and elite Africans who wore European suits and called villagers "bush niggers." By the time Lowell arrived, tanks lined the stifling-hot beaches, while everyone else took shelter from the heat.

By jeep, Lowell went to the special Pan Am camp three miles from the airfield, overlooking the ocean. Barracks and officers' quarters had running water and electricity supplied by three power stations and a thirty-thousand-gallon elevated water tank. They had sewer service and a septic system, refrigeration for food, eight warehouses, a laundry and boiling house, workshops, two mess halls with kitchens, a recreation building, and a fifty-bed hospital—all constructed out of gorgeous mahogany wood.[5] The rooms felt exotic, filled with African carvings and masks, and ornaments from hunting adventures that included elephant tusks and antelope horns.

Few Americans saw Africa. Yet here was Lowell, a farm boy from Indiana, now among the privileged few. Malaria expertise did, indeed, bring great riches, as Samuel Darling had promised. Not in money and gold, but in travel and worldliness. And it came with something money couldn't buy: the satisfaction of being genuinely useful. Looking all around him, Lowell eyed the stagnant rivers, swamps, water-filled buckets, and catchments baking in the morning sun. His public health calculus immediately ran the numbers: Millions of jerking larvae would soon morph into malarial mosquitoes that would feed on thousands of infected people and create another widespread outbreak.

If Lowell trained his whole life to fix just one major problem, this was it.

AS Pan Am Africa's new medical director, Lowell didn't look like much—all elbows and knees under his oversize sun helmet. He spoke in scientific jargon and spent much of his time wandering around the air base, behaving more like a guest than a boss. He didn't bark orders or demand results, like Trippe's other corporate managers. Instead, he talked to everyone, and concentrated most of his efforts on the rank and file. He described to the men the nature of their beast and involved the entire airfield in his work.

This was typical Lowell, playing down his status as a Pan Am director even as he had more authority than any man there. Trippe had given him executive status to deal with the malaria problem, which he used as a broad brush, not a big stick. He knew exactly what needed to be done and he acted quickly, using "brains rather than brawn," as one Pan Am employee put it.[6]

He brought with him newly hired nurses and doctors to care for the sick while he kindly asked carpenters to retrofit all buildings with dual spring-operated, outward-swinging doors set six feet apart. The resulting vestibule, he explained, would catch mosquitoes as they tried to fly in. He also had eighteenth-of-an-inch mesh screening flown in from the States on a high-priority ticket, bumping bullets for space on aircraft. Then he had maintenance men fasten it over every door and window of every building in Pan Am's camp. Workers were put in charge of doing nothing but inspecting screens, repairing holes, and ordering more screens when needed. Screens, screens, and more screens were the centerpiece of defense.[7] Everything else was backup, including drugs.

The British were skeptical of his approach, so he proved his point; he ran a side-by-side study. The British barracks remained as they were, with no screens and poorly maintained bed nets. But the men took five grains of quinine daily as a prophylactic. Next door, Pan Am employees followed Lowell's instructions and meticulously maintained screens and bed nets, and sprayed nightly with insecticides. They took no quinine. Between June 1 and October 30, 1942, of a hundred British air force officers, eighty-seven had to be hospitalized for an average of ten days for malaria. During the same period, only one of the sixty Pan Am employees contracted malaria. Lowell learned that that one man had slipped out after dark to another camp, probably for sex, "where he did not observe personal protective measures," Lowell later wrote in a report.[8] Of the British forces, he wrote: "There was a minimum of eight hundred and seventy man days lost because of malaria alone in the unprotected group, compared to 10 days in the carefully protected group. As the disabled men were highly trained personnel in the air force, it can readily be seen that the money lost by their inactivity would have paid for an intensive program of malaria control."

To further prove his point, he took heavy equipment and construction teams off runway maintaining and put them into service to drain swamps, and build dikes of concrete and steel to prevent seepage. Next he hired villagers to poison with Paris Green all ponds within a mile of the base, in every direction, once a week. Then he had teams nail plywood mosquito bars over every Pan Am–occupied bed, and from them hang soft white netting that flowed like Roman drapes from ceiling to floor. Lowell trained villagers to spray a mixture of pyrethrum and kerosene on all indoor structures, including their own huts. Mosquitoes landed on the greasy film to digest blood meals and died within hours.

Lowell also ran malaria orientation courses, appealing to the men's intellect. They weren't just coming down with fevers, he told them; they were housing microbes that chewed on their blood cells and damaged their organs. He explained that all native quarters were moved a mile from the air base because the mosquitoes he worried about didn't fly much farther than that for blood. Placing African workers in a different "mosquito community" at night, when biting occurred, segregated malarial microbes to anophelines that couldn't fly far enough to reach the men of Pan Am. He needed them to understand that evening trots to village girls for sex risked more than just venereal disease. "The doctor got much results simply by explaining that to the boys," observed a Pan Am manager.[9] To make sure the lessons sank in, he appealed to their wallets. Violation of the rules, including strict adherence to wearing long trousers and shirtsleeves after dark, and sleeping under mosquito nets, was punishable by docked wages.

Then he set out to find remaining mosquito breeding sites. According to a Pan Am employee, Lowell "caught some mosquitoes in traps, dyed them red and set them free. Then he went mosquito hunting in the fields and bushes to try to find them."[10] He found thousands of larvae growing in pools of rainwater caught by hundreds of gasoline barrels that cluttered the airfield. So he hired "African boys" to paint the barrels with old engine oil. He also ordered them to dust seepage areas, footprints, tire tracks, and other waterlogged divots with Paris Green at least once a week to eliminate the last remaining larvae. "By such means he dried up the mosquitoes right around the airports and didn't bother with the rest of Africa. . . . It was an achievement of permanent value because of the ease with which it was done," observed another Pan Am employee.[11]

Lowell lavished spreadsheets with drug treatment data. He brought from home supplies of experimental doses of atabrine and sulfa mixtures, measuring for the first time how these different doses worked against wild African malaria—

versus controlled lab-raised malaria in the psych wards. He created cohorts, some getting quinine, some getting atabrine, and some getting the sulfa mixtures. The atabrine performed poorly, relative to quinine, until enough was administered to turn the men yellow; then it suppressed symptoms about as well as quinine. His big surprise was that one of the sulfa mixtures acted more quickly and without side effects. This mirrored what he had seen in his monkeys back in Michigan, but not in his patients there. He concluded that *falciparum* microbes were far more vulnerable to sulfa than the lab-controlled St. Elizabeth strain of *vivax*. He saw that different types of malaria require different treatments; that malaria was, in fact, not just one disease but many different diseases; and that drug-treatment strategies to protect soldiers would need to vary depending on where they were sent for battle and which parasites attacked their blood.

His findings he wrote up in reports that he sent back to the NRC.

LOWELL dried up malaria, which impressed everyone, even British leaders who believed Africa's main malarial mosquito, *A. gambiae*, to be the impossible phoenix of the insect world. No matter how often they sprayed, she rose from the ashes to infect again. *A. gambiae* had tormented European colonizers for centuries. Many thousands of mercenaries and colonial officers barely penetrated the coastal zones before meeting their fate with deadly *falciparum* carried by this flitting murderer. Yet here was a skinny professor from America beating the beast.

Americans en route to Africa heard the horror stories. But no one, until Lowell arrived, fully explained why British officers called Africa the "white man's graveyard." Lowell explained that *A. gambiae* was actually a complex of mosquitoes. Some sought blood meals from animals, while others, the worst of them, would take it only from humans. These so-called anthropophilic *gambiae* were (and still are) the world's most efficient malaria carriers and, by far, the deadliest insects on earth.

To look at, they were plain Janes in Africa's stunningly diverse and gargantuan insect world. Small, brown, and silent, with an almost pain-free bite, they often went unnoticed. The continent's inch-long biting flies, hand-size orb spiders, giant mantises, goliath beetles, and frighteningly ugly centipedes scared even Pan Am's grizzled construction teams. But it was the ubiquitous featherweight *A. gambiae* that threatened their lives, Lowell explained.

Lowell trained his eye on this one mosquito not just to protect Pan Am's employees. The National Research Council had asked him to assess the threat of an *A. gambiae* invasion of North America, given the frequent flights between Miami and West Africa, via Brazil. More than a decade earlier, *A. gambiae* had sneaked onto cargo ships steaming from Dakar to Brazil's port of Natal and flitted ashore to lay eggs in irrigated hay fields. In 1930, a horrified Rockefeller entomologist, Raymond Shannon, spotted their larvae while doing a mosquito survey involving yellow fever. He immediately warned Natal officials, who understood the grave danger and sanitized the city with antilarvae agents. But they chose not to flood the surrounding hay fields with tidal waters, because the move would have destroyed the hay, not just larvae.

Entomologist Gerry Killeen explained what happened next: "The African mosquito slowly and inconspicuously spread westward to the wetter, more populated valleys of the Assú, Apodí, and Jaguaribe rivers where widespread agricultural activity presented diverse and abundant opportunities for propagation."[12] By 1938, *A. gambiae* inhabited eighteen thousand square miles of Brazilian countryside and triggered an explosive epidemic. Conservative estimates suggest that at least a hundred thousand people shook and sweated through fierce infections, resulting in some twenty thousand deaths. This was the worst malaria outbreak in the history of the Western Hemisphere. That year, according to mosquito expert Andrew Spielman, "[Seventy] percent of the cotton crop was never picked because the harvesters were too sick. Hospitals ran out of beds. Pharmacies exhausted their supplies of drugs. Power companies lacked the necessary workers to maintain electric service."[13] Some areas were so infested that mosquito catchers assessing the size of the *A. gambiae* population trapped well over a hundred specimens per room, per dwelling.

Families healthy enough to travel fled the region in a panic, while everyone else fought for their lives. The dead lay unburied for days, because few people were well enough to dig graves. It was a national disaster and an international public health spectacle.

In 1939, Dr. Barber, maker of Paris Green, summed up the situation: "There is no doubt that this invasion of *gambiae* threatens the Americas with a catastrophe in comparison with which ordinary pestilence, conflagration, or even war are but small and temporary calamities. *Gambiae* literally enters into the very veins of a country and may remain to plague it for centuries."[14] The U.S. Public Health Service feared the invasion would soon reach Florida.

With encouragement from American diplomats, the Brazilian government in 1939 joined forces with Rockefeller entomologist Fred Soper and launched an unprecedented assault on the alien mosquito. Soper, using cartographical maps, divided the eighteen thousand square miles of *gambiae*-infested northeast Brazil into parcels small enough for one hired hand—a larvae inspector. The inspector's daily schedule had every hour planned in which he, by foot or on a bike, patrolled his parcel and rooted out *gambiae* from homes, trash, ditches, and puddles. He spilled Paris Green on potential *gambiae* breeding spots and climbed under joists and rafters to kill adult mosquitoes where they rested. Over him was a district supervisor, typically responsible for checking the work of five larvae inspectors. And over him was a manager responsible for several districts. Supervisors earned bonuses if they found a mosquito their larvae inspectors missed, and inspectors' wages were docked if they were caught veering from their regimented daily schedule. Soper ran a rigorous, autocratic, hierarchical, and thus airtight operation.

During one of Soper's surprise inspections he famously called a supervisor from the field to sit with him. The man hesitated because his clothes were covered in dirt and sweat. But Soper, always tidily dressed in woolen suit, vest, and tie, told the man that a sector chief with a clean shirt in the middle of the workday would be fired for laziness, no questions asked. Malcolm Gladwell, writing about Soper for the *New Yorker* in 2001, recounted another famous story. It involved an accident at a munitions dump: "Soper, it was said, heard the explosion in his office, checked the location of the arsenal on one of his maps, verified by the master schedule that an inspector was at the dump at the time of the accident, and immediately sent flowers and a check to the widow. The next day, the inspector showed up for work, and Soper fired him on the spot—for being alive."[15]

Soper was pious and hard driving, and achieved the impossible. A year before Lowell landed in Accra, Soper had eradicated *gambiae* from Brazil; he wowed the world of public health with a feat that remains perhaps *the* greatest victory against mosquitoes, given the odds.

In 1942, the NRC asked Lowell to use his position with Pan Am Africa to work with officials on a strict containment policy for *A. gambiae*. It included aerosolizing insecticides on every aircraft departing from Accra, Fisherman's Lake, Roberts Field, and the new route coming up from the Congo—with the top-secret cargoes of yellowcake uranium headed for New York. Once the planes landed in Natal, no one was allowed off the flight until a team came aboard and exploded another round of "bug bombs." For ten minutes, everyone sat still while a thick fog engulfed the fuselage. It burned passengers' throats, and caused some people severe nausea and occasional claustrophobia-induced outbursts. But no one dared leave before the allotted time under threat of dismissal.[16]

The efforts worked. Lowell reported to the NRC that Brazilian health officers had counted more than nine hundred dead *A. gambiae* on Pan Am Africa flights landing in Natal. The bombs also killed an abundance of other stowaway insects, including aggressive jungle flies—the dreaded tsetse—that carried deadly African sleeping sickness.

MALARIA around Fisherman's Lake, in Liberia, proved the most difficult to tame, and not just because of the thick surrounding jungles. The site's importance grew in stature as Clippers belly-flopped daily onto the lagoon. There, the biggest malaria threat came from villages within a mile of employee barracks. Unlike Accra, where British rulers wrote the laws governing local people—making it politically easy to burn down villages and relocate them a mile away, or to impose mandatory weekly pyrethrum spraying—Liberian villages were organized under laws written by the central Liberian government in Monrovia and executed by local chiefs. The people fished the lakes, farmed the land, hunted small game, and lived in long-established communities. Neither Pan Am nor the U.S. Air Command could simply order them around. Lowell preferred to truck workers a full mile away to make all Africans sleep with their own mosquito colonies, segregated from the Pan Am compounds. But with that not an option, he retrofitted his programs, which meant working with local chiefs. By March, village elders allowed the use of pyrethrum canisters for hut-to-hut spraying. And they allowed medical treatment for the many VD-infected village girls–turned-prostitutes.

Lowell taught teams to screen barracks and drain swamps, and he ran his malaria-orientation courses. Strategies here had to be adjusted, given the dense jungles, heavy humidity, and close-by local women who kept sneaking onto the compound to earn cash for sex. He contended with oppressive heat that discouraged the use of bed nets, because they stopped airflow. And *gambiae* breeding sites were everywhere. So he added to the regular uniforms helmet nets and mosquito boots, which looked like canvas button-up gaiters that protected the ankles and calves from bites. He also insisted that everyone wear long pants and sleeves, and lots of bug spray. He treated everyone with either atabrine, sulfa, or quinine, entering data into his spreadsheets as nurses helped him keep track of outcomes. The infirmary, packed with malaria and VD patients, soon held just the latter. Nurses instead began focusing on the intense heat and humidity, the nightmarish insects, and the insanity of boredom. Eventually they all had to be sent home. This, to Lowell, was success.

Most of the remaining air bases were smaller and easier. Lagos, about 250 miles west of Accra, wasn't too bad. It had nineteen Pan Am employees in poorly screened British barracks. So Lowell set up screening regiments; enforced bed net and protective clothing standards; ran orientation sessions; and warned of the double jeopardy—VD and malaria—lurking in the villages after dark.

Fort Lamy, the next stop in the Free French–run country of Chad, was more difficult. It had an intense rainy season that made roads impassable and mosquitoes unbearable. Pan Am's five employees all caught malaria. The airfield was on Lake Chad, which was the size of an inland sea, bordering three countries. This stop started out as an emergency airfield only, and was so inconsequential to the battles that raged in Libya less than a thousand miles to the north that it was attacked only once, by a lone German pilot who dropped a few bombs and buzzed off, never to return. This was hippo and crocodile country, frequented by herds of elephants, zebra, and buffalo. Pan Am pilots, if they could help it, flew over Fort Lamy for the next stop. That is, until a thousand cases of whiskey arrived by barge with no orders for its next destination. Overnight, Fort Lamy's well-stocked shebeen became a favorite stopover for pilots.[17] And that meant its malaria had to be tamed. There was no way to sanitize this vast waterway ruled by intensely territorial hippos—Africa's most dangerous animal to humans because of their violent attacks on boats. So staff hung screens and bed nets, and used insect repellent, even in bed.

The remaining airfields, most of which were small repair-and-refueling stations located in increasingly arid climates on the way to Khartoum, added color to the tedium of antimosquito work. For example, in El Geneina, Sudan, a half-domesticated giraffe met pilots face-to-face as they climbed from the cockpit. And two hundred miles farther east, an airstrip in El Fasher had a young lion mascot named Leo living with the RAF pilots.[18]

Another five-hundred-mile jump from El Fasher was Pan Am's secondary hub on the banks of the Nile River, twenty miles north of Khartoum. The crews built grass tennis courts and a swimming pool. The region got barely six inches of rain a year, had little if any malaria, and, in the summer, had temperatures that reached 125 degrees Fahrenheit. There, Pan Am kept nearly a hundred employees. Virtually every aircraft using the route stopped here for refueling and repairs, and opportunities to buy ivory carvings and other African art.

From Khartoum the air route forked in two directions. One turned north along the Nile to Cairo, with one emergency airfield in Luxor, near the temples of Karnak and Thebes, and the tombs of the Valley of the Kings. In Cairo, Pan Am leased luxury two-story homes with uniformed Egyptian servants. Pilots rode camels to the pyramids and drank in a rooftop cabaret on the Continental Savoy with belly dancers and prostitutes. The region had light, seasonal malaria. But this was about to change, as no containment policy was in place for aircraft out of West Africa heading east; only planes flying west to the Americas were bombed for bugs. In the spring of 1942, as Lowell made his way across the Sahara, *A. gambiae* sneaked onto at least one flight and colonized the northern stretches of the Nile—the effects of which would not be felt for another year. Meanwhile, the route expanded over Saudi Arabia into Iran, where fifteen Pan Am engineers tuned up bombers slated for the Russian front.

The other fork from Khartoum went due east to Salalah on the Arabian Sea, and across the Indian Ocean, to Karachi. Then continued east to New Delhi and Calcutta, with plans for extending over the Himalayan "hump" to China.

While Lowell worked across the African continent, malaria transmission dropped from nearly 50 percent across all Pan Am employees, and nearly 100 percent at Accra and Fisherman's Lake, to less than 1 percent.[19] Much relieved, Pan Am managers stopped plans to move their hub to the wickedly hot deserts outside of Khartoum. It would remain in the oceanfront colonial paradise of Accra.

Some forty-five years later, during an interview with an oral historian, Lowell said of his time in Africa: "I never worked so hard or felt so great about anything in my life."

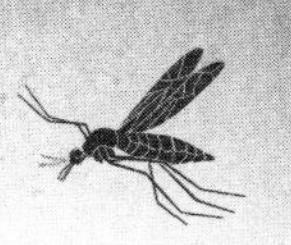

CHAPTER 14
Guadalcanal

Two weeks after the Bataan surrender, Lowell's former boss from the Leesburg Malaria Station, Paul Russell, gave a lecture at the Army Medical School in Washington, and afterward stopped in on Colonel Simmons in the surgeon general's office.

Russell could have predicted the malaria disaster on Bataan; the studies he did for Rockefeller in the 1930s were unequivocal. Most people on the peninsula carried the disease. Of course it would spread to troops. The math was pretty clear and simple. So he volunteered to help the army avoid another Bataan. At nearly forty-eight, he was too old to serve as a battlefield doctor. But he wanted to serve, because Russell believed in service. He had become a public health crusader to help mankind; he proselytized the virtues of hygiene and sanitation to help bring everyone into the temple of good health, a church that belonged to all people. These were skills of an evangelist that he inherited from his preacher father. But Russell's calling—and he did hear a calling—was not to save souls but to save people, as his son described decades later. Russell went to medical school to be a medical missionary and from there learned the religion of public health. Rockefeller provided the perfect outlet for him. But with the war on, he felt he might be useful to the army, maybe as a volunteer of some sort.

Colonel Simmons, in his usual oversize manner, said volunteering was all very fine, but what he really needed was to have Dr. Russell in uniform, working with him inside the War Department. So Russell took leave from Rockefeller and on May 20, 1942, was sworn in as a lieutenant colonel in the Army Medical

Corps. The rank understated his position as chief of the army's Tropical Disease Section, which didn't exist yet. Positioned directly under Colonel Simmons, it would be a catchment for in-house malaria expertise where strategies would be developed to prevent another Bataan.

Dr. Russell rented a tiny house on Klingle Street in northwest Washington, muscled through the administrative headaches of setting up a tropical disease operation within the protocol-laden hierarchy of the U.S. military, then got to work. One of the first things he did was to think like a mosquito, as Dr. Darling had taught him. How would he stop them from spitting malaria into troops sleeping under trees in forward areas? Repellents and bug bombs were an obvious starting place. In July, he flew to Camp Blanding in Jacksonville, Florida, to work with Westinghouse Electric & Manufacturing Company on a new pressure cylinder for "bug bombs" that used Freon to aerosolize the insecticide pyrethrum, made from African chrysanthemum flowers. They weighed a pound and did the work of a gallon of insecticide. These bombs would reduce mosquito infestations in hospitals and fallback zones.

This fit a broader plan worked out before Russell joined the army, in which antimosquito work would be done by navy Construction Battalions, or the CBs, also known as the Seabees. These skilled builders would move into occupied territory and, while constructing airstrips, roads, hospitals, housing, warehouses, gasoline storage facilities, and bridges, they would also do sanitation work. With some guidance, military planners saw no reason the Seabees couldn't operate as mosquito brigades, dousing the rear with Paris Green. This, in theory, would create malaria-free zones for infected troops to fall back to. Once on the mend, they would rotate back to the front lines to replace the next round of fever-stricken men. Troop strengths were calculated with an assumption that roughly 10 percent always would be sick or recovering from malaria—not perfect, but an improvement over the 20 percent loss of strength during the Great War.

The plan was deeply flawed, as Lieutenant Colonel Russell well knew. Burdening the Seabees with antimalaria work was akin to solving linear algebra with no known variables. The Seabees were building tradesmen, not entomologists, so expecting them to flesh out important biological factors for malaria sanitation was unrealistic.

In time, Russell would be proven right, as not even worst-case scenarios foresaw the malaria nightmare to come.

. . .

THROUGHOUT the first few months of 1942, Allied-held territory in the Pacific diminished rapidly. Guam and Wake Island fell with Bataan. Then the Japanese steamed southward to annihilate the American, British, Dutch, and Australian command (ABDACOM). More than sixty thousand Allied prisoners of war were taken, including several hundred Americans fished out of the seas after the sinking of the USS *Houston*. Nearly all of them were shipped to the mainland to build what would later be fictionalized as *The Bridge Over the River Kwai*, based on the true story of how these POWs suffered terrible malaria and malnutrition while forced at gunpoint to build a railroad linking Japanese supply depots in Thailand to the Rangoon rail line in Burma.

By May 1942, Japan held Manchuria, parts of Shanghai, Hong Kong, Burma, Thailand, Malaya, Singapore, French Indochina, Sumatra, Java, Borneo, and, of course, the Philippines—basically all of Southeast Asia and everything in the seas between the Asian continent and Australia. With the destruction of ABDACOM, Japan also occupied the resource-rich Dutch East Indies, taking control of stores of oil, rubber, and tungsten. And the Japanese also very strategically took Java. The Dutch destroyed cinchona bark already stripped and processed for quinine, to deny it to the enemy, then fled. With them went 90 percent of the world's quinine supply. This was anticipated, but nonetheless devastating news to malaria planners.

A ferocious naval battle in the Coral Sea permanently blocked Japan's attempts to land forces on the islands of Fiji, New Caledonia, New Hebrides, and Tonga, thereby creating a buffer for New Zealand—and providing staging areas for the battles to come. It also robbed the Japanese of two aircraft carriers needed for their planned June attack on Midway—a spit of coral in the middle of the Pacific with an American air base. The Japanese needed the victory at Midway to finish what they had started at Pearl Harbor: the complete destruction of America's strategic power in the Pacific. No more air hubs and no more protected naval bases meant no more Americans. But American code breakers learned of the plan and turned the outcome around. Instead of another Pearl Harbor, the U.S. Navy ambushed the Japanese and won its first victory in the Pacific. In a war whose outcome could have been won by naval battles and aerial dogfights launched from these floating runways, Midway turned into a disaster for Japan. The battle diminished the Imperial Navy's frightfully superior sea power to about that of the Americans. The playing field had been leveled. For the first time, American planners felt they had a fair shot at taking back the Pacific, which translated into an eager and aggressive look at the possibilities—still in need of focus.

THAT focus came in July 1942, when Allied spies caught the Japanese building an airstrip large enough to service heavy bombers on the Solomon Island of Guadalcanal. If completed, the Japanese would have the means to launch air raids on U.S. naval advances steaming in from the United States. This had to be the first target of attack. Navy commanders planned for the entire 1st Marine Division—sixteen thousand men—to storm the beaches, capture the airfield, and evict Japanese forces from there and the nearby islands of Tulagi and Gavutu. While the marines fought on land, U.S. warships would best the Imperial Navy at sea, preventing it from landing replacements. Once secured, Guadalcanal would become the staging area from which Allied forces would launch an island-to-island assault to take back the Philippines and reach all the way to Japan.

The 1st Marine Division's units, scattered around the Pacific, all steamed for either Wellington, New Zealand, the Fiji islands, or Samoa for training. Some came on well-stocked navy vessels with tight berthing compartments and well-run mess halls. Others were stuck on filthy cargo ships covered in soot. And others traveled on well-appointed luxury cruise liners. The bulk of them, however, rode aboard the MV *Ericsson*. This conscripted Swedish cruise liner's 1,544-passenger capacity was stretched to fit some five thousand marines. The mess hall took four hours to serve meals made from rancid stocks of oil, eggs, and reprocessed butter. "The bakery ran out of proper shortening ten days from San Francisco" and an oil substitute triggered rampant diarrhea that struck half the crew, according to navy records.[1] Just when morale couldn't sink lower, a spoiled roast beef dinner hit the men so fast some relieved themselves on deck, unable to wait in long lines for toilets. By the time the ship docked at Wellington, the men, on average, had lost between sixteen and twenty-three pounds.[2] Many wondered what kind of putrid chaos lay ahead.

Wellington temporarily allayed their fears. Troops camped in screened tents with cots. They ate three meals a day, regained weight, and got medical attention. Those with a venereal disease or found physically or mentally unfit for intense tropical heat were separated from their companies and handed over to the quartermaster for stocking duties. Everyone else received inoculations for smallpox, yellow fever, typhoid fever, and tetanus. On July 22 they all boarded ships for the Fiji island of Koro to practice night landings, which turned into a "disaster."[3] Transport ships miscalculated the shoreline and dropped anchor far out at sea. Marines climbed down rope ladders to inflatable rafts with outboard motors that broke down in the salt water, while coral reefs tore through the rubber. The consolation was that rafts were ruled out for landing on Guadalcanal. Instead, a limited number of Higgins boats would shuttle waves of marines from ship to shore, back and forth until everyone landed.

Two weeks later, the operation commenced.

REVEILLE sounded at two forty-five a.m. on August 7, 1942, just as the American's eighty-nine-vessel convoy broke through low-hanging clouds to enter the waters off Guadalcanal. By four a.m. the assault force was topside, loaded down with a sixty-day supply of ammunition, eyes fixed on the horizon as the convoy approached the dark shadow of Guadalcanal's mountains.

Dawn soon brightened a silent shoreline of waving coconut palms and six-foot-tall kunai grass underneath an ancient volcano—a quiet paradise about to be blown to bits by a sea of gunships cocked for battle. At six fourteen a.m., the sound of waves sucking at the hulls broke to loud concussions of cannon fire. A moment later, the shoreline exploded in fireworks. Then came the fighter planes, showering the grasslands and foothills with sheets of bullets. This went on, round after round, for three hours in an assault designed to scare the enemy and secure safe landing for the marines. But the salvos also shredded the ecosystem; the smooth, spongy earth became pocked with mosquito-friendly puddles.

When the cannonading finally ended, the 1st Marines swung their muscular legs over the ships' railings to climb down cargo netting that dropped them onto bobbing Higgins boats some sixty feet below. Fighter planes from the accompanying carrier fleet buzzed overhead, knitting a protective cover as wave after wave of packed boats spilled troops onto Guadalcanal's gray sand beaches. Here, five degrees south of the equator, thousands upon thousands of heavy boot marks further churned up the landscape.

More than six thousand marines landed on the northern shores of Guadalcanal that day, as did mountains of supplies, including jeeps, cases of canned food, boxes of spare parts, tractors, electric generators, carts of shells, antiair-craft guns, and tanks. Troops, surprised and energized by the lack of resistance, worked all day and night in bucket brigades, heaving materials from the beaches, through the island's belt of coconut palms, then the sharp, tall grasses, and finally to the shaded jungles. All the while they tore up whatever pristine grounds survived the salvos.

That first night was mostly quiet, broken by occasional rifle shots or rounds of machine guns—mostly friendly fire—and the rumbles of heavy cannonades and artillery twenty miles north on Tulagi and Gavutu. After dark, the sweaty and exhausted men bathed in the salty sea, laid out ponchos for bedding under palm trees, and drifted off. Many barely noticed as Guadalcanal's anopheline mosquitoes fed on their blood. They assumed the annoying stings were the least of their worries. No one watched as the biting females rested in shrubs to digest the stolen blood. Nor did the men see these insects later squat in countless puddles created by the invasion—to lay two to three hundred offspring each.

The next day, marines awoke to scout the area for Japanese soldiers. They found shacks with tipped chairs, half-empty drinks, and chopsticks still leaning on bowls of rice and meat. The Japanese air base—the number one target of the assault—was taken without gunfire. It too had been abandoned, along with a large booty of supplies, including new wooden barracks, vast tent encampments, shacks with iron beds, a bathtub, bottles of saki and wine, crates of canned food, cases of soda and beer, bicycles, trucks, jeeps, medical supplies, radios, and bed nets. Clearly, the U.S. assault had taken the Japanese by complete surprise. The enemy had set no trap and left in a rush. Evidence of the past months' activities abounded, while the infirmary held a single hint of the future. There, on a grass mat, an emaciated, listless enemy soldier tossed with fever, waging his own personal battle—at a cellular level—against malaria.[4]

ABOUT midday on August 8, a large formation of Japanese torpedo bombers filled the skies over Guadalcanal. The U.S. naval convoy "sent up a murderous fire" to shoot them down.[5] One crashed on the deck of the USS *George F. Elliott.* A huge explosion and belching inferno sent the navy's camouflage-painted workhorse to the bottom of the sound, taking with it all the medical supplies and equipment of E Company. Also lost were medical supplies of H and S companies, which were loaded at the top of the hole in Wellington—to be first off—but during maneuvers in Koro ended up at the bottom. These critical supplies "thus trapped, were never unloaded."[6]

The combined loss of vital equipment and medicines was barely noticed at first. Fewer than a hundred men occupied hospital beds, mostly for severe sunburn, dehydration, and heat exhaustion, plus a few bullet wounds caused by trigger-jumpy marines spooked by stories of Japanese snipers hiding in trees and bushes. By day four, a hearty perimeter around Lunga Point engulfed the newly captured airfield, which was converted for Allied use and renamed Henderson Field.

On day six things took a turn. Colonel Frank Goettge, an intelligence officer, set out with a patrol to scout the shores west of the perimeter but got pinned down on a beach by Japanese soldiers, who spent all night picking off the men of his patrol. Only three managed to slip into the dark waters and swim four miles back to the perimeter. Three nights later a fierce attack on Allied vessels sank three American cruisers and damaged a fourth. Fear of further losses, especially the aircraft carriers, compelled the commander of the carrier fleet to withdraw from Guadalcanal, "leaving land forces to care for their casualties alone," one medical officer complained.[7]

Still, casualties on Guadalcanal remained light, mostly because the enemy underestimated American troop strengths. This became clear on August 21, when Japanese colonel Kiyonao Ichiki landed nearly a thousand soldiers on the beaches near Lunga Point with instructions to take back the airfield. In what came to be called the Battle of the Tenaru, more than eleven thousand U.S. Marines massacred all but about a hundred of Ichiki's men. Over nine hundred mangled bodies littered the mouth of the Ilu River.[8]

Americans suffered so few casualties that medical officers wondered whether they overestimated their needs. The division field hospital had only 172 patients; the mobile medical units of E and A companies had another ninety or so—out of eleven thousand men.[9] By early September, however, admissions rates spiked. Horrid sanitation problems allowed flies to infest slit trenches and then swarm the mess halls, triggering outbreaks of dysentery. Food supplies ran dangerously low, so rations were cut. Undernourished and exposed to food-borne diseases, the men suffered from weakened immune responses, making them vulnerable to opportunistic microbes, like fungi that turned the men's flesh into festering, open sores they called jungle rot. Nightly rains and daytime humidity kept their socks and underwear forever soaked, making the infections particularly severe around the feet, groins, and buttocks. Units patrolling the perimeter remained on the move, marching through mangroves and over muddy hills. Whenever they could, they rested in water-filled tents or open fields, sweating, rotting, and swatting mosquitoes.

With the U.S. Navy's carrier fleet gone, Japanese transports easily landed reinforcements for their starving and diseased soldiers, who'd been living in caves and jungles since the Americans landed. Meanwhile, few additional supplies arrived for the marines, and limited U.S. aircraft protected the skies. Every night, when exhausted and hungry Americans fell into their wet holes to sleep, Japanese nuisance raiders dropped bombs, fired shots, and then disappeared into the black. Systematic sleep deprivation worked like torture and sent dozens over the edge. But overall casualties still remained lower than expected. The marines were tough and killed far more Japanese than they lost in battle.

They also collected prisoners—a move that would be their downfall.

These soldiers and laborers had spent months on the island building an airfield. Guadalcanal's hypervirulent strain of *vivax* had long since passed to them, a hundred of whom were now held captive inside the American perimeter. This proximity combined with an enormous production of mosquitoes created perfect conditions for a perimeter-wide outbreak. The first signs of it appeared in mid-September when forty-eight marines checked in with fever. A handful had dengue (a mosquito-borne viral infection aptly nicknamed break-bone fever); the rest shivered with malaria.[10]

IN the late 1800s the Solomon Islands became part of the British protectorate. Independent companies leased land and pressed local farmers into hard labor on plantations of rubber and copra, a coconut product used in soap. Over time, local ways broke down. And new malaria strains surely came and went with each group of European overlords.

But never did the Solomons see the kind of malaria that was about to develop on Guadalcanal. The more each side lobbed bombs and carved waterlogged tire tracks into rain-saturated soil, the more nature conspired against both. "Once a trail became impassable because of rutting, a new trail parallel to the old one would be started, adding to the miles of ruts," observed one medical officer.[11] Roads were built hastily, creating hard-to-eliminate water catchments. All contributed to mosquito production.

It wasn't possible for the Seabees to establish a malaria-free zone in which sick marines received treatment and convalesced—as envisioned by medical commanders—partly because of lost supplies and poor planning, but mostly because of the devastated ecosystem. Mosquitoes bred virtually everywhere: in blown-apart mangroves, splintered jungles, clogged rivers, debris-filled foxholes, bomb craters, boot marks, tire tracks, trash piles, discarded cans, and empty barrels. The destruction was nowhere near over. Japan kept landing troops. And the Americans kept digging in. The ninety-by-twenty-five-mile island, with its razorback ridge and sprawling wetlands, became a stepping-stone toward world domination. Both sides conscripted the local people into labor, burned their villages, and destroyed their fields. The old African proverb "When the elephants fight, it is the grass that suffers" precisely fit the scene. But these elephants of war weren't just killing the grass; the destruction was so over-the-top that it gave nature a chance to fight back.

American medical officers asked unit commanders to insist on long pants and shirts at night, and begged to have malaria-seeded POWs and conscripted laborers removed from the perimeter. But these officers had no authority and couldn't convince commanders that antimalaria standards were at least as important as combat readiness. The mere few dozen cases of malaria led one high-ranking officer to declare: "We are here to kill Japs and to hell with mosquitoes."[12]

CHAPTER 15
A Malaria Manhattan Project

Just as the Manhattan Project recruited the country's top physicists to work on an atomic bomb, the malaria project recruited top medical investigators and organic chemists to build a chemical bomb against malaria. No more quinine from Java meant the War Department needed something else, fast. The British should have led the research, but couldn't because of air raids on their chemical labs and severe shortages of everything needed to make anything, even flour for bread. So Americans led the way.

But without enough malaria experts in the United States to do it all—many were recruited into the military, like Paul Russell—Surgeon General Parran reached into other fields to compel researchers into the project. These efforts brought in heavy hitters, including Dr. James Shannon, a kidney expert and medical professor (and future head of NIH) at New York University; Dr. Eli K. Marshall, a physiologist, kidney expert and trained organic chemist at Johns Hopkins University; Louis Fieser, a chemist, author, and expert in steroids and opium alkaloids at Harvard; Dr. Robert F. Loeb, a pioneer in blood electrolyte research, professor of medicine at Columbia University and associate director of its neurological institute; and Lyndon Small, an organic chemist and expert in morphine and opium alkaloids, head of the Drug Addiction Laboratory at the University of Virginia, U.S. technical representative to the League of Nations, and editor in chief of the *Journal of Organic Chemistry*. All these men were giants in their fields and members of the National Academy of Sciences. They brought in top-rate assistants, including Kenneth C. Blanchard, an MIT biochemist and mathematician with expert knowledge of sulfa drugs; Leon H. Schmidt, a medical investigator with expertise in primates and a director at the Christ Hospital Institute of Medical Research in Cincinnati; and soon, Dr. Alf Alving, a Swedish renal expert at the University of Chicago, whose thick accent, sharp wit, and excellent investigative skills were well-known to and deeply liked by Drs. Shannon and Marshall—the two who would soon take charge of the malaria project.[1]

These malaria rookies traded sterile labs with petri dishes and protective hoods for humid insectaries buzzing with different species of anopheline mosquitoes, or laboratories stacked with cages of chicks and ducklings, all of which had to be infected with malaria by hand using a syringe with parasite-filled blood. Or they had to learn how to give malaria to lunatics in state hospitals and psych wards. All had to quickly absorb the deep complexity of malarial microbes. What they brought to the table, however, was huge. The blood experts offered insights on what the microbes actually did to the immune system and were masters at tracking absorption rates of chemical compounds in tissue—which signified the peak of activity against invading germs. This invaluable technical know-how came with strong personalities, and a lot of bickering and complaining about one another.

Marshall at Hopkins used blustery language and expletives to make important points that often were at odds with the parasitologists and malariologists, whose investigations, he felt, lacked quantitative rigor. He found their use of bird malarias as a means to understand human malarias preposterous. These were alien classes, he would say. Bird malaria and human malaria looked nothing alike.[2] But these alien bird microbes were all the malariologists could offer—human ones couldn't be grown in vitro or in *any* lower animal. Initial screenings on live chicks sick with bird malaria were all they had.[3] So Marshall cursed and complained, then ramped up the number of chicks and ducks used in studies to create quantitatively significant data that showed how and when bird studies could be extrapolated to actually mean something with respect to the human studies. Johns Hopkins's malaria lab soon had cages of chicks infected with *P. gallinaceum* piled high, almost to the ceiling. Their delicate down feathers wafted in the air as Marshall blew by on his way to some important meeting.

Meanwhile, Lowell worked his job in Africa, knowing the group he helped launch was ballooning in size and complexity—and, hopefully, efficiency.

. . .

THE number of American insectaries mushroomed. Everyone wanted their own mosquitoes, so new insect labs opened in Chicago, Washington, New York, Memphis, Atlanta, and Boston. Larvae were raised in shallow enamel-coated pans inside screened rooms. Nets covered the pans to catch the insects as they molted into flight. The captured mosquitoes were then put in jars with screen lids and pressed against the bare arms, legs, and bellies of an increasingly large number of deranged patients strapped to their beds. As the insects used needle-like proboscises to slice into the patients' flesh and either suck in the sexual-stage microbes from the already infected or spit newly hatched offspring into the still well, these unwitting patients contributed to the war effort. The insectaries weaponized millions of mosquitoes contained in small rooms that, if allowed to escape, would surely infect residents of the adjacent cities, at a time when the world's quinine supply belonged to the Japanese. These hothouses were a biohazard and had to be kept off rooftops to protect "against air raid damage" by order of federal authorities.[4]

Many scientists for the first time, ever, stepped into asylums and tried their hand at infecting with malaria helplessly confused patients.

Shannon just a year earlier had set up a clinical investigation unit on renal diseases at Goldwater Memorial Hospital. When he got the call to do malaria, he converted his clinic, brought in paretics from other institutions, and eventually absorbed Bellevue and Manhattan State hospitals into his project. Lyndon Small, the master organic chemist, moved to the South Carolina State Hospital in Columbia to work alongside easygoing Martin Young, and the more cocky and difficult, although highly competent, G. Robert Coatney. Between the three of them, they ran a program in which Small made the drugs, Young raised the mosquitoes, and Coatney studied patient blood for parasites. Dr. Mark Boyd, meanwhile, used his Rockefeller-run lab at Florida State Hospital in Tallahassee to train newcomers in the intricacies of manhandling lab-raised strains of human malaria, both in patients and mosquitoes.

These cooperating state hospitals struggled financially and welcomed the government-sponsored support. In many ways, patient care improved because the malaria doctors were so attentive, constantly taking temperatures, blood samples, and urine samples, and controlling patient fevers with different drugs and dosages. Patients also were fed well to avoid life-threatening symptoms that malaria often triggered in the malnourished. Some were even cured of their deadly, demoralizing syphilis infections.

From today's perspective, the whole enterprise was fraught with troubling ambiguities. Jauregg's Nobel Prize legitimized malaria as a treatment for syphilis. Here, however, scientists launched a large-scale project that used paretics in drug studies *and* to warehouse microbes that were then shared with other researchers. The work was kept under a veil of secrecy, as the public couldn't be expected to understand the science of it all. Every able body was needed to win the war, including the insane.

And so an organized, secretive project designed to build a chemical bomb against malaria-causing microbes grew in size and stature, with Surgeon General Parran as its chief political cheerleader and head recruiting officer.

THROUGHOUT the first half of 1942, the work was paid for and supervised by the National Research Council and the U.S. Public Health Service, which got funding through the Committee on Medical Research, the medical arm of the White House's Office of Scientific Research and Development—an umbrella organization that oversaw all war-related, science-based work, including the Manhattan Project. OSRD was a unit of President Roosevelt's Office of Emergency Management, where moneys were made available for secret projects not subject to public transparency rules.

An alphabet soup of offices and committees formed a confidential funding stream to support this antigerm weapons program. The resulting string of acronyms and abbreviated titles ran a line from the White House to the researchers, with the OEM funding the OSRD's CMR, which commissioned the NRC to dole out contacts to the NIH, the U.S. PHS, and Rockefeller's IHD.

A small cluster of scientists—a company by military measurement—formed the hierarchy of this medical weapons program, which would soon eat up a quarter of all funding available for war-related medical research. They developed their own fiefdoms. Each took on a different aspect of the problem, hired their own people, and spoke a jargon-laden language that few on the outside understood. Subcommittees to bring in different talents for the different contracts were formed on an as-needed basis.

The conceptual framework was ideal for research—a bottom-up approach in which scientists steered the course. The feds were there merely to pay the bills.

Federal overseers didn't understand malaria well enough to act as directors. But they saw the effects of malaria in Bataan and braced for more outbreaks in the Pacific, and committed to keeping the money spigot open.

This hands-off approach wouldn't last long. By the fall of 1942, the structure of the Malaria Project grew rigid and the NRC no longer decided who got money; the Committee on Medical Research did. Lead investigators used to talking to Lowell or one of the other central figures for money to explore a research question now sent in formal proposals and had to appear before a review panel, if necessary, in the Carnegie Institution's offices at 1530 P Street in Northwest Washington. The new structure ruffled some feelings. For example, Carl Johnson at the Gorgas Memorial Laboratory in Panama City, who worked with Lowell and had ties to Rockefeller and the U.S. Public Health Service, saw his project zeroed out—the committee refused to give him more money because they couldn't see the value of his work, which included giving native laborers a type of malaria carried by spider monkeys, called *P. brazilianum*, to measure their immune response to it. Johnson had been working in tandem with Lowell, who had been doing similar work with different monkey malarias at the Battle Creek asylum. But the OSRD's medical committee stamped Johnson's work "no good," while it called Lowell's "good but subject to improvement"—largely because he was in Africa, running his project in absentia.[5]

Other vaccine proposals were also rejected—as Lowell and others had provided enough evidence to make those studies appear naive or misguided. The committee refused many ideas, like one to reopen large studies of Paul Ehrlich's methylene blue because the drug had already failed many decades of testing.[6] Fruit enzymes for possible antimalarial properties didn't sound plausible. But studying California plants and extracts for activity against malaria did, so a small grant went to that. Even though the oxygen-choking German drug plasmochin was "out of the realm of any therapeutic promise," the committee approved just about every proposal that examined analogues of this 8-aminoquinoline. There was just something about this drug that on paper suggested it should work—maybe even be a "cure" *and* a transmission blocker (that is, work against the tissue stage of the parasite and the sexual stage). If only the side chains could be rotated to make one that was toxic enough to kill the microbes but not so toxic as to also kill the lab dogs, because dead dogs meant a compound was too poisonous for human studies.[7]

. . .

BY late 1942, almost half of the project's overseers were in the military, with a typical meeting consisting of twelve civilian scientists, ten military representatives, at least two staffers from the United States Public Health Service, and a high-ranking member of the OSRD's Committee on Medical Research (someone who judged the malaria work in the context of all medical projects linked to the war effort).

Meeting notes went to the chair of the Committee on Medical Research, A. N. Richards, who was famous in his own right and a distinguished professor of pharmacology at the University of Pennsylvania. He answered to the head of the OSRD, the brilliant inventor and technology genius Vannevar Bush—who one year after the war would conceptualize the infrastructure for the World Wide Web. The renowned Vannevar Bush had been President Roosevelt's top science adviser before the war, and convinced him to create the OSRD to augment military research. Soon, the OSRD led the way in radar technology, submarine tactics, secret devices for the OSS, nuclear weapons, and, of course, medical research. Bush was a man so firm in his positions that he once openly disagreed with Winston Churchill during a top-secret meeting on the atomic bomb project—and prevailed. A 1942 issue of *Collier's* described Bush as the "man who may win or lose the war."[8] His OSRD created the scaffolding for today's government support of science. On the medical front, it would lead the way for a major expansion of the National Institute of Health, breaking it into many separate institutes.

Those disgruntled by the OSRD takeover of the Malaria Project were appeased by concessions, which included more money and the power to draft trained medical professionals directly into their projects. Under the OSRD umbrella, the malaria work was presumed important, which encouraged cooperation across all sectors. More than fifty chemists at American university labs, including UCLA and Cal Tech, volunteered to help make compounds. Additional drug companies also signed up, including DuPont, American Cyanamid, Abbott, Eli Lilly, and Dow Chemical Company. These universities and commercial labs pumped out new compounds for testing on patients at seven state hospitals—Lowell's in Michigan, Rockefeller's in Tallahassee, Shannon's three hospitals in New York, the U.S. Public Health Service's two labs at St. Elizabeth's (in cooperation with the army), and the South Carolina State

Hospital in Columbia (run by Lyndon Small's group). Others, like Boston Psychopathic and Manteno State Hospital, would soon join.

The number of compounds, lab studies, clinical studies, and toxicity tests outgrew the small organization's administrative structure, so a clearinghouse for information was set up at Johns Hopkins University, under Eli Marshall's muscular control. This created a second hub for the project, which Marshall called "the high command in Baltimore."[9] There, a humble, hardworking physician named Frederick Wiselogle cataloged all incoming compounds, as well as those already tested by Lowell and his colleagues in the previous two years. All were assigned survey numbers, or SNs. The compound Winthrop had sent Lowell in 1940—which worked well against bird malaria and was never heard of again—entered the ledger as SN-183, indicating it had been the hundred and eighty-third compound cataloged. Kenneth Blanchard, the MIT biochemist recruited by Marshall, asked if he could explore this SN-183, as it looked a lot like the 8-aminoquinolines but different enough to be important, perhaps. But he was ignored.[10]

Wiselogle's catalog system came to be known as the "survey" of antimalaria drugs, which created a mainframe of information for administrators to work from. Surveyed compounds shelved for, say, toxicity were redlined to avoid further testing. The unique properties of the incoming compounds were known only to top administrators to protect the intellectual property of each participating drug company. But they all received details on failed compounds, to avoid duplication. This kept chemists focused on new possibilities and away from the old ones, like SN-183—which should have been reexamined, as Blanchard suggested.

DR. Small's project was the largest and most comprehensive. He and a small team made the compounds; tested them on chicks infected with bird malaria; ran toxicity tests on dogs, rats, and monkeys; and conducted trials on people. As could have been predicted, he stepped away from Jauregg's dictates. Martin Young raised mosquitoes that physicians then used to transmit *vivax* and *falciparum* to more than five hundred male patients. Half had dementia praecox (schizophrenia) and were used for drug testing, and half had syphilis and were used as controls. His studies produced promising results with the sulfa drugs, especially sulfadiazine.

With Small as a model, James Shannon ramped up his project. His first chore was to increase his access to people, or "clinical material," which he referred to simply as "material." In an October 26, 1942, "Dear Coke" letter to Dr. E. Cowles Andrus, a famous cardiologist from Johns Hopkins who oversaw several medical projects for the OSRD, Shannon mapped out his plan to double his human subjects by using two new groups. In addition to syphilitics, he proposed using conscientious objectors and schizophrenics. He surmised that the New York metropolitan area had enough of this type of "material" to meet his needs. So he sought permission from William Tiffany, the commissioner of mental hygiene for New York, to concentrate "it" all into one malaria ward at Manhattan State Hospital. He felt localizing patients into a single place would increase the quality of care and data collection. On November 4, an assistant of Tiffany's wrote to Richards approving Shannon's plan.[11]

This would be the first in a series of project expansions, giving scientists greater access to more needed "material." No one openly questioned the ethics of using nonsyphilitics; perhaps project scientists thought fever therapy might actually help schizophrenics. Jauregg had passed away in 1940 and was no longer around to guilt malariologists into using his therapy as conservatively as possible. No studies had definitively ruled out other classes of the mentally ill—although Jauregg had provided ample evidence suggesting they would not benefit. And the war effort needed bodies. As a group, the Malaria Project took a giant step away from the principles set out by Jauregg and marched forward with new ones—ones that made a broader range of mentally crippled patients fair game for the malaria work.

SUPERVISORS, including Richards and Vannevar Bush, were told the malaria problem would be solved quickly. Henry E. Meleney, the aging president of the American Society of Parasitologists, a special adviser in the program's chain of command, and secretary pro tem of the malaria conferences, in January 1943 wrote the following to Richards: "It is impossible to predict time required for the discovery of a satisfactory substitute for quinine or atabrine, but it may be accomplished within a short period."[12] Richards sent the memo to Dr. Bush, no doubt to assure him that malaria would soon be under control and that he need not divert his focus from the OSRD's number one priority, the Manhattan Project.

CHAPTER 16
Better to Be Yellow

By the invasion of Guadalcanal in August 1942, the Medical Corps had a rough estimate of quinine supplies for the Allies: enough for about a year, based on an assumption that the Southwest Pacific and China-Burma-India regions would be the only areas complicated by malaria. Medical planners presumed the scheme for Europe hadn't changed: that American and British infantry would cross the English Channel to broadside Hitler in northern France, where malaria would not be a problem.

However, the surgeons general of the army and navy knew otherwise; the new front against Hitler would involve malaria-seeded regions of North Africa and Italy. For this, war efforts on both sides of the world would need a quinine replacement—soon. Dreams of a vaccine lived on. But most in the chain of command just needed a chemical-based drug to replace the lost quinine. Everyone agreed on this. The debate, then, was over where to concentrate energies, with one faction pushing the need to find a radical "cure"—to keep at the 8-aminoquinolines until they came up with a safe analogue. Others wanted to explore whether a prophylactic could be derived from already known fever suppressants, especially aspirin. They quickly learned it couldn't.

By the time the first cases of malaria trickled into field hospitals on Guadalcanal, the Malaria Project had tested more than five hundred potential chemicals, with particular attention to three groups: sulfa-based compounds, variations of the unusual yellow drug atabrine, and derivatives of plasmochin—all originally German-made.

Good outcomes with the sulfa drugs were reported out of Africa, in Lowell's studies, and out of Dr. Small's project at the South Carolina State Hospital. But they all showed the compounds worked well against *falciparum*, not against *vivax*. So chemists worked on different sulfa analogues to see whether they could make one against *vivax*. Nothing definitive turned up yet. Meanwhile, plasmochin toxicity kept disappointing everyone, as human studies confirmed its dangers.

That left atabrine. The army surgeon general already had been told in July that this bad drug damaged the liver. "In spite of the streams of propaganda poured out by the German Chemical Trust and by its subsidiaries in the United States," its toxicity made it useless to the army.[1] But it was all they had.

American chemists used three names for this bitter yellow pill: quinacrine, atabrine, and 9-amino acridine. Quinacrine was the name given by its American manufacturer, Winthrop Chemical Company; atabrine was the German trade name; and 9-amino acridine was its chemical name. To make it, chemists boiled down several thousand barrels of the industrial fabric dye acridine yellow to be used as the base molecule, with a long side chain ending in an amino group attached at the ninth position. The work required specialized glass-lined boilers and ingredients made only by I.G. Farben in Germany. Making it became a big problem once the United States entered the war.

Other options were explored. The U.S. State Department launched an emergency project to grow quinine in South America, involving a dramatic rescue of cinchona tree seedlings just after the Bataan surrender. Army intelligence officer Colonel Arthur F. Fischer had previously served as chief of the Philippines Bureau of Forestry and had commandeered seedlings from Java to start a plantation on the southernmost island of the Philippines chain, where the humidity helped the finicky trees grow. President Herbert Hoover had ordered the creation of this cinchona tree farm in 1917, amid the malaria nightmares of the Great War. By 1936, the plantation was producing modest amounts of totaquine—a weaker version of quinine that worked well when mixed with pure quinine. When Bataan fell, MacArthur sent three Flying Fortresses to the island to pick up key personnel. Two of the planes were shot down, but the third made it. That plane carried Colonel Fischer and a supply of his cinchona tree seeds to Australia. Then Fischer flew to Washington.[2] By the summer of 1942 his seedlings were sprouting on a farm in Costa Rica set up by the State Department with emergency funds. This plan, however, offered no guarantee. "Cinchona trees are delicate and difficult to grow," Fischer warned. Their need for specific humidity, temperature, elevation and rainfall "are very exacting."[3] Full yield would take seven years. No one knew when the war would end so Fisher's project became a high priority. But for the war's immediate needs, it was useless.

The National Research Council, meanwhile, began studying the alkaloids of lower-quality quinine to measure whether the United States should buy as much of it as they could from South American cinchona plantations, which the United States had always snubbed because of their inferior quinine. Now researchers wanted to know whether totaquine and another product called cinchonine, also derived from these inferior trees, could be used to prevent malaria. This was considered a sidebar to the main goal, which was to find new drugs even better than the best quinine. So the assignment went to a new doctor, George A. Carden, a Boston physician with no malaria-related experience, but keen administrative and problem-solving skills. He took the task seriously and quickly solicited Merck, S. B. Penick Company, New York Quinine and Chemical Works, and Inland Alkaloids Company to process any suitable cinchona bark imported from South America.[4] During this work, federal officials uncovered a Nazi-led effort to buy up all South American bark, denying the Allies access to it.[5] This raised an important question: Why did the Germans want quinine if they were using atabrine? Was atabrine as bad as everyone thought, or were they just trying to deny the Americans access to a quinine replacement? No one knew.

IN May 1942, the NRC summoned Lowell Coggeshall home from Africa to brief project scientists on his successes using atabrine along the air route. Lowell caught three cold, rattling, all-night military flights over the Atlantic, up the coast of South America, and into Miami, then a commercial flight to Washington, DC. On June 5, at the NRC's headquarters on Constitution Avenue, near the Lincoln Memorial, he presented his findings. In side-by-side studies, using quinine as a control, he reported that both drugs acted equally in protecting pilots from acute symptoms—though the pilots still carried malaria in their blood. Neither drug cured the men. Atabrine, like quinine, generally took about ten days to suppress the worst symptoms—not ideal. But it saved hundreds by preventing cerebral malaria. Pan Am lost no employee to severe malaria, which showed how important it was to have something, be it quinine or atabrine.

But because neither prevented malaria—they only mitigated symptoms—Lowell reiterated his faith in the sulfa-based compounds. He reported the results of a study in Sierra Leone in which the British used sulfa to cure RAF pilots of African *falciparum*. He theorized that the right sulfa mixture could also be used as a prophylactic. Then he stated the obvious to military leaders still not ready to hear it: Screens would be their best weapon, then bed nets, long sleeves and pants, and education—tell troops the truth. Educate them on malaria and how it gets into mosquito spit. Warn them emphatically of the dangers, teach them how to prevent mosquito bites, and threaten them with demotion if they got sloppy. *Then* have drugs as a backup, because that's all they're good for.

That was his best advice. A few days later, Lowell brought this same message to the War Department, where he briefed key staff on malaria prevention. Then he returned to the West African jungles, and flew across the Sahara and over the Arabian Sea for his next assignment: sanitizing malaria from airfields in India so fever-free pilots could fly supplies over the Himalayas, for Allied forces about to fight Japan for Burma.

AT the June 1942 NRC meeting, Colonel Simmons made the military's instructions clear: They wanted the project to figure out how to make atabrine workable. No one liked this drug. But it was all they had. The FBI pressured executives at the Winthrop Chemical Company to open up licensing rights to other companies.[6] The agency was working from a 1941 investigation by the Justice Department that accused Winthrop of maintaining illegal business ties to Bayer in Germany after the American blockade of German products. Those records also showed that Winthrop relied on Germany for key ingredients and that Winthrop, trying to make the drug on its own, had produced flawed stocks that the company sold to the military. Evidence of this was coming in from the field. For example, secret reports from Panama described U.S. Army doctors attempting to use atabrine to fend off malaria during summer maneuvers. The doctor in charge reported poor results, attributing the problem to Winthrop-made pills found undissolved in soldiers' bedpans. They were covered in a thick shellac that even stomach acids couldn't dissolve. The lots of atabrine made in Cologne, Germany, however, dissolved quickly and worked well in patients. The canal zone's commanding general sent back Winthrop's atabrine and requested that the U.S. surgeon general ask the company to try again.[7] Another report from

Australia revealed in a secret study using atabrine that the drug was not working at all, although none was found in bedpans. Drs. Shannon and Small also had a hard time establishing adequate atabrine levels in blood when they used stocks made by Winthrop. All this raised suspicions and led a few vocal critics to accuse Winthrop of being pro-Nazi.

Military leaders wanted no part of this larger political debate of Winthrop and its legacy as a partly German company. The Medical Corps just needed Winthrop and other drug makers to get the formula right and begin producing 0.1-gram doses by the millions for use overseas. Winthrop executives pushed back, insisting that the company patents be respected—even though they were written when Winthrop still partnered with Bayer. That partnership was severed by the outbreak of war. The U.S. Custodian had seized Bayer's holdings in the United States. But that seizure gave the government only one vote of three on Winthrop's executive board. So the custodian couldn't legally force the government's hand without making the dramatic move of seizing Winthrop's patents. Such an extreme measure would have alarmed other American companies that held patents written through partnerships with German companies, prewar. But above that was the military demanding that nothing slow production of atabrine. The Medical Corps just wanted the atabrine problem worked out.

During this time, the military secretly asked Allied Chemical & Dye Corporation and Pharma Chemical to make bootleg versions of the 0.1-gram pills, to see if they could. Company chemists had to improvise on ingredients, because they lacked complete patent information—which Winthrop claimed not to have. Nonetheless, these companies managed to make the pills, which, for the most part, worked about as well as the German-made originals. So the War Department commissioned them to make millions more, packed for shipment overseas.

In the meantime, Shannon used reserves of German-made atabrine on his paretics in New York. In large doses of the drug, the side effects turned volatile, from gastro disruptions to paranoia. Lower doses produced uneven results. No pattern explained why some patients soiled their beds or doubled over in pain while others remained fine. But no one died or became dangerously ill.

He also measured atabrine levels in patient blood and urine to track absorption and pinpoint when enough was present to start killing off parasites. This also proved tricky, because absorption rates varied greatly. But his general impression was that the drug started working around the time the dye saturated tissue and turned patients yellow. Military leaders had no other option. They couldn't afford another Bataan. And Shannon's studies showed atabrine to be relatively safe. So, in early September, the navy approved the use of atabrine in the Solomons.

BY late September 1942, the first shipments reached Guadalcanal. Orders spelled out in a circular letter instructed medical officers to make sure each man took two 0.1-gram pills twice a week, on nonconsecutive days, with a meal. This was the established dosing recommended by Winthrop and shown by Shannon to sufficiently saturate tissue. The Medical Corps set up distribution in the mess tents or with rations out in the field. Enough of the men complied for medical officers to observe trends. They reported that some pills did nothing—presumably because the active ingredients degraded during the trip overseas—while others caused reactions. The most hair-raising were rashes that, in a few men, progressed grotesquely, with skin falling off in sheets, creating open sores that attracted flies and oozed with pus. Another few developed erratic mood swings, violent anger, and deep depression. More than half experienced the standard diarrhea, vomiting, and cramps.

Many of the marines stopped taking the pills. "Lack of supervision by the responsible line officer became apparent when hundreds of tablets were picked up by messmen," recorded one navy medical officer.[8] Malaria rates quickened, with more than three hundred cases in the hospital. So commanders formed choke points in the mess tents. Unit commanders placed the pills in each man's mouth, but many, and perhaps most, simply stepped outside and spit the pills out, turning the sands yellow.

These men weren't casually insubordinate. They just couldn't tolerate the drug. It made many vomit up rations that were hard enough to swallow (mostly Spam mixed with worm-infested rice left by the Japanese). The food situation was bad, really bad. Rations had almost run out. The men were hungry, sunburned, and dehydrated—and unwilling to take another hit against their health.

In late September, the rains fell, filling every divot and rut, every bomb crater and boot mark. Mosquitoes squatted in a million little puddles to lay up to three hundred eggs each. The more the two sides lobbed artillery shells, sank boots into the grasses, and carved waterlogged tire tracks into the saturated soil, the better the terrain for mosquito breeding. Never before had the island supported *this* kind of mosquito production.

When the insects rose up by the millions, no entomologist was there to distinguish among them. All six of the South Pacific's malaria-carrying anophelines bred on Guadalcanal. But only one could potentially cripple forces: *A. farauti* Laveran, named for the French military doctor who first spied malarial parasites under a microscope.[9] This mosquito preferred humans to animals and could fly more than a mile to hunt down half-naked marines fighting in the woods. *Farauti* also bred just about everywhere—sunny *and* shaded waters, tidal- *and* river-fed swamps and lagoons, streams *and* wells, trash piles *and* slit trenches. The larvae developed into mosquitoes as quickly as seven days—whereas others took up to two weeks. Once in flight, with the right weather conditions, it hunted for blood as early as three p.m., long before expected and many hours before troops tucked themselves under mosquito nets—the few who had them.

By October, the number of hospitalized Americans on Guadalcanal exceeded two thousand—out of eleven thousand.[10]

Washington responded by sending another circular letter, which arrived in early November. It ordered everyone to take atabrine or risk court-martial. Commanders whose men failed to turn yellow would be held responsible and risk demotion.

Unit commanders begged and cajoled and threatened the men, with some success. But then a rumor spread that the drug caused impotence. It didn't matter that commanders and medics denied the rumor; the men shut down. They refused to follow orders, evident in the thickening yellow sands around the mess tents. So the rules changed again. Unit leaders had to place two pills in each man's mouth, observe his Adam's apple glide up and down in a swallow, and then check the tongue and cheeks to ensure the pills went down.[11]

More men finally turned yellow, but the infection rates continued to soar—more than thirty-two hundred hospitalized in November.[12]

Farauti just kept on breeding in every discarded tin can; every footprint, tire track, and bomb rut; every slit trench, foxhole, and drainage ditch. Her larvae kept molting. She kept feasting and spreading malaria while the stench of decaying coconuts and human waste hung in the steamy air. As higher-ups debated the pros and cons of atabrine, Guadalcanal turned into hell on Earth—a wet, bug-infested, trash-strewn, rotting hell.

When the division and company hospitals filled, medics placed the sick shoulder-to-shoulder under coconut palms, the fronds long since blown off.

There for days they convulsed through chills, then roasted with fever, desperate thirst, and pounding headaches.

For every battle casualty, ten men lay sick with malaria. The landscape was of bald trees, rutted fields, flowing mud, and cot after cot of sick marines. The official calculation of the malaria rate was expressed as nearly three thousand infections for every one thousand soldiers, per annum—and that was just hospital admissions. At that rate, every man would have malaria three times by the end of a year. By comparison, the VD rate among marines before they embarked for Guadalcanal was forty-five per thousand, per annum—which had alarmed navy medical officers. Now those rates were nothing. Marines fell to the ground by the dozens every day. Officers nicknamed the island Operation Pestilence.

Overwhelmed medics told headquarters that heaps of bullets and firearms lay unused. Navy medical corpsmen needed more medicine, preferably quinine. They also needed guidance on what to do about the mosquitoes. And replacement troops were needed to relieve the demoralized marines now racked with disease.[13]

Malaria also devastated the Japanese. But a nightly flow of transporters sneaked through the dark waters off the island's northwestern shores to land fresh soldiers—a waterway the Americans called the Tokyo Express. These new arrivals remained healthy for three or four weeks before feeling the effects of the island's near–100 percent malaria transmission rate. The Express landed roughly sixteen thousand Japanese troops, who quickly dug in to the west and east of the American perimeter, determined to take back their airfield. Hirohito's military strategists decided Guadalcanal was *the* battleground to determine the fate of the war, and they threw his remaining navy and infantry at taking it back. His forces were amassing on the nearby island of Rabaul, readying to overwhelm American defenses and kick the marines back to New Zealand and Australia.

The American carrier fleet that had fled shortly after the landings floated somewhere near Australia preparing for a return to Guadalcanal, but no one on the island knew that. For all they knew, they'd been abandoned there, starving, sick with fever, and wrung out. They had the distinction of harboring strains of malaria that relapsed as often as the mosquito populations bred—a mysterious synergy rooted in the evolution of the germ and its insect host. Mosquitoes picked up new strains of *vivax* and *falciparum* the Japanese brought from other islands, then bit the marines, launching new infections just as their initial infections rebounded for a second attack. The men were doubly loaded with different versions of the same blood-destroying disease.

Relief came in November, when the American carrier fleet returned from Australia and gave cover to land replacement troops, food, mosquito nets, and more atabrine. The men dreaded the arrival of more of the gut-rotting pills. But they loved the nets. They had slept on the ground, usually atop their ponchos to keep the wet earth from saturating their clothes, and used blankets to stop mosquitoes from biting their faces and interrupting needed rest. Now they had nets to stop the biting. "But the nets really came too late. We were full of malaria," observed one marine.[14]

The replacements troops were no panacea. The 8th Marines came seeded with a nasty microbe of their own, which they picked up during training on the island of Samoa. This other mosquito-borne parasite caused a horrible disease called filariasis, discovered a hundred years earlier by Ronald Ross's mentor Sir Patrick Manson. These microscopic worms, once in the blood, were known to grow several inches long, clog the lymphatic system, and, in textbook cases, cause severe swelling of the limbs and scrota—a condition called elephantiasis. (A famous photo in the annals of public health is of a barefooted Asian man carrying his freakishly enlarged testicles in a wheelbarrow.) But this long-term degenerative disease wasn't like malaria; it didn't cause acute attacks, nor rob men of their ability to fight. For the time being, the men of the 8th were a godsend. They helped tired and diseased troops hold on to Henderson Field and root out enemy snipers. Then they, too, contracted malaria, on top of the yet-to-be-diagnosed filariasis.

By December, the navy evacuated more than four thousand marines, mostly for relapsing malaria.[15] Those remaining were finally replaced later that month with units of what came to be known as the Americal Division—combined U.S. infantry units under MacArthur's command made up of the army's 164th, 182d and 132d Infantry. These young soldiers, with fleshy cheeks and full waistlines, relieved the emaciated marines.

Once off the island, all marines were ordered to their unit medical officers for physical examinations. Most had survived at least one bout of fevers and now complained mostly about rotting flesh around their groins, buttocks, and armpits. They'd lost fat and muscle, which made them look old, even though, on average, they were younger than the fresh army soldiers coming ashore.

Some ten thousand sickly marines—who just months earlier used thick, muscular bodies to haul millions of pounds of supplies onto Guadalcanal's volcanic beaches—were shipped to Australia. There, doctors examined their skeletal bodies, only to find 34 percent unfit for further duty, mostly because of relapsing malaria.[16]

Their fate was still undetermined. They couldn't be sent home, because they were malaria carriers and the U.S. Public Health Service feared they would bring the Pacific's vicious strains of malaria back to the United States, where they would spread rapidly throughout the South. These men had endured unspeakable conditions on Guadalcanal, but couldn't return to the arms of their mothers. For the same reason, they were unwanted in Australia, New Zealand, Fiji, and other nonmalarious islands. They were a burden to already overtaxed commanders focused on the logistics of war. Rehabilitation of spent troops wasn't possible. For a short-term solution, "convalescent" camps were set up around the theater on the promise that the soldiers would be isolated from population centers and confined to screened compounds to prevent their malaria from seeping out.

These troops had it made, compared to two thousand of their fellow marines who passed their health examinations and were loaded onto transports headed for Milne Bay, along the malaria-endemic southern coastline of New Guinea. In December, these 2nd Marines relieved an Australian force that was 80 percent infected with malaria, half of which was *falciparum*, with some turning into dreaded blackwater fever. The Americans, in taking control, immediately tore up the ecosystem by hauling supplies, building roads, and digging trenches. Mosquitoes bred like crazy, including the dreaded *A. farauti*, which passed parasites from the Australians to the Americans.

These Americans in New Guinea skipped the convalescent camps in Australia and Fiji because their fevers hadn't relapsed. For unknown reasons, the parasites they picked up on Guadalcanal burned out in their blood, which made them look immune to the paralyzing relapses. For this strength they were sent back to the front lines. But New Guinea's mosquitoes carried a different strain, the most virulent encountered. Three weeks after the soldiers landed, it hit hard. First the chills, which the men called "seizures." Then the "oven." Then the "bone-cracking rack."[17] Then the chills again. They wrapped blankets over their heads, cinched at the neck, and shivered with chattering teeth.

After a month in Milne Bay, the American malaria transmission rate measured in at thirty-three hundred cases per thousand per annum—everyone would soon get it at least three times in the coming year.[18]

Guadalcanal and New Guinea veteran Robert Leckie, of the 2nd Marine Division's E Company, described his attack on New Guinea:

> To lie on my back was torture, to lie on my stomach a torment. I tried to lie on my side, but even here my bones ached as though they were being cracked in the grip of giant pliers. I could not eat, I could not drink—not even water. They fed me through the veins, intravenously . . . [for] ten days, two weeks. All the time I lay baking—not burning or flaming, understand, but baking, as though I were in an oven . . . hearing people alive and talking around me . . . but comprehending nothing, lying there, only a rag of aching bones slowly shrinking in the glowing oven of malaria.
>
> Then the fever broke. Sweat poured from my pores like a balm . . . and then came the chills. I kept on shivering and they piled the blankets higher over me. It was well over one hundred degrees outside, but they covered me with blankets . . . And I still shook.[19]

Several days later he sat up and rejoined E Company.

BACK on Guadalcanal, a majority of soldiers in the Americal Division quickly fell sick, and many had to be evacuated to Fiji. But unlike the situation on Bataan, the U.S. Navy fought fierce sea battles—probably the fiercest in history—to provide ground forces with backup supplies, medical services, and air cover. Both sides lost valuable vessels, sunk in the deep waters off Guadalcanal's northern shores, which came to be known as Ironbottom Sound. The Allied fleets, however, sank more battleships than the Japanese. By the end of January the U.S. Navy had assembled a fifty-ship blockade. This shut down the Tokyo Express. Some forty-five thousand U.S. replacements grew to vastly outnumber the enemy and overran their bivouacs, which were filled with starved and diseased Japanese soldiers, some hiding in dugouts with dead, decomposing comrades. Malaria and beriberi had decimated them so completely that thousands couldn't march to the shoreline for emergency evacuation. By February the Axis enemy was captured, dead, or gone.

"Those who suffered from malaria acquired on the island have never been counted, but it is a safe assumption that almost every man who served on the island during the period of Aug 7–February 9 fell victim to the disease sooner or later, and in a vast majority of cases the attacks were recurrent over a long period of months," wrote a Marine historian in summarizing the Guadalcanal campaign shortly after the war. Had the naval battles gone differently and replacement troops for the men on Guadalcanal been delayed, the Japanese could have easily reversed momentum and taken control of the island, he wrote.[20]

Guadalcanal now became a staging area and airport for U.S.-led battles to come. But the military victory over the Japanese solved that one problem. Mosquitoes still ruled the island. And atabrine wasn't working. The whole stinking place required sanitization from anopheles and flies. Given the near-complete environmental destruction, and the perfect ecosystem for mosquito breeding, the first step had to be protecting the men from bites.

No more relying on drugs. The great yellow hope of atabrine had failed.

CHAPTER 17
The Other Side

Brigadier General Paul Hawley commanded the medical preparations for a massive buildup of American forces in Great Britain, which was code-named Operation Bolero. Bolero was to culminate in a cross-channel assault on Germany's forces in France—code-named Operation Sledgehammer—to break through Hitler's Great Wall. Sledgehammer was to commence sometime in late 1942. Hawley, the European theater surgeon, oversaw an army of consultants on everything from psychiatry, dentistry, and dermatology, to infectious diseases like tuberculosis and VD. He worried about upper respiratory disease and demanded that soldiers be given more than the thirty-one square feet allotted to them in the drafty barracks thrown together in the chilly English countryside. There the men "sneezed in the next guy's face," spewing infectious germs.[1] Hawley also scrounged for extra motion-sickness pills to save the men from severe nausea during the anticipated all-night trip across the hectic and choppy waters of the English Channel. He sent desperate letters to supply services for lab equipment and medicines—the most important being morphine for pain and sulfa powder to prevent war wounds from going septic in a theater that would soon swarm with more than two million American soldiers.

Hawley, a square-shaped, sturdy man of immense opinions, had too many worries. Countless conferences, meetings, and top-secret negotiations went into preparations for taking back Europe. Hawley was in on all of them to report on medical preparations, which portended the obvious. Surgeons learned to operate under mock battle conditions amid "huge warehouses filled to the roofs with bandages . . . for men then healthy who would soon be cripples," described Ernie Pyle, a United Press correspondent sent to England to report on the buildup there.[2] Pyle wrote his descriptions to remind Americans of the sacrifices made by their sons and husbands. But Hawley needed no reminder. Even when he lay buried in piles of paperwork so necessary to acquiring vast supplies for prolonged battles, he worried about the men's health. He even wrote personal letters to medical corps and quartermaster colonels, beseeching them to give his supply requests the highest priority. His letters held the tone of both commanding general and doting mother, and supply officers loved him for it. They tried their best to help him prepare for the carnage to come.

Then, in July 1942, just as U.S. Marines landed on Guadalcanal, the plan changed. It came seemingly out of the blue, but really after heated negotiations between Roosevelt and Churchill. In the end, Churchill won the argument. Instead of hitting Hitler head-on in northern France, U.S. and British forces would invade and occupy Vichy-controlled North Africa and use the continent's Mediterranean coastline to amass troops for an attack on Hitler's underbelly—from Sicily, to Italy, to southern France and eventually to Germany.

The new war plan satisfied many needs, the most important of which were to distract Hitler from his assault on Russia and clear the Mediterranean of Germans. But for those working on logistics, the change was a nightmare. Overnight, Hawley's basic equipment list underwent "radical changes."[3] On short notice he added to his worries a host of new potential afflictions, malaria chief among them. U.S. troops would encounter it at every stop on the way to France. Malaria maps showed intense summer infections across Algeria, Tunisia, Sicily, and Italy. He had no supplies for malaria, and the quartermaster offices were ninety days behind in getting medical orders filled. No greater challenge threatened the success of the North African campaign than the gargantuan matter of shifting supply chains from England to Africa. It ate up his time and completely overwhelmed his staff, which consisted of only eight officers and fifteen enlisted men.

To address the malaria problem, his deputy, Colonel J. F. Corby, placed an "urgent" order for atabrine.[4] Hawley had no experience with the drug, but intelligence reports said the Nazis used it, and his circular letters said it had replaced quinine on the list of available medicines for his theater. Requesting twenty-five thousand tablets per company of soldiers, he asked that they be sent by air transport. He also cabled the National Research Council for copies of all reports and meeting minutes for the different medical committees, including those from the specialized Malaria Project, to be sent to him in London by diplomatic pouch.

Meanwhile, he badgered the office of the army surgeon general in Washington to track down supplies he'd ordered months earlier. For the African campaign, he warned that he had just two stethoscopes, only pieces of mobile hospital units, no sulfanilamide (for dressing wounds and treating VD), and insufficient stocks of just about everything.

On October 18, just three weeks before the North African landings, Hawley got an uncharacteristically colorful reply from a deeply regretful supply officer. Colonel Francis Tyng of the Medical Corps explained that Hawley's requests from July, sent to the quartermaster embarkation headquarters in New York, had only just arrived. "Where they were between the time you dispatched them and the time they arrived in New York is unknown to anyone in America," wrote Tyng. He described swamped procurement officers drowning in paperwork required by the War Production Board. "The confusion that exists in Washington . . . over critical materials is so great that no one seems to be able to work his way through the maze. . . . The Medical Supply Program is on the verge of collapse. I believe we are very close to a major scandal." He complained that he had only five officers and seventeen civilians managing $220 million of medical supplies for lend-lease, U.S. military, and overseas civilian needs. Amid calamitous confusion, all Tyng could say was that a great number of Hawley's materials "have been floated" across the Atlantic and "the number of sinkings have been relatively small," which led Tyng to conclude that the supplies would either arrive shortly or had been pilfered along the way, maybe in Greenland, Iceland, Ireland, or Scotland.[5]

Hawley wrote back on November 7, the day before the landings. "My dear Tyng . . . I had to go to General Eisenhower and tell him that the stage is all set for the biggest medical scandal since the Spanish-American War." To add to his nightmares, the Army Air Corps had used his name to cable the U.S. for the same stuff he himself had requested, then used their airfreight capacity to deliver the goods to their pilots and support staff. Hawley cornered the Air Corps surgeon, Malcolm Grow, and called him on it. But he bowed to Grow's ingenuity. "The supply situation being what it is, one can't blame him."[6] (Grow went on to win many medals for ingenuity, including the development of special body armor, helmets and suits that protected pilots from bomb blasts and extreme temperatures. The Malcolm Grow Medical Center at Andrews Air Force Base in

Ohio is named in his honor.) Hawley knew the army had many persuasive men vying for supplies and services, and respected the ones, like Grow, who worked the system for the sake of their men, not their own comfort.

Hawley, still in Great Britain, finally received some of his long-awaited medical supplies. Atabrine was not among them. Nor were the medical books, journals, and bulletins he requested, or sulfa supplies to manage an expected rash of VD in U.S. soldiers about to hit the brothels of Casablanca, Oran, and Algiers. Many of the provisions that had arrived were unusable. Large pieces of medical equipment looked like they'd been dropped on the docks, as they were smashed to bits. Hawley complained in one letter to the Quartermaster Corps that he received ruined jeeps and motor cars and "shells so dented they won't fit in guns."[7] He became so humbled by his supply-related torments that he felt grateful for anything that arrived intact.

ON November 8, 1942, the North African landings commenced. Within two days, three U.S.-led task forces occupied Morocco and the Algerian ports of Oran and Algiers. In the next two weeks, tens of thousands of additional troops landed with the hope that they would quickly move into Tunisia and secure the seaports there. Many Americans arrived by way of the Pan Am air route through Liberia to avoid German submarines menacing the seas around Casablanca and Gibraltar. From Fisherman's Lake they headed north to the Mediterranean coast, where they joined a sea of tent cities in the arid, dusty plains near Oran.

When Hitler learned of the landings, he sent as many forces as he could spare across the Mediterranean to Tunisia, and ordered them to form a blockade around Tunis and Bizerte, seaports that sat just a few hundred miles from Sicily and the toe of Italy's boot. Hitler knew the Allies coveted these ports, perched atop the African continent, perfectly located for staging an assault on Nazi-held Europe. He aimed to occupy them first.

American negotiators worked furiously to secure Vichy French capitulation so Allied troops could move quickly to Tunisia and stop Germans from amassing an army there. But the French waffled, buying the Axis powers time, which they used to land enough tanks, big guns, trucks, munitions, supplies, and troops to set up a fearsome defense. Then they fanned across the countryside to brace for an Allied attack. Their orders came straight from Hitler: "North Africa, being the approach to Europe, must be held at all costs."[8]

Vichy leaders in Tunisia eventually received orders from their supreme leader in Algeria, François Darlan, to fight the German buildup. But by then Nazis were pouring into the Tunisian ports. They threatened to execute six thousand French sailors if the French resisted at all. Within days, the U.S. consulate in Tunis was turned into Axis headquarters, with no protest from the French.

Put in charge of the whole affair was one of Hitler's favorite commanders, Field Marshal Albert Kesselring—nicknamed "Smiling Albert" either because of his eternal optimism or his unusually large teeth, or both. He enthusiastically planned a counteroffensive, sure he and his men would prevail—or at least delay the Allied advance on Europe long enough for Hitler to pummel Russian forces in the East.

As Allied troops marched two thick columns of thirty thousand men into Tunisia, inexperienced and overpacked Americans slowly tossed off personal gear to make room for more ammunition. Onto the dirt roads fell overcoats, field jackets, mess kits, extra clothing, canteens—and mosquito nets.

The stage was thus set for a massive collision of forces in the malarious foothills of Tunisia's Atlas Mountains.

A French doctor in Tunisia's green pastures heard the thunder of war and quietly vanished, abandoning a secret project.

Dr. Jean Schneider had arrived there in August, completely unaware that Americans would soon invade North Africa. He'd been sent to test a top-secret drug made by Bayer chemists in the mid-1930s. All he knew about it was that they had discovered it after years of turning molecules around a quininelike centerpiece in search of a structure that acted like quinine, only better. By the time their country went to war, these German scientists rearranged molecular scaffoldings a thousand different ways, finally stumbling onto a series they referred to as 4-aminoquinolines. They apparently brought their work to the Société Parisienne d'Expansion Chimique, better known as Specia, in Paris, perhaps because of constant Allied bombings of I.G. Farben's factories—including Bayer's in Elberfeld. At Specia, chemists made a 4-aminoquinoline called sontochin that they deemed ready for a large human trial. Schneider later said he was "induced" to run a big study on this drug. In July 1942, around the time Churchill and Roosevelt quietly plotted the North African invasion, he left France.

Once in Tunisia, Schneider traveled by mule along gravel roads, lugging some fifty thousand 0.10-gram tablets of sontochin in three varieties, labeled C, R, and M, plus several boxes of a liquid version in vials prepped for intramuscular injection.[9] He aimed for seven whitewashed adobe villages between the Medguda and Rarai rivers. This picturesque corner of Tunisia's northern plains Ernie Pyle later described as "ripe country" of "violently green" pastures "rich in nature's kindness."[10] Schneider landed there because it was also highly malarious.

He arrived two months after the start of Tunisia's vicious malaria season—which ran June to December. Many of the twenty-two hundred Arabs in his selected villages were taking turns with fevers. Most had lived with malaria and experienced only flulike symptoms. For their young, who often died from it, they fell to their knees and prayed. Schneider promised relief from the misery in exchange for cooperation in the drug study. It probably helped that these peasants had endured a hundred years of French rule and knew compliance was easier than fighting. Whatever the reason, all seven villages cooperated, for the most part. Schneider stayed among them for the next three months, running a study in which two villages received no drug, as controls; one got a weekly dose of atabrine; and four took varying doses of this new drug, sontochin.

Conditions for the study were less than ideal. His late start meant he couldn't be sure of the new drug's prophylactic potential. Only one group continued the drug through Ramadan, which ran from the second week of September to the second week of October that year. Then, a month later, the Allies invaded. With battles about to ravage the countryside, Schneider fled the villages for Tunis and the protection of German troops there.

He left a final round of data collection to his hired technicians, who kept poor inventory of the facts. One of the four groups on sontochin, however, had a particularly resourceful nurse living among them in the village of Belgacem. That nurse, and probably an assistant, dutifully knocked on 350 doors every night to deliver a small 0.1-gram pill. The results were "remarkable."[11] After two months on sontochin, the village's 25 percent infection rate dropped to less than 1 percent. Of that 1 percent, not a single person developed symptomatic malaria—their parasite levels stayed so low they never got sick. This meant sontochin probably worked as a prophylactic, even amid heavy swarms of malarial mosquitoes. Better still, it possibly cured those already infected. All other groups had breakthrough infections and, in general, higher rates of parasites in their blood—easily explained by the respite in drug use during Ramadan. On toxicity, Schneider ran a test on children and found no bad effects, even in them.

Schneider was thrilled. These were the most promising results he'd ever seen against malaria. Under normal circumstances he'd publish his results and be the talk of his profession, with invitations to speak at international conferences and new funding to advance his career. But war bore down on everything. No one trapped in North Africa could be sure they'd still be alive in the coming months. So he sat on the results and helped prepare Ernest Conseil Hospital in Tunis for the expected crush of battle casualties, ready to accept the authority of whichever victor took charge—praying to survive the salvos.

THE American-led advance on Tunisia slowed, largely because of inadequate supplies. The abysmal medical situation was a small part of the problem, but significant to corpsmen. Stores of their equipment that landed with troops were looted or severely damaged. Essential medicines, syringes, and X-ray machines disappeared from the ports, or arrived so smashed-up that they were unusable. Meager medicines and bandages not stolen were quickly used up on severe burn victims and battle casualties from the two days of fighting against small but tenacious Vichy French forces in Algeria and Morocco. General Dwight D. Eisenhower, commander of Allied forces in Africa, personally asked Hawley to "step in and straighten things out."[12] Hawley flew to Oran shortly after the landings to do inventory and hold meetings. His trip resulted in major changes, including a shift to hard plastic packaging for essential items, supply codes on boxes instead of content labels, and armed patrols assigned to depots. More medical supplies were ordered to replace those stolen and wrecked. And he started calling in consultants to help with specialized problems, most prominently troop medical needs.

Colonel Simmons, Hawley's one contact back in the United States on the malaria question, promised him that atabrine stocks would arrive in North Africa in time for the spring malaria season.[13] Severe shortages of the drug and unanswered questions about dosing and side effects were still being worked out. But Simmons assured Hawley's staff that the problems would be resolved in the next few months. The theater would have sufficient stocks in time to suppress Africa's deadly cerebral malaria, which could kill up to 30 percent of the afflicted if allowed to go untreated.

Courtesy of Ambassador Theodore "Ted" Russell, from Paul Russell's personal files

Courtesy of Ted Russell, from Paul Russell's personal files

The Wilbur A. Sawyer Papers, National Library of Medicine

Top left: During his first trip to do malaria work in 1924, graduate student Lowell T. Coggeshall holds a water moccasin he killed while collecting anopheline larvae in a swamp near Leesburg, Georgia. Fifteen years later, he helped mastermind the Malaria Project.

Top left: Dr. Samuel Darling (*right*) led the malaria research in Leesburg with his lead physician, Paul Russell (*left*). This photograph was taken in 1924.

Left: Wilbur Sawyer, a director of the Rockefeller Foundation's international health programs, uses a dipstick to take larvae samples from "Swampville" near Leesburg, September 1924.

Left: Julius Wagner-Jauregg pioneered "fever therapy," the use of induced malaria to treat brain damage caused by late-stage syphilis.

Below: Jauregg (*in black crosstie and long mustache*) supervises a blood transfusion from a malaria-infected man (*left, background*) to a psychiatric patient with neurosyphilis (*center, foreground*). Jauregg's treatment for the madness caused by late-stage syphilis was successful roughly 30 percent of the time.

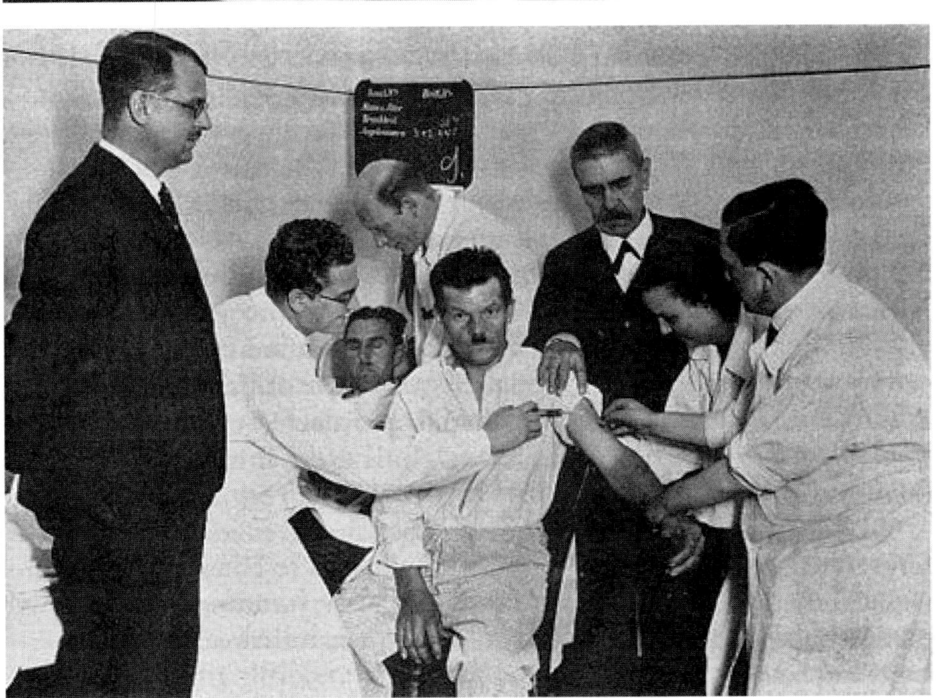

Medical University of Vienna

National Archives

Americal Division soldiers disembark at Red Beach on Guadalcanal in December 1942, replacing malaria-crippled divisions of the 1st Marines.

National Archives

An infantryman suffers through the furnace-hot fevers, icy chills, terrible nausea, and pounding headaches of malaria in the 7th Portable Hospital in New Guinea, March 1943.

Courtesy of Ted Russell

Paul Russell (*second from left*) lands in Melbourne for a meeting regarding malaria discipline with General Douglas MacArthur, May 1943.

Photo by Paul Russell, courtesy of Ted Russell

The officers' camp on Guadalcanal that Paul Russell called a "mud hole," May 1943.

National Museum of Health and Medicine

Malaria survey "dipstick" troops in training, circa 1943.

National Archives

A malariologist lectures new arrivals to Guadalcanal on the difference between anopheline mosquitoes biting with their abdomen pointed upward and regular culicine mosquitoes.

National Archives

Dipstick soldiers of the 18th Malaria Survey Unit dip for mosquito larvae in a stream near Kamaing, Burma. Larvae are brought back to a lab and examined under a microscope.

National Archives

Locals clear a ditch during the rainy season of 1943 on Guadalcanal to prevent water from clogging and creating mosquito-friendly breeding pools.

National Archives

Workers clear debris from the Lunga delta on Guadalcanal, which was followed by dragline and dynamite ditching, then oiling to kill remaining larvae, October 1943.

National Archives

Left: A pharmacist mate on Guadalcanal peers through a microscope, searching for evidence of malaria, March 1943.

Below: Infected red blood cells viewed under a microscope confirm the presence of malaria. Blood-stage parasites, called merazoites, used hemoglobin in each red cell to multiply into another thirty or forty parasites, eventually infecting billions of cells.

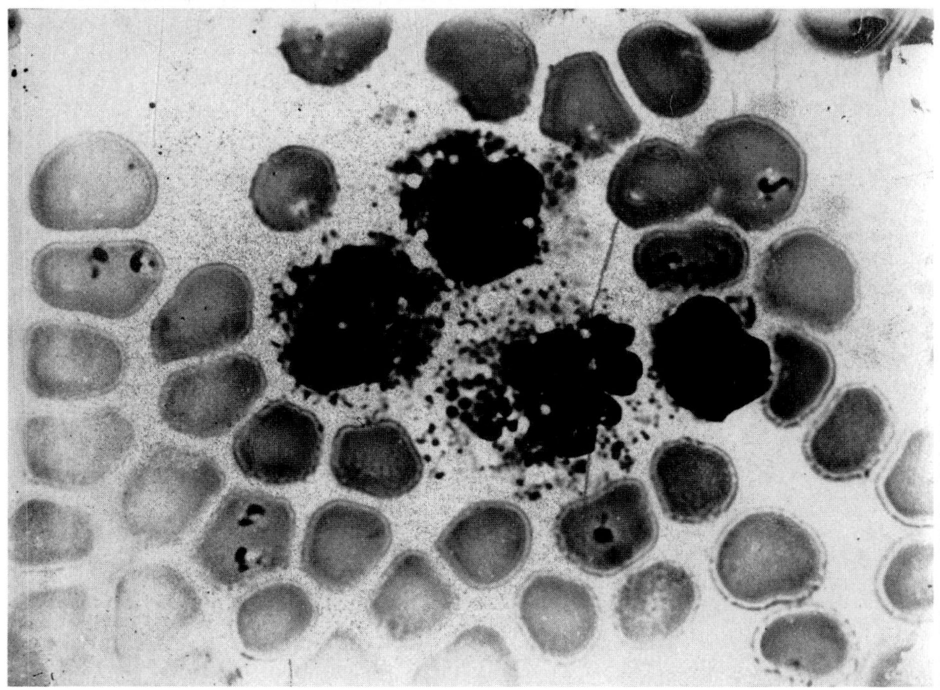

National Museum of Health and Medicine

Courtesy of the George and Katy Abramson Papers, Cornell University

The villages of Shangri-La and Paradise in Liberia were built by the U.S. Army to each hold three hundred prostitutes, with a third village, Idlewylde, to hold those infected with venereal disease. In addition to VD, these camps controlled malaria by keeping soldiers inside the mosquito-sanitized perimeter of the Army reservation, instead of seeking sex in off-limits villages.

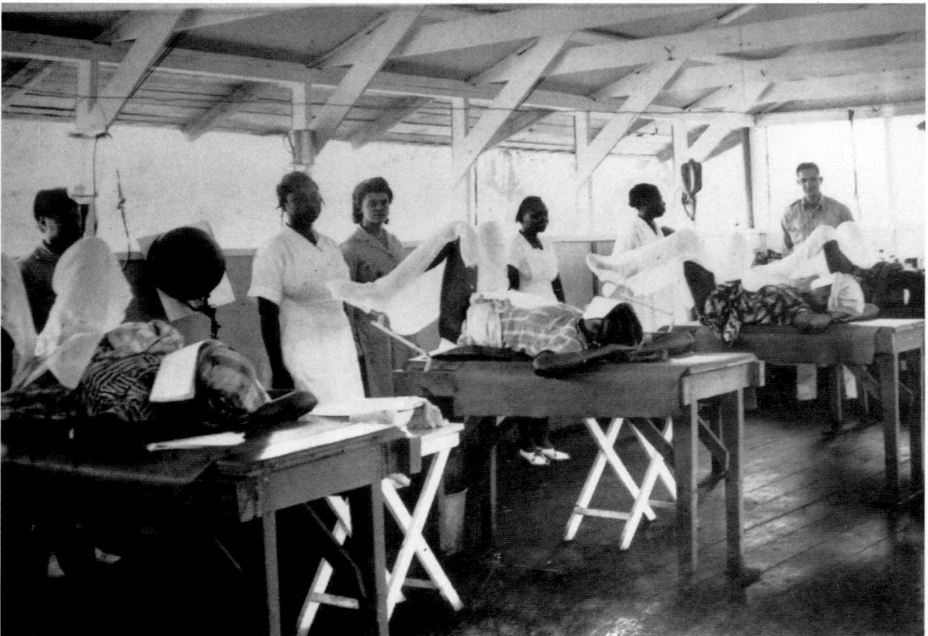

Courtesy of the George and Katy Abramson Papers, Cornell University

U.S. Army doctors and nurses inspected prostitutes of Shangri-La and Paradise every Sunday, sending those "peppered" to the MP-guarded VD prison camp, Idlewylde.

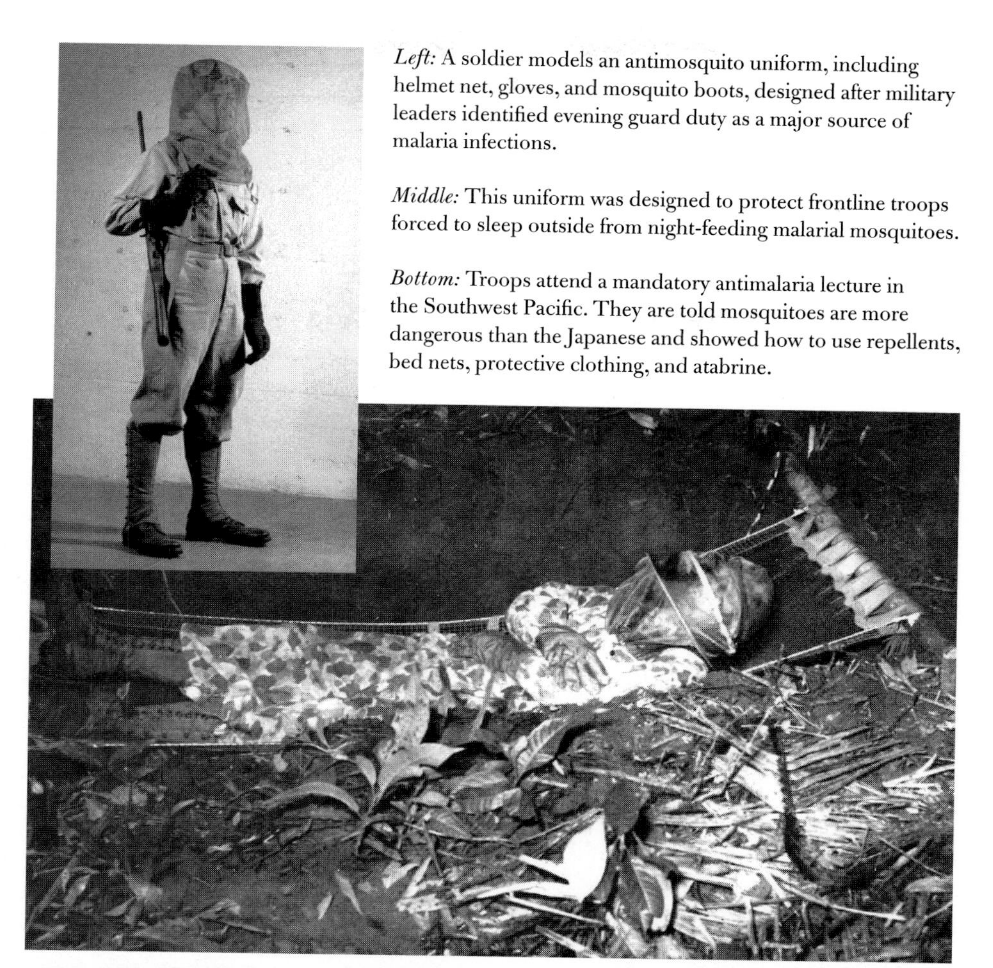

Left: A soldier models an antimosquito uniform, including helmet net, gloves, and mosquito boots, designed after military leaders identified evening guard duty as a major source of malaria infections.

Middle: This uniform was designed to protect frontline troops forced to sleep outside from night-feeding malarial mosquitoes.

Bottom: Troops attend a mandatory antimalaria lecture in the Southwest Pacific. They are told mosquitoes are more dangerous than the Japanese and showed how to use repellents, bed nets, protective clothing, and atabrine.

All photos on this page courtesy of the National Museum of Health and Medicine

National Archives

A Medical Corps officer explains malaria suppressive therapy charts to General Douglas MacArthur (*far right*) in the 101st Station Hospital in Queensland, Australia, December 1945.

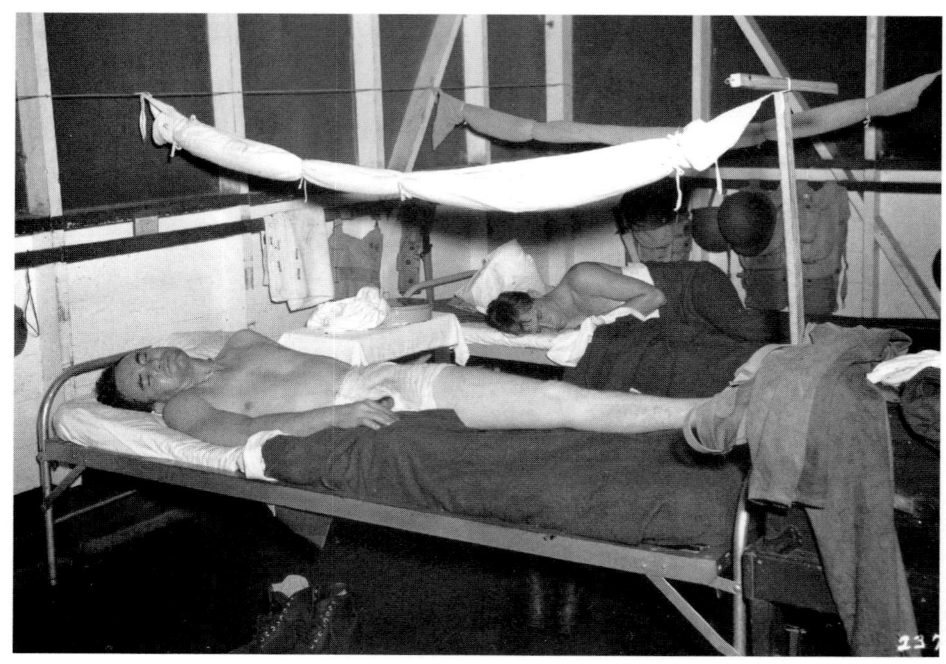

National Archives

Patients suffer through malaria symptoms at an outpost in the Panama Canal Zone, February 1943.

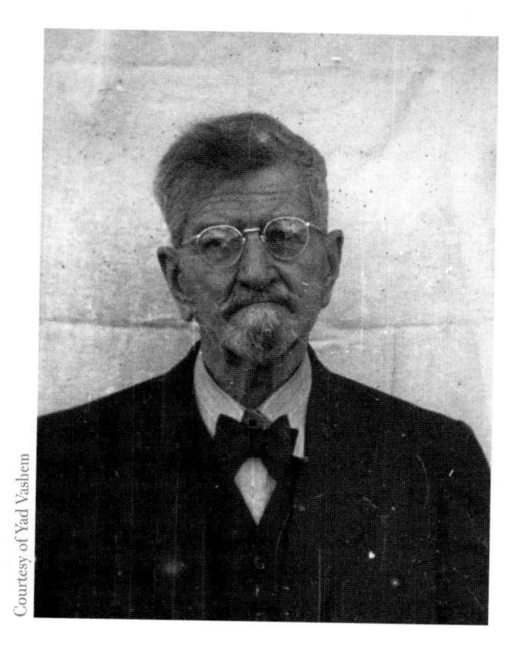

Courtesy of Yad Vashem

Claus Schilling, a world-renowned German malaria expert before the war, was recruited by SS leader Heinrich Himmler to run malaria experiments on prisoners at the Dachau concentration camp. Schilling was convicted of war crimes and hanged in 1946.

Adolf and Benito antimosquito poster.

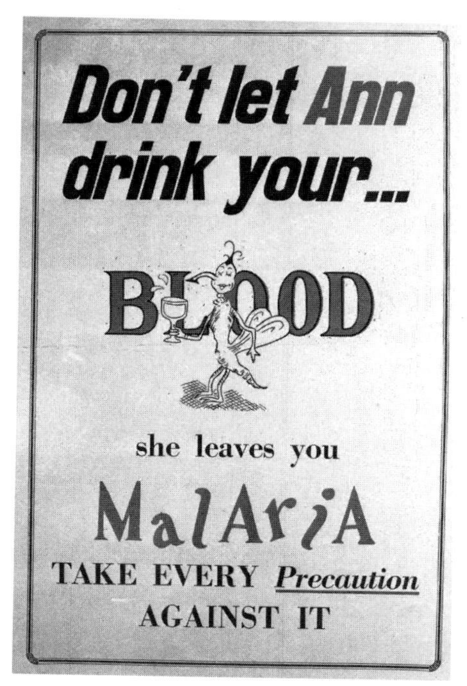

A propaganda poster featuring Ann the mosquito, designed by Captain Theodor S. Giesel, later known as Dr. Seuss.

All posters on this page courtesy of the National Museum of Health and Medicine

National Archives

The 6th Malaria Survey Unit, assigned to MacArthur in the Southwest Pacific Area, used eye-catching billboards to promote atabrine use early in the war. This strategy was later deployed in all malarious theaters.

National Archives

Locals mix 5 percent DDT in kerosene, which was then poured into backpack sprayers and applied around military camps and nearby villages, under the direction of the 31st Malaria Control Detachment, in Assam, India, June 1945.

National Archives

T/4 Frank A. Lee, of Gainesville, Florida, inspects mosquito breeding in pools around empty barrels at a dump site in New Guinea, July 1944.

National Archives

Malaria control headquarters in New Guinea, March 1944. Technicians and officers studied mosquito larvae to pinpoint and destroy anopheline breeding sites, and examined blood slides to find and treat the infected. Reducing human and insect carriers contained outbreaks.

National Archives

A soldier reads a malaria control message over a radio broadcast in India, August 1945.

National Museum of Health and Medicine

Natives spray tire tracks filled with rainwater on Guadalcanal.

National Archives

A building in Battipaglia, Italy, marked as treated with DDT by the 134th Malaria Control Unit, August 1944.

National Museum of Health and Medicine

Workers mix Paris Green in Italy, circa 1944.

National Archives

Aircraft dust anti-larvae chemicals over areas around Naples flooded after the Nazis sabotaged pumping stations and created massive anopheline mosquito breeding. The germ warfare failed to produce crippling malaria in Allied troops because of DDT and fast-acting malaria units led by Paul Russell.

National Archives

A C-47 of the 5th Air Force sprays DDT over the ruins of Manila, April 1945. This highly effective method of malaria control replaced difficult drainage and ditching work.

Courtesy of Northwestern University Library, Leopold and Loeb Papers

Top: Clarence Darrow sits by Nathan Leopold (*left*) and Richard Loeb (*right*) during their murder hearing in the fall of 1924. Darrow famously saved them from the gallows and, instead, won them sentences of life, plus ninety-nine years. Loeb was killed in prison in 1936; Leopold survived and volunteered for the Malaria Project, which helped him win parole in 1958.

Right: Swedish kidney expert Alf Alving was recruited into the Malaria Project in 1944 to give malaria to patients at Manteno State Hospital and experiment on Stateville prisoners. He performed among the most valuable malaria research of World War II and continued making important discoveries until his death in 1965.

Courtesy of Carl Alving

Not completely satisfied, Hawley sought more information. He heard that a malaria expert was already on the continent: Lowell T. Coggeshall. So he dispatched an attaché to locate the doctor and bring him to Algeria.

BY the time the Americans landed in North Africa, the Pan Am African resupply route had flown more than a quarter million miles under two contracts with the War Department: one for ferrying supplies and aircraft from the United States to Liberia and the Gold Coast, the other to transport those supplies and aircraft over the Sahara Desert to Khartoum and Cairo, then India. Lowell had sanitized the airstrips, taught Pan Am Africa employees how to protect themselves from mosquitoes, and, where necessary, managed irrigation projects specifically designed to reduce mosquito populations. He found that propaganda and even direct orders to wear protective clothing, use bed nets, and stay inside screened barracks after dark weren't enough; the men needed to understand the concepts behind the orders. When they understood the disease, the mosquito, and the consequences, they fell in line and cooperated with his program. His experience gave him far greater trust in education than in atabrine's ability to prevent malaria. He planned to bring these lessons to the National Research Council and recommend that soldiers be given a course in malaria prevention—not just be ordered to do as they were told.

Lowell's multipronged attack on malaria allowed Pan Am Africa to operate unfettered by disease. More and more flights dipped down to the Belgian Congo to move more and more uranium ore to Staten Island for the Manhattan Project. Many daily flights now flew over the Arabian Sea to India, Burma, Russia, and China. By the time Hawley landed in Oran, the Army Air Corps had taken control of the air route. Juan Trippe, Pan Am's president, had resisted. But the takeover was inevitable and would have happened sooner if the Air Transport Command were better organized.

On October 1, 1942, Major General Harold L. George spoke to Pan Am staff. He said the air route "must be welded into a bigger organization, which the leaders of this nation are trying to create for only one purpose and that is to make those yellow-bellied sons of Hirohito and sons of Adolph Hitler regret the day they ever armed themselves against the peace of the world."[14] He said nothing of how Pan Am pilots had frequently frustrated the army chain of command on intelligence matters. In one incident, Major Thomas Dawson, assigned to the

Pan Am air route by the War Department, reported that information mishandled by the airline had prevented opportunities to "drop depth bombs on Axis submarines which we have been seeing and knew were in locality for the past 10 days."[15] The grave consequence was the torpedoing of *West Irmo* off the Gold Coast, and the loss of essential supplies—including screens for barracks.

In another case, a German spy at Fisherman's Lake sent signal flashes to an offshore U-boat, which army intelligence intercepted and asked Pan Am managers to investigate. They came up with nothing. German spies also had been seen taking pictures of Americans building the airfield at the Firestone plantation, creating great concern "over the possibility of an air raid."[16] Then a German broadcast was intercepted indicating detailed knowledge of the air route and promising to "eliminate shipping" from the United States to the West Coast of Africa "by submarine warfare." This happened just as three thousand U.S. troops were en route to Liberia. The War Department was forced to provide additional escorts to all cargo and transport vessels leaving Brazil for Africa. A temporary solution had cargo transported to Nigeria, instead of Liberia or the Gold Coast, and then shipped by land via an ancient road that cut a straight line to Egypt. But the first attempt ended in disaster, as malaria casualties along the thick jungle route forced the team to return to port.

Major General George said none of this as he addressed Pan Am employees. Instead, he invited all pilots, radio operators, weathermen, air base managers, construction managers, and medical staff to remain on the job as commissioned officers. Lowell, who ostensibly worked for Pan Am's transport division while running top-secret drug experiments for the NRC, received specific orders to return to the United States. The U.S. Army surgeon general wanted him either in Australia under MacArthur's command or in China.

Lowell, who was personally commended for his malaria work by Churchill's chief medical adviser, Lord Moran, was in Accra when Hawley's attaché caught up with him. He was getting ready to board a small aircraft headed to Liberia, where he planned to catch the next Clipper over the Atlantic. Lowell missed his wife and young children and was ready to return home to them. When Hawley's messenger introduced himself and told Lowell of his mission, the doctor didn't complain. He hated not going home. But by now he was used to malaria running his life. He took his name off the flight ledger and had Hawley's pilot lead the way. As nightfall consumed the landscape and the sweet smell of day turned musty and chilled, the two walked to an unarmed cargo plane sitting on the air base's vast runway. At ten p.m. they took off into the black of night to avoid enemy fire, made a brief stop at the air base on Lake Chad, then pointed north to the Mediterranean coast, soaring ten thousand feet above a dark mass of disputed territory. The flight was longer than the one that would have taken him over the Atlantic toward home.

Shouting over the roar of the engine, the pilot told Lowell that North Africa belonged to the Allies, but if they were shot at they'd have to turn around, as the plane was unarmed. Many hours later the plane snarled and shook over the Atlas Mountains, and the temperature dropped. Lowell "almost froze to death" in his khaki shorts and elephant hat. The pilot noticed and shouted: "Here, you've got to have some clothes," and tossed Lowell a major's jacket, long trousers, and a brigadier general's overcoat, with one star. He warmed up as the plane approached a landing strip some fifteen miles outside of Oran. As the sun rose, they set down.

On the tarmac the pilot flagged a Catholic chaplain speeding by in a jeep and asked if he could give Lowell a lift into the city. "He'll take you in," the pilot assured Lowell. But the chaplain stopped at an army hospital at the edge of town and gave Lowell directions from there. "You just keep walking right down this street and eventually you'll see a building with an American flag on it and that's the headquarters. That's where General Hawley will be."

Sunshine burned the morning dew as Lowell walked stiffly down a tree-lined street cluttered with U.S. Army flotsam. His elephant hat, so appropriate for steamy West Africa, was out of place here. But so was everything else. This, Africa's most European city, with broad sidewalks, outdoor cafés, terraced buildings, and grand churches, also had poor Arabs dressed in rags, garbage in the gutters, shockingly gaunt dogs, and more horse carts than autos. Each city structure stood as an architectural clue to the city's tumultuous past—ruled by Moors, then Spaniards, then Turks, then Spaniards again, then Turks again, and, in 1831, the French, who held the city until surrendering it to the Americans in the weeks before Lowell landed there. The momentous shift in power was palpable; the city's streets were void of French and Muslim residents, who were still unsure of their safety. While the dusty roads bore the clutter of a second wave of U.S. infantry, who had landed in the Oran port just days before, they were clean compared to the mess they would soon become, once hundreds of thousands more arrived to rest up for the long trip to Tunisia.

This part of Africa was new to Lowell. He carried no weapon and had gleaned only scraps of information about the American invasion forces. Tired and sore from his all-night flight, he walked on, following the chaplain's instructions. But not two blocks in, the silent skies cracked with gunfire. Men in parachute pants appeared from nowhere, "running around corners, popping their rifles." Lowell ducked up alleys and crouched for cover, "trying to avoid these fellows," he recalled decades later. They were French Zouaves loyal to the Germans, shooting at God knew who. Lowell was scared and thought about the University of Michigan identification in his pocket. The ID would at least get his body back to the family he had almost gotten home to see.

Soon enough, however, the riflemen moved on. He continued making his way to the U.S. headquarters. As he turned one corner, he ran smack into Ernie Pyle. "My god, Cogg!" was the first thing Pyle said. "What are you doing over here?" Lowell's old friend from Indiana University was now the famous syndicated war correspondent. They hadn't seen each other in years.

Pyle had arrived in Oran with the second wave of American soldiers. He knew U.S. leaders ordered leniency toward the kind of rebel resistance Lowell had just survived. Because the occupation of North Africa was as much about winning over Vichy leaders as it was about opening a front along Hitler's underbelly, all soldiers were ordered to not engage unless it became absolutely necessary. Roosevelt and Churchill agreed the United States had to be the face of the mission, because the French *so* hated the English. Americans also were a better bet for winning over Arabs, who hated the colonial empires of both France and England. This would be tough, as many Arab ethnic groups liked the Germans. The Nazi occupation thus far had allowed them more freedom and prosperity than the French ever had, especially in the countryside, where Germans paid good prices for fruits and vegetables they bought for troops and to ship home to Munich. That changed when the Americans took over; oranges and other crops were left rotting in the fields.[17]

Making matters worse, the Americans' attempts at diplomacy came off as soft, and U.S. troops appeared sloppy. This lost them respect and allowed German propaganda to easily convince Algerians that the Nazis would prevail, which encouraged the bold display of armed resistance by French Zouaves Lowell had just witnessed.

When Ernie Pyle and Lowell said their good-byes, each walked in a different direction to make his own indelible mark on the war. Pyle set out to tell millions of American readers about the brotherhood of battle. And Lowell walked on to find Hawley and, ultimately, to help the War Department lodge a counterattack against its number one medical enemy: malaria.

ARMY officers in North Africa had no idea how to manage their inevitable malaria problem. For starters, medical reports in Hawley's possession, as of the landings, pertained only to North Africa's problems with typhus, smallpox, and the plague. He was also looking at woefully inadequate evacuation policies that would require the sick and injured to be removed from rugged, moonlike terrain on mule-back—far from ideal and possibly not even doable.[18] Lowell stood among a cadre of consultants invited to meet with the general. He soon made his points, emphasizing education, personal protection, mosquito nets, and, in frontline situations, drugs—whatever Hawley could get his hands on—be it quinine, atabrine, or sulfa mixtures. Hawley just needed something to stop Africa's virulent malaria from killing troops.

Lowell also helped straighten out conflicting information on the use of blood plasma for injured soldiers. The Office of the Surgeon General warned that before plasma from malarious regions could be used in transfusions it first had to be examined for malarial parasites. Hawley had neither the equipment nor the manpower for such work. Lowell brought assurances that he wouldn't have to. Lowell had done studies that showed malarial parasites didn't survive longer than ten hours in centrifuged blood, making plasma transfusions safe after that time.[19]

Lowell left Hawley's office to find his way back to the airstrip and grab a seat on an Air Transport Command flight to Accra, now running daily from Oran. The North African winter chilled the night as he departed for the sunny Gold Coast, then the steamy jungles of Liberia. On December 10 he finally caught a Clipper home.[20]

CHAPTER 18
Taking Tunis

While Lowell returned to the United States to rejoin the NRC's Malaria Project, Hawley's medical supply nightmares played out to varying degrees. Each of the North African task forces—based in Casablanca in the west, Oran in the center, and Algiers in the east—ran its own show. The Central Task Force for Oran managed to acquire its own stock of atabrine and included it in ration kits for individual soldiers. The atabrine bottles had a sharp warning: to be taken only with a doctor's instructions. But most of the men, including Ernie Pyle, needed no warning. They had heard the horror stories about atabrine and decided to brave malaria. Had they known anything about the disease they might have chosen the remedy. "Personally I decided never to take mine," Pyle wrote. "I had talked to one doctor from the South, a malaria specialist, who took his and thought he was going to die. He said he'd rather have malaria and get it over with."[1] Neither Pyle nor a growing sea of young American soldiers bothered about mosquitoes yet. They had much bigger worries.

During this time, Pyle came down with a chest infection that soldiers called the "African flu." He embedded within a growing sea of U.S. troops camped in enormous tent cities in the arid plains around Oran. They had few blankets and insufficient protection from the winds that kicked dust into their throats. In many, like Pyle, the quick cold of the desert nights turned a raw throat into a chest infection. These respiratory problems, along with the wounded from the previous month's landings, occupied the base hospitals. Malaria was barely an issue, except for the few cases that trickled in with soldiers exposed to it in Liberia and the Gold Coast. The locals didn't bother stocking quinine until March, at the start of malaria season. The only talk of malaria, from a planning standpoint, was in the assembly of medical supply kits, which included morphine, sulfa pills and powder, water-purifying pills, and, for some, atabrine tablets.

MEANWHILE, the initial clash of forces along the Algeria-Tunisia border went horribly wrong for the Allies. Many thousands of young men perished in tragic battles with now-famous names. It was a brutal, bitter winter in which green American soldiers were picked off like props in rifle practice as they naively walked upright along ridges; or were ambushed and slaughtered by seemingly retreating panzers; or were blown apart in groups formed randomly in the trenchless rocky terrain.

But with spring came reinforcements, new supplies, and a better battle plan. General Eisenhower, quickly realizing he wouldn't sweep in and take Tunisia with thirty thousand troops, called in more than a hundred thousand additional soldiers. The continent shook with moving armies. They crossed into Tunisia and by sheer force of numbers and airpower pushed their way through the Atlas Mountains into the sun-soaked pastures and floodplains on the other side. The Germans quickly dug in around rugged hills just fifty miles from Tunis, ready to pummel the advancing wall of Allied troops and machines.

Robust mosquitoes rose from the waterways, while the Allies descended on the countryside. Even though peasant farmers avoided interactions with soldiers, this "had no effect on preventing malaria, as it was virtually impossible to avoid setting camps and bivouac areas within mosquito flight range of Arab villages," observed a medical officer.[2] Warm weather and full rivers produced the usual bumper crops of larvae. As March turned to April, these larvae molted into mosquitoes that fed on infected Arabs and would soon be delivering mouthfuls of microbes to Allied and Axis soldiers throwing artillery at one another in battles that would create even more perfect breeding grounds for even more mosquitoes.

DURING this phase, Rommel quietly slipped away to Germany, ostensibly because of a malaria infection, but presumably to form a new battle strategy. Major General George S. Patton also slipped away to begin planning a secret attack on Sicily.

A few dozen feverish U.S. soldiers in the Tunisian hillsides trickled into field hospitals. The new theater surgeon, Brigadier General F. A. Blesse, assumed the soldiers had picked up malaria in Liberia, because Tunisia's malaria season was still at least a month away. But he knew if the campaign dragged into the summer, hyperendemic malaria would knock nearly everyone off their feet, and the Allies would lose important ground. Without Tunisia, hope of quickly attacking the Axis in Italy would dissolve. With Hawley working back channels in England, and Blesse making his own pleas, large shipments of atabrine finally arrived in North Africa. Immediately, they were added to all ration kits.[3]

On April 22, everyone in the campaign, from frontline soldiers to supply officers in the rear, received orders to begin taking 0.2 grams of the bitter yellow pills twice a week. A week later, on April 28, the bloody battle for Hill 609 ensued. In the middle of it, on April 30, everyone swallowed his third pill. For the men on 609 it went down just hours before fierce fighting. Historian Rick Atkinson described the scene: "All day the valleys rumbled with artillery fire; the crack of shells splitting rock carried from the hilltops. Hundreds of men fell sick in apparent reaction to atabrine. . . . Weak and nauseated they vomited down the front of their uniforms and fouled their trousers with uncontrolled diarrhea before rising on command . . . to stumble forward again."[4]

Similar problems developed elsewhere in the campaign. In the 59th Evacuation Hospital, nurses measured the effects of that third pill. Seven hours after taking it, staff and patients felt a sudden and intense wave of nausea and "immediate bowel evacuation." Over the next nine hours came "copious vomiting and diarrhea." Recovery began about seventeen hours later. This was the common experience of 76 percent of the hospital's bedridden patients, and more than 50 percent of staff.[5]

Reports across the theater poured in, from battlegrounds and transport ships. "Very few units escaped toxic manifestations," lamented Brigadier General H. L. Morgan.[6] He reported that the British "ordered the First Army to abandon the treatment temporarily (until the end of hostilities in Tunisia)." An estimated 30 to 50 percent of soldiers and staff on atabrine experienced debilitating cramps and diarrhea. Several epidemiological investigators tried to determine whether the problem was food-borne. But all the evidence pointed to atabrine, always after taking that third pill in that second week.[7] And all of it was made by Winthrop.

. . .

AS in the Pacific, troops in North Africa stopped taking atabrine. They opened their ration kits and tossed out the little pills that General Hawley and others so valiantly struggled to provide them. The stuff tinged the soil a sickly yellow, advertising the fact that the men had chosen malaria over uncontrolled diarrhea. Whispers that atabrine robbed men of their manhood by making them impotent and sterile didn't help. U.S. commanders suspected German POWs of starting the rumors to discourage atabrine use by the Allies. The Germans' one major medical advantage was that they understood proper dosing of atabrine to keep malaria attacks and side effects to a minimum. In November 1941, German malariologist Gerhard Rose—working for the Robert Koch Institute for Infectious Diseases in Berlin, and later to be convicted at Nuremburg—teased out the difficulties of this drug and proved that smaller, more frequent doses worked as a prophylactic, while large, less frequent doses did not. It had to do with two things: natural variability in individual dosing by manufacturers and variability in the rate at which each person absorbed the drug. Rose demonstrated that "shock doses" of 0.2 grams twice a week, which Bayer recommended and the Americans prescribed, was less effective and possibly the cause of reported explosive diarrhea in German troops occupying the Balkans. His work instructed the Reich board of health to establish a regime of 0.06 grams daily for prophylactic use in troops.[8] Atabrine would never be a cure, but in small daily doses it eliminated enough parasites in soldiers' blood to keep them on their feet and fighting. Thus, while German troops carried plenty of malarial parasites, those parasites had no measurable impact on troop strength. If the campaign dragged into the summer, this would be a major advantage for the Nazis.

By May 7, however, the Allies broke through the lines around Bizerte and Tunis. As U.S. and British commanders bickered over the course of future operations, especially who would lead whom into Sicily and Italy, more than 275,000 German and Italian soldiers surrendered. Allied troops questioned German, Italian, and Vichy French citizens and soldiers, separating friend from foe. The former were left to fend for themselves, with their homes in ruins, and the latter were rounded up and put into camps.

Captured German documents, signed by Kesselring, described a multi-tiered program to protect soldiers from malaria: screened buildings, bed nets when possible, long pants and shirts at night, strict use of repellents, education for foot soldiers, and a daily hit of 0.06 grams of atabrine.[9] But for the atabrine dosing, the German strategy paralleled the plan pushed by Paul Russell and Lowell Coggeshall. The similarities were a product of prewar antimalaria collaborations between scientists from the United States, Germany, France, England, and Australia. The German dosing, however, was different. And important. The papers were sent by pouch to Colonel Loren Moore, the U.S. theater malariologist in Algiers, on high priority, because battle commanders assumed they held great value. They were received, no doubt, with a big "no kidding"—but for the dosing. That was sent directly to the Malaria Project.

On May 10, the Allies entered the Ernest Conseil Hospital in Tunis. Jean Schneider approached an American infantryman and whispered that he had in his possession a cure for malaria. Schneider asked that the medication and his study results be sent to the American theater surgeon—not the British one. He included a letter addressed to Colonel Moore that outlined his case. Schneider said he had been "induced" into conducting research on a compound discovered by Bayer. He said he had what he believed to be a new, highly valuable drug, one that no one else knew about. He described the work he'd done the previous summer and fall, testing this compound on Tunisian villagers. He enthusiastically offered his data and thirty thousand tablets to the Americans—again, *not the British.*

Despite having no leverage with the new lords of North Africa, Schneider boldly wrote: "I submit all of the results of my experiments concerning the prophylaxis or the curative treatment of Malaria, herewith. In exchange for which, I ask Colonel Moore to send me a letter constituting a receipt for the samples and data, and a guarantee that the commercial rights of the Specia Society [Société Parisienne d'Expansion Chimique] will be safeguarded. . . ."[10]

The letter, data and tablets were promptly delivered to Colonel Moore. A month later, with Schneider's documents finally translated from French to English, Colonel Moore packed the data and pills and sent them, via Brigadier General Blesse, to the U.S. Army's Office of the Surgeon General, Preventive Medicine Division, in Washington, DC. They arrived there in July.[11]

CHAPTER 19
Mosquito Brigades

Long before the world divided into the Axis and Allies, U.S. military planners mapped war scenarios that included attacks from Germany, Japan, *and* Great Britain. The world of the late 1930s was volatile and increasingly expansionist. Anything seemed possible. So Congress passed the Selective Training and Services Act of 1940, requiring men between twenty-one and thirty-five to register for the military draft, leading to an unprecedented expansion of U.S. military forces. Hundreds of new training camps, bases, and military-related industrial plants were created or expanded, many located in the South to take advantage of cheap land and labor, and a climate conducive to year-round training. But these places still harbored pockets of endemic malaria in the poorest people. The influx of workers and draftees created a potentially explosive numbers game in which local mosquitoes would suddenly have thousands of malaria virgins coming down from Northern states for training. More people to infect would create a higher basic case reproduction rate. The combination of densely populated servicemen training in and around shredded ecosystems with infected populations nearby would trigger outbreaks in troops. It happened during the First World War, when Southern training camps had summer infection rates that exceeded 17 percent—nearly two ruined soldiers for every ten trained.[1] Their fevers broke through on the front lines in France, making them useless to the war effort.

To prevent this from happening again, the War Department in 1941 effectively sanitized camps and industrial sites. But the military lacked manpower and authority to do the surrounding areas, where malaria-seeded people and malarial mosquitoes lived. In early 1942, the U.S. Public Health Service re-created a new subordinate agency to do the work, called the Office of Malaria Control in War Areas (the predecessor to today's Centers for Disease Control and Prevention). Both the War Department and the MCWA worked from principles developed by Ronald Ross and used by malariologists ever since—oil to stop mosquitoes from laying eggs, larvicide to kill already developing larvae, fill to eliminate wetlands, and hard labor with heavy equipment to dig drainage ditches.

These new efforts in the early 1940s were more efficient than those of the previous decades because of the technical skills amassed by public health professionals during the intervening years—skills created by men like Samuel Darling and developed further by men like Paul Russell and Lowell Coggeshall. These public health preachers showed the world the importance of identifying and targeting precise mosquito species, which required a fair level of technical know-how.

Guiding the MCWA in the gargantuan effort of sanitizing a mile radius around more than five hundred Southern military installations was none other than Dr. Louis L. Williams, the man who had taught Lowell how to hand-crank Paris Green out the back of a de Havilland while soaring over the treetops of Quantico. Williams was now training hundreds more Lowell Coggeshalls to spew dust over huge swaths of the South.

The work paid off. In 1942 and 1943, the malaria rate among troops trained in these camps went from a high of 3 percent—with an expectation that it would climb quickly in the absence of sanitation work—to just 0.01 percent (almost none). To accomplish this, the War Department and MCWA spent more than $14 million to dig 18.6 million linear feet of ditches; clear 178,000 acres of scrub and brush; flush debris from fifty million linear feet of clogged channels and streams; line with concrete more than one million linear feet of drainage systems; eliminate well over forty-two thousand acres of wetlands; spew 614,000 pounds of Paris Green and other larvicides; and pour 5.9 million gallons of oil into U.S. waterways.[2] All told, by the time the war ended, 11.7 million gallons of oil would be spilled in Southern waters—to stop malarial mosquitoes from breeding. (For a point of comparison, the 1989 *Exxon Valdez* spill in Alaska's Prince William Sound was conservatively estimated at eleven million gallons of crude.)

Paul Russell, the army's new chief of the new Division of Tropical Medicine, had to figure out how to do what the MCWA was doing, only in militarized zones overseas. This was a seemingly impossible prospect. How would he manage mosquito control in the middle of war zones, with troop movements and artillery fire and air raids as constant but unpredictable variables he had no control over? No one, not even Russell, had an immediate answer. But he had become one of the world's foremost malaria experts for good reason. He had been at it since the 1920s, and understood better than anyone the variables he could control: infected people interacting with malarial mosquitoes. As malaria decimated troops on Guadalcanal, he sat at his new desk in Washington thinking like a mosquito, as Darling had taught him to do.

He knew their bites would be deadlier to U.S. troop strengths than Japanese snipers and German panzers, but that the kind of sanitation work so effectively scrubbing malaria from U.S. military installations would not work overseas. While mulling the possibilities he formulated a solution: Why not create specially trained, portable units to attach to different infantry battalions as needed? These units of roughly twelve to fifteen men each—under the authority of a theater malariologist and including an entomologist, a parasitologist, medical doctors, and enlisted lab technicians, plus a driver or two—could swoop into overseas theaters and sanitize fallback areas, bivouacs, hospital sites, airfields, launching areas, and all other strategically important grounds, but for the front lines, where the work would be impossible to do. The units would collaborate with sanitary engineers and Seabees on safely locating permanent and semipermanent bases away from infected populations, and consult with both on where to point their heavy machinery for filling or draining anopheline-breeding waters. Survey maps made by these specially trained units would show construction teams where and how to destroy malaria carriers, with no time wasted on mere nuisance mosquitoes.

His units would bring with them screens, insecticides, and bug bombs. They would teach foot soldiers how to protect themselves at night from bites, and show them how to flatten ration cans into tin pancakes—so the cans wouldn't fill with rainwater and give enemy anophelines a safe place to breed. These units would be invaluable to commanders, and generals would call them a godsend.

Troops would be sent to a base somewhere in balmy Florida or Louisiana, or maybe Alabama, along the gulf, where they would sleep in barracks, eat in mess halls, and awake to reveille. But instead of learning to shoot weapons to kill soldiers, they would learn how to use dipsticks to catch mosquito larvae; use microscopes to count malarial egg sacs in the insects' guts; and use magnifying glasses to study the different characteristics of anopheline mosquitoes and larvae. Then they'd learn how to hunt the insects down and kill them. Some could even learn how to take blood smears from people and check spleen size to count infection rates in villages around bivouacs and staging areas. Everything Russell learned from Rockefeller could be taught to troops in this boot-camp format. They would spread the Rockefeller know-how across war zones, creating huge advantages for the Allies. The project was doable. To Russell the need was obvious.

But it wasn't to the military.

His senior officers said no when Russell first lobbied for the program. The army sanitary engineers were already scheduled to sanitize for mosquitoes, as the Seabees were doing for the navy. "No need for duplication," was the initial response of the surgeon general's office.[3]

Then marines fell by the tens of thousands on Guadalcanal, despite the Seabees' efforts to sanitize for mosquitoes. Reports arrived in headquarters that thousands of men in the North African campaign had been infected with malaria during stopovers in West Africa, where mosquito controls put in place by Lowell Coggeshall had broken down in his absence. Russell sat as the lowest-ranked officer in high-level staff meetings explaining why. He had been to the army camps and bases in Puerto Rico and Panama, where malaria was still a problem, and explained why efforts there failed, too. By December 1942, either his persuasive style or the sheer number of fallen troops, or both, finally convinced Army Chief of Staff General George C. Marshall that this Lieutenant Colonel Russell made sense. By Christmas, Marshall had signed off on Russell's malaria boot-camp plan.[4]

But Marshall added a rather large caveat: Malaria units would be activated for training only after theater commanders formally requested them. To Russell, this meant the number of units would always be four to six months behind the need, as that was how long they would take to train. Nonetheless, General Marshall blessed Russell's mosquito brigades, a huge victory for this public health missionary.

Administrative work piled on. Russell needed a location, curricula, instructors, and a process for finding appropriate recruits—ideally, enlisted men with at least some training in biology, preferably at the college level. While intercepting these men from infantry training, Russell worked the chain of command, encouraging theater generals to request the units well before they moved into malarious areas.

In the middle of it all, a chemical substance landed on his desk. Its arrival in the U.S. had been facilitated by the U.S. Military Attaché in Bern, Switzerland, where Allen Dulles served as America's chief spy.[5] U.S. intelligence officials had received reports that the Germans used against potato bugs this "very effective pesticide" that they called Gesarol, or dichlorodiphenyltrichloroethane. They arranged for samples made by the Geigy Company to be sent to the United States, some of which were routed to Russell.[6] Russell quickly shipped the samples to his former Rockefeller colleague, Dr. M. A. Barber, who had turned Paris Green into a larvicide in 1921. Barber, by now, was testing insecticides for the war effort in Orlando, Florida.[7] He and other investigators tried this crystalline powder against a range of pests and found it worked beautifully on a type of body lice causing terrible typhus outbreaks in North Africa. They shook this odorless, tasteless substance like talcum powder inside the clothing of test subjects, and found it harmless to the skin. All the while they marveled at how effectively—almost miraculously—it worked against lice. The army hired Fred Soper, who had led the assault on *A. gambiae* in Brazil, to run field studies inside lice-infested POW camps in Oran and Morocco.[8] Meanwhile, researchers in Florida studied other uses for this excellent new substance. When the War Department stamped it "top secret," they listed it under a generic name: DDT.

Russell had no idea that this powder would, in less than a year, revolutionize his work and help solve the malaria problem. But here, in late 1942, he still had only old chemicals like Paris Green, and a few new weapons, like Freon bug bombs, that were good but in short supply.

AS malaria nightmares in the Pacific continued to ruin troops, requests for Russell's mosquito brigades trickled in at first, then flooded his office. In the first two months, December 1942 to January 1943, commanders in the Pacific, North Africa, West Africa, China-Burma-India, Middle East, and Caribbean theaters requested a total of eight malariologists, fifteen assistants, sixteen survey units, and thirty control units (the survey units to study and map anopheline breeding sites, and control units to sanitize them). In the next five months, the requests would more than double, and then double again by the end of the year, so that by December 1943 more than two hundred malaria units were requested to help troops in every theater.[9]

The officers—malariologists, assistant malariologists, parasitologists, and entomologists—went through regular medical corps officer training, then headed to Mark Boyd in Panama for specific training in malariology. The sanitation officers went through a separate course run by the Tennessee Valley Authority at the Wilson Dam in Alabama and, during the winter months, went to a different location run by the Rockefeller Foundation and the Florida State Board of Health. Or they went to Costa Rica to train in a special facility attached to the Pan American Highway Project. Eventually, all these efforts were brought together in the new Army School of Malariology at Fort Clayton, in the Panama Canal Zone.[10]

As for the enlisted men, they trained at Camp Plauche on the western outskirts of New Orleans in Jefferson Parish, at the base of the Huey P. Long Bridge, on the banks of the Mississippi River. There, more than six hundred whitewashed, single-story barracks lined a grid of streets contained within barbed-wire fencing. In addition to malaria units, the camp trained enlisted troops and officers of the Army Transportation Corps, and served as an animal remount station for overseas needs. The malaria units were small, by comparison, and served a practical purpose. While training in antimalaria work, they also sanitized the camp. One recruit noted in a letter home: "Today I noticed an accumulation of water in front of the barracks and investigated. It was teeming with mosquito larvae. What a paradox—here with malaria survey and control units and mosquito larvae by the front door!"[11]

The coursework was disorganized and "so darn repetitive," according to students plucked from the undergraduate labs of the country's top universities. These were Russell's handpicked recruits who would do wonders overseas, as soon as he could get them there. But the military had procedures. At Camp Plauche, they sat in the same courses as other fellow recruits who didn't come from fancy universities but maybe liked biology in high school. Together they were the dipstick recruits, and together they learned the basics of malaria work. In outward appearance, they were 100 percent military, dressed in fatigues and subjected to bed checks, guard duty, calisthenics, drills, inspections, and even infiltration courses—which meant living on K rations while spending fourteen-hour days crawling in the mud under barbed wire, amid live machine-gun fire and exploding dynamite. But the best of them were, on the inside, true naturalists. Their barracks showed it, with shelves weighted down by heavy biology textbooks and taxonomy dictionaries. During leisure time these men caught and identified the region's wildlife: green snakes, puff adders, chameleons, ground skinks, narrow-mouthed toads, tree frogs, lubber grasshoppers, and snapping, mud, and box turtles, many of which decorated the barracks—some stuffed, some live in cages—giving the quarters the feel of a science museum.

For work, they spent the days hunting down larvae in the marshlands near the levee on the opposite banks of the Mississippi, home to large seabirds, six-foot-long alligators, freshwater crustaceans, and prolific fig trees and bamboo. But the larvae were their focus, which they brought back to base and studied under microscopes. At night they took turns catching mosquitoes while sitting bare-chested under the hackberry, oak, and cypress trees, always with a buddy to help catch the biters before their proboscis bayonets sliced into flesh. They then studied and identified the different mosquito species, looking for and counting the anophelines.

To the other battalions, these units were a bit odd—too wholesome and earnest. They loved their microscopes and binoculars, and spent hours floating down the Mississippi—birding, of all things—when the French Quarter and prostitutes were within walking distance! Denton Crocker, who had just graduated in biology from Northwestern University, landed at Camp Plauche in August 1943 to join the 31st Malaria Survey Unit, which would be shipped to New Guinea.[12] He paired up with Cornell graduate Bob Roecker, among others, to become one of the most effective malaria units sent overseas.

Of New Orleans, Crocker said: "It is a wide open town and the men have wild stories to tell. One of the unit sergeants came in very drunk last night. If he isn't careful he'll lose his rating. I think I'll wait until Sunday when I can go in during the day . . . and can see more of the old architecture and perhaps take a boat trip on the Mississippi. There is a lot of immorality in the army. I'm thinking particularly of sex immorality. . . . Most distressing of all is that married men are far from exempt. The women that I saw as we passed through the city and that I've seen around camp have been a poor bunch and many of them are doubtless infected with disease."[13]

While at Camp Plauche, he wrote his girlfriend at least twice a week, describing his dips in Lake Pontchartrain, and all the gulls, terns, and plovers, and locals fishing for crab and shrimp, and green bananas hanging from trees over "negro shacks." He described kerosene lanterns strung along the dark shores of this giant lake, needed for evening shrimping, and the moss-covered cypresses with "knee" roots reaching out of the swamps for air. His letters described exotic scenery, relative to the tame flora and fauna of his home state, Massachusetts.[14]

Of the culture around New Orleans, he described dances with "beats of savage intensity" and girls "of the gum chewing variety." He asked his parents to send him books like Julian Huxley's *Evolution: The Modern Synthesis*, and Theodosius Dobzhansky's *Genetics and the Origin of Species*. He spent his nights with books like Thomas Wolfe's *You Can't Go Home Again*, and Bertrand Russell's *A Free Man's Worship*. Then, in a rush of romance, he wrote a fabulously romantic letter, asking his girl to marry him.[15]

New Orleans got into his blood. After a few short months he loosened up, even faked a pass and joined his buddies for jazz at the Vieux Carré, where black musicians performed—which he described as if it were a novelty. He seemed mesmerized by the culture, and by black people, and particularly by one rotund African American woman he watched for hours while she played the piano "with a large glass bowl for tips and a glass for drinks, which always filled up when she emptied it," he wrote.

The steamy streets swarmed with servicemen, many falling onto the cobblestones from noisy bars. He looked into one and saw "a nearly naked woman . . . gyrating wildly up on the bar, her high heels clicking out the beat of the music. Her only clothing was a flap of cloth, about the size of a 50 cent piece, which with an occasional thrust of the hips she could flip up, rendering her completely naked."[16]

The next day Crocker wrote to his fiancée professing his bad mood, and declared: "how terribly much I need you—both physically and mentally."[17]

More than twenty-five hundred enlisted men trained in malaria worked here. The reception they would receive overseas would not be nearly as friendly as in the French Quarter.

THE U.S. capture of Guadalcanal marked a huge victory back in Washington; Americans had kicked in the first door to Japan's captured archipelagoes. But the island's shredded landscape and prolific mosquito production threatened to make the capture meaningless. If the Americans couldn't beat the mosquitoes, they couldn't keep the island. Surrender, however, was not an option, because the process of moving the staging area from New Caledonia (a thousand miles south of the Solomons, halfway to New Zealand and well outside the malaria belt) to the newly captured Guadalcanal had already begun. Perhaps more important, this hellhole of an island couldn't be abandoned to stage elsewhere because the Japanese wanted it. To leave would be to surrender it to the enemy, who would certainly use it to launch attacks on Allied forces.

Everyone agreed on the importance of keeping Guadalcanal.[18] Area commanders would work from here toward their next objective: to take back New Guinea for staging an attack on the Port of Rabaul, on the island of New Britain—the linchpin of Japanese defenses and about five hundred nautical miles from the huge Imperial Japanese Navy base on Truk, in the Caroline Islands. Allied island hopping would begin from Guadalcanal to another eleven malarious islands scattered across 1,550 nautical miles of the South Pacific area, then into the Southwest Pacific, starting with New Guinea, driving farther and farther westward and northward, all the way to Japan.

If Paul Russell's mosquito brigades could beat Guadalcanal's mosquitoes, they would create the template for malaria control throughout the region.

The first Malaria Control and Survey Units arrived to a complete disaster. Throughout February, U.S. infantry counted more than nine thousand dead Japanese soldiers killed by disease, mainly malaria and beriberi. "This was nearly half their dead on the island," observed one American medical officer.[19] Meanwhile, American casualties from malaria—nearly all of the replacements sent to relieve the 1st Marines, including the 2d Marine Division, the Americal Division, and the 147th Infantry Regiment—were quarantined. Americal troops went to Fiji and the marines went to Australia, where they were studied and used in drug tests. The first experiment was suggested by the Malaria Project, back in the "Zone of the Interior"—the military's name for the United States. Project scientists wondered whether the infections could build immunity against subsequent infections, so the quarantined men were all taken off atabrine. Then doctors recorded symptoms to see if suffering through relapses would build the men's defenses against subsequent attacks from the parasites hiding in their soft tissue. As it turns out, it didn't. "The rehabilitation period of units who were subject to this procedure was approximately double the time estimated for this purpose and meant a serious decrease in the number of available combat troops."[20] The men just kept getting sicker and sicker, stuck in their military purgatory, unable to go home, unable to go back to their units. Severe depression set in. Commanders kept requesting more troops to replace these spent forces.

Russell's malaria units on Guadalcanal had to stop replacement troops from meeting the same fate. To do this, the units had to sanitize a hundred and ten square miles of occupied territory along forty-five miles of the island's northwest coastline.[21] These "skeeter chasers" arrived in March 1943, led by Major John Rogers. This young doctor had trained in New York with James Shannon and then in Panama with John Maier, who had assisted Lowell Coggeshall in New York in the late 1930s. After Rogers's time in Panama, the army labeled him a malariologist and shipped him out. He was smart, reliable, focused, and ready.

His team of twelve men arrived with standard-issue bedrolls and knapsacks, and multiple copies of circular letters describing their mission, in case commanding officers didn't have copies. But the units had no official attachment and lacked specific orders; they didn't know whom to report to. They weren't doctors there to treat servicemen, so they didn't fit in with the Medical Corps. And they weren't soldiers, so the staff sergeants ignored them, calling them "skeeter chasers" and "dipshits," a wordplay on their chief weapon, the dipstick.

To make matters worse, they had very little equipment and supplies. This was in part a function of the paperwork that required all requests for repellents to go to the Quartermaster Corps and all requests for antilarvae agents like Paris Green to go to the Corps of Engineers, with no coordination. Plus, pyrethrum and Freon, the main ingredients in bug bombs, were in short supply.

But this was an emergency. Several tons of African chrysanthemums, from which pyrethrum was derived, arrived in the United States from Kenya and was set aside for Guadalcanal. Industrial air-conditioning systems in New York City were stripped of Freon and, with the Kenyan pyrethrum, hastily refashioned by Westinghouse into antimosquito bug bombs. The officers of the Quartermaster Corps worked furiously to prepare everything for shipment to the Solomons. To the best of their knowledge, the ship launched from San Francisco and was due to arrive at about the same time as Russell's malaria units—in March 1943—provided the cargo vessel survived the voyage.

The bug bombs were especially needed to keep mosquitoes out of the field hospitals, barracks, and labor camps. Millions of virgin mosquitoes kept rising up from the shell holes and bomb craters, trash piles and slit trenches, tire tracks and boot marks. Infected men in quarantine might as well have been pools of poison seeping into the water supply, for from them the island's mosquitoes became infected and then passed parasites to every replacement who stepped foot on the island.

First and foremost, the malaria units needed to create a barrier between the sick and the rising mosquitoes, and bug bombs were the key. Screens, which still hadn't arrived, would help. But they would be of limited help here where anopheles started biting well before dark.

Guadalcanal had competition for antimalaria supplies; Australian and Fijian authorities demanded bug bombs to prevent local mosquitoes from biting American marines and soldiers sent there for isolation. Other island authorities wanted ships and planes previously in the Solomons to be bombed with bug spray before landing in their ports, placing even more pressure on meager supplies. All the bombs designated for Guadalcanal had to arrive; there were no replacements for them just yet. Production was still nowhere near catching up with demand.

Weeks went by. Cables flew back and forth. South Pacific supply officers demanded to know what was going on. Finally Russell investigated. He was the one who commissioned Westinghouse to make the bug bombs on an emergency basis. He needed to know why they hadn't arrived. So he sent two officers from Washington to San Francisco to track the lost supplies. They reported in late March that a clerk had confused the code name for Guadalcanal and accidentally sent the vital cargo to Hawaii, where there was no malaria. "This unfortunate occurrence will delay the arrival of the much needed Freon aerosol dispensers . . . for about six to eight weeks," the report concluded, as more than 57,600 pounds of material sat unopened and possibly degrading in the ship's hot hold, at the wrong port.[22] This was not the biggest mistake of the war. But to Russell and his malaria units on Guadalcanal, it was a disaster.

Russell made desperate calls to get the materials designated as a number one priority for shipment. This would force quartermaster officers to send the misplaced supplies by Air Transport Command, getting them there in just days. But he was rebuffed by red tape. Only theater commanders could make such designations. So Russell made more phone calls. Even after Major C. Case of the quartermaster branch received cables from the offices of General MacArthur and Admiral Chester Nimitz, declaring bug repellent and bug bombs their top priority, over explosives and bullets, Russell still couldn't get the antimalaria supplies on a flight. A cable from the quartermaster in charge said the problem was "due to the enormous volume of highest priority supplies now awaiting shipment by air from the San Francisco Port of Embarkation."[23]

In the end, the supplies ended up on a slow boat to the Solomons at the same time a million bottles of repellent and another fifty thousand bug bombs reached

San Francisco for shipment to the South Pacific. A million linear feet of screens were also ordered. Everything was scheduled to arrive sometime in May. In June, malaria units on Guadalcanal were still waiting.

These units did what they could with what they had—atabrine being their only real tool. The malariologist heading the units, Captain Roger Page, made a bold decision to try more frequent, smaller doses, to see if that would be easier on the men's stomachs and encourage them to take their pills. Instead of two 0.1-gram pills twice a week (total 0.4 grams), he cut the pills in half and gave the men one 0.05-gram pill every day, with a double dose on Sunday (total 0.4 grams). Then he reported to his superiors that the smaller doses caused fewer gastrointestinal disruptions, and seemed to absorb better, as evidenced by the men's yellow skin and low sick rates. But instead of being congratulated, he was reprimanded for veering from established War Department policy. His report eventually reached Russell just after the army captured Nazi malaria-control instructions that showed German soldiers used atabrine in a similar way. Russell wrote Page immediately to encourage him to continue thinking creatively on the malaria front, and to commend him for figuring out how to use atabrine on his own. In Russell's letter, between the lines spoke a scientist frustrated by the dunderheaded problems created by chain-of-command dogmas.

MEANWHILE, the number of malaria cases mounted and the Pacific battles wore on with dramatically reduced troop strengths. At least fifty-seven thousand U.S. sailors and marines and roughly two hundred thousand U.S. soldiers contracted malaria in 1943, alone.[24] In all theaters, the total number of infected would reach half a million by the end of the war.[25] MacArthur already had the equivalent of six divisions down with the disease.

On March 18, 1943, he wired General Marshall with the following message: "Find Dr. Russell. Send him to me." Three days later Russell was on a red-eye to Hamilton Field, north of San Francisco, where a dental chart was taken. When he asked, "Why?" the dentist said: "Your teeth might be all that's left!" On March 24 he boarded a C-47 for Hawaii that touched down for three hours before taking off for a night on Christmas Island, then onto Tutuila and to Fiji by dinnertime, then over the international date line to Nandi and deep mud in New Caledonia—at which time he had to stick his arm out the plane's window to wipe it clean with a newspaper. Finally, on the night of the twenty-eighth he landed in Brisbane, Australia. The next day he met General MacArthur.

Russell said he had never seen so much decoration on a uniform. But Russell wore his theater ribbons from Panama and Puerto Rico, where he worked on the malaria problem. And to that expertise MacArthur graciously deferred. He called Russell "doctor," not lieutenant colonel, and wanted to hear the doctor's ideas for fixing his terrible malaria problem. "I have signed and distributed all the antimalaria directives prepared by my health officers," he told Russell. None of them produced results; his men were still falling by the thousands. He knew he had a public relations problem as much as a disease problem. The War Department's circular letters all warned his men of the dangers of malaria. But they didn't listen and got sick. Then they learned firsthand that malaria was nothing compared to the hell of battle.[26] The symptoms were intense. But then came severe fatigue and days of sleep—heavenly sleep.

Pretty nurses fed them medicine in clean facilities with fresh water. If they suffered multiple relapses, they might be evacuated home, or at least to a civilized place where there were women and hot showers—free of bombs and bayonets and blood and maggot-filled rations.

This disease was now the War Department's number one medical problem, and not just because of fevers; it was ruining the men's will to fight, a danger far more insidious. By March 1943, none of Guadalcanal's hardened commanders would have dared to repeat the flippant "to hell with mosquitoes" comment made by the ignorant unit leader in September 1942. A major change had occurred in just five months.

While MacArthur and Russell squatted on a divan, the general made a declaration that is today famous in the annals of malariology. He told Russell in a tone that grew in power and purpose: "This will be a very long war if for every division I have fighting the Japs, I must have one division in hospital with malaria and another recuperating in New Zealand!"[27]

This day's diary entry was as long as any Russell had made, and closed with the following: "He ordered me to make a complete survey of malaria in every part of his theater and hand the report *directly* to him."

CHAPTER 20
The Jump

Russell started in Port Moresby, New Guinea, arriving there by flying boat on May 8, 1943—the same month Jean Schneider surrendered his secret pills in Tunis, on the other side of the world. On this mysterious, continent-size island, Russell would soon demonstrate about as well as any event in history what happens when an unstoppable force hits an immovable object—one gives. In this case, at this time, the immovable object of malaria appeared far more powerful. But Russell had only just begun.

MACARTHUR needed to secure New Guinea to launch an assault on Japan's nearby fortress in the port of Rabaul on New Britain, where the Japanese had multiple airfields, a huge naval force, and an estimated ninety thousand troops. Rabaul was a beehive, not to be disturbed until American forces had comparable strength in the region. To take it from the Japanese would be to force them out of Melanesia (the Solomons, the Bismarcks, and New Guinea) and northward to Micronesia (to Truk in the Carolines, and to other tiny specks of islands from Palau to Kinini, Eniwetok, Makin, and Tarawa). Neutralizing Rabaul was *the* strategic imperative of Melanesia, code-named Operation Cartwheel. It would clear the way for U.S. forces to move northward and take back the Philippines.

To make this happen, MacArthur needed airstrips and staging areas on the northeast coast of New Guinea. And for that he needed to expel the more than three thousand Japanese troops occupying the area.

This would not be easy.

The Melanesian chains were the least explored islands of the Pacific. Europeans called them "the black islands," because very little was known about them. New Guinea stood out as the largest and most mysterious of them all, shaped like a potbellied dragon in flight, and stretching more than a thousand miles from nose to tail. Malaria ruled its coastline. But just inland, as the foothills shot up into huge uncharted mountains, malaria disappeared in elevations too high to support anopheles mosquitoes. In these jungle-covered highlands, explorers and naturalists had reported seeing warring tribes that practiced cannibalism.

To own this island as an occupying force was to dominate the coastal tribes, leave the warring mountain tribes undisturbed, and set up camps and weapons in strategically defensive positions along the beaches.

Both Japanese and Allied forces had tactical reasons to occupy New Guinea—the Allies for their push to Rabaul and the Japanese to prevent that push, and to create a buffer around their newly acquired, resource-rich islands of the East Indies, especially Java. Both sides ignored the hulking central highlands, which had been heaved upward over millions of years by plate tectonics and ferocious volcanic activity. There, tribal people farmed yams and spoke a thousand different languages, completely isolated from the world. As war consumed the planet, the highland people remained virtually untouched.[1]

Battles for New Guinea were confined to the dragon's tail, a narrow stretch of lower mountains that ended in a place called Milne Bay at the tip of the tail. The northern and southern coastline of this long peninsula had beaches and lowlands for occupying forces, and tamable locals easily pressed into labor. MacArthur's forces held precariously on to Milne Bay, and on to the underside of the tail at Port Moresby. This port looked toward Australia, just a few hundred miles to the south. The Japanese had sent troops over the mountains to try to capture this seaport in July 1942 while malaria burned through MacArthur's forces. He asked for more of everything—troops, transports, and airpower to repel the Japanese. But commanders in Europe said no; they could spare nothing in their attack on Hitler. Plus, their analysis of the Pacific theater was positive. Melanesia preoccupied Japanese forces, which slowed the assault on China, allowing Roosevelt to keep his promise to help Chinese leader Chiang Kai-shek.

But when Guadalcanal fell to the Americans, the Japanese concentrated on holding New Guinea. MacArthur dug in while the enemy amassed forces, and mosquitoes and malaria infected everyone.

. . .

RUSSELL landed on New Guinea just as an army doctor there was threatened with court-martial for a report that mentioned soldiers "fucking sheep."

The doctor was really just trying to get help with the island's astronomical malarial rates, which had reached four thousand per one thousand men, per year. "That meant everybody was ill with malaria four times a year," remembered Roger O. Egeberg, the doctor who wrote the letter. "In six months, we had to evacuate a third of our soldiers to Australia. I had tried to get help—work crews—locally, but since every outfit was short-handed, the answer would be, 'Look, Doc, I have to unload these ships. I can't spare you any men to dig a ditch.' Or, 'Sure, Roger, I'd like to help you drain that swamp, but I have to man this ack-ack battery and I'm down to half strength. You were glad to hear our noise last night, I bet.'"[2]

So Egeberg wrote letters up the chain, begging for help. He needed more oil, Paris Green, and labor for ditching. And he needed this drug atabrine he kept hearing about but had none of. He also asked for "a little authority" to take charge of the problem, to instruct commanders on where to locate camps and how to protect their men from mosquito bites. But the replies he received focused on his venereal disease rates, with no acknowledgment of the malaria problem. "Our V.D. rate was 8 per 1,000 per year," acquired by men on furlough, he wrote in his memoir. He fumed over the stupidity of supply officers sending him condoms instead of insect repellent, and ordering him to write reports about VD when he had out-of-control malaria. The final straw arrived from U.S. Army Services of Supply headquarters in a letter that said: "You will reply by endorsement hereon. . . . Do you visit your prophylactic stations late at night? . . . Have you obtained cooperation of the civilian authorities so you can apprehend and treat the contacts?"—"contacts" meaning prostitutes. This reference to the sex trade sent Egeberg over the edge, because Milne Bay had no prostitutes. All they had were sheep.[3]

In his reply he mapped out the men's choice: "eating the fucking sheep or fucking the bleating sheep," which he called a choice that "would work itself out, though somewhat on the order of killing the goose that lays the golden egg." He ended his report with a poem:

"Out Milne way, no amateurs, no whores to take our semen. But forms galore, reports, and charts we have to manufacture so men in higher headquarters can have the facts they're after. If men in higher headquarters could go into the field, we're sure that forms and charts, etc., would be severely 'pealed."[4]

Three days later his CO received a telegram. It was from HQ in Port Moresby demanding the arrest of Major Egeberg "for the use of obscenity in official communications."[5] His CO, Colonel Burns, had been quietly trying to get the arrest warrant recalled when Russell arrived at Milne Bay with Colonel Maurice Pincoffs, also from the army's tropical disease division. According to Egeberg, they "looked at the statistics and saw in every dipperful of pond or puddle water the countless mosquito larvae wriggling there. They examined children in one village and found they all had malaria. They went back and reported I had been desperate."[6]

Russell probably saved Egeberg. He parachuted in, did a survey of the problem, and used data, not expletives, to show Services of Supply just how wrong they were in assessing Major Egeberg's medical program. That data provided indisputable evidence that Egeberg needed an antimalaria sanitation campaign, not VD directives. Soon thereafter Services of Supply withdrew the arrest warrant.

From there, Russell moved to other military sites in and around New Guinea, including Dobodura, Buna, Gona, Cape Kellerton, Sananda, Mukawa, Oro Bay, Goodenough Island, Waga Waga, and Gili Gili. Then he flew back to Brisbane to debrief MacArthur.

On May 25, Russell reported that not even officers carried out the general's malaria directives. He told MacArthur that medical officers lacked authority to do their work, yet took responsibility when it wasn't done. Better to give those officers authority and, if combat commanders refused to comply, make them responsible for malaria rates. Motivate unit commanders by requiring a separate report for each case of malaria. The threat of more paperwork alone would be a potent catalyst for change.

Russell also advised MacArthur to immediately request another dozen malaria control and survey units, to get them queued up for training in New Orleans. Last, he recommended that malaria personnel and supplies be given top priority for movement. This no doubt came from Russell's infuriating experience trying to get his specially made bug bombs and pesticides from Hawaii to their intended location on Guadalcanal. As far as he knew, those supplies still hadn't arrived on the island.

MacArthur set in motion all three recommendations, and the results were immediate. Egeberg said that a new directive making malaria control a command function arrived in Milne Bay shortly after these HQ doctors left. "It was just fourteen or fifteen words, but they changed the whole picture," Egeberg wrote. "The 'eephus' was now on each C.O., the technical advice was in my bailiwick, oil and Atabrine came, and we were allowed to move the inhabitants of three villages." He said within six months, the malaria rate dropped to just 140 per thousand per year—from four thousand per thousand—and "Milne Bay became the healthiest base in the Army."[7]

Then, to Egeberg's utter amazement, General MacArthur invited him to serve as his new personal physician. He had been handpicked for the job *because* of the obscene letter he had sent to the SOS.

AFTER leaving the general, Russell island hopped on navy PB4Y patrol bombers, fitting in behind the pilot, under the turret gunner's feet—flying in torrential rains, always "loaded for bear." On Guadalcanal he found the malariologist sick with dengue at Koli Point and the officers' camp a complete "mud hole" that everyone called Mildew Arms. In the villages, he examined spleens and pricked fingers, and found infection rates from 60 to 90 percent. He witnessed troops working near *A. farauti* breeding sites, as if she weren't the enemy.

War debris clogged the rivers like dams, creating giant bogs of stagnant water. A high water table filled foxholes and shell and bomb craters. The destroyed landscape was pocked with water-filled ruts. The enemy *farauti*—in her distinct uniform of spotted wings and tiger-striped legs—lived everywhere. He noted it all in triplicate reports, then packed for home.

On June 12, when he was supposed to leave, the Japanese staged an air raid against Henderson Field. But American, Australian, and New Zealand air defenses completely destroyed enemy aircraft. "No bomb fell on Field," Russell wrote in his cryptic syntax for that day's diary entry, adding a flash of detail: "Intelligence Tent broadcast all stages of the thrilling air battle. This was the last air attack by Japs on Guadalcanal." If only he could get these expert gunners to take mosquitoes as seriously as they took Japanese planes.

He left later that day, island hopping toward home on all manner of aircraft—including an old de Havilland, nearly having to ditch in the Pacific's shark-filled waters between Palmyra and Pearl Harbor, and sleeping on the floor with mail bags from Pearl Harbor to San Francisco.[8] Finally, on the summer solstice, June 21, he arrived in Washington. Total miles traveled during this two-and-a-half-month tour: 26,434.

INFECTION rates at base sections on New Guinea, Guadalcanal, Efate, Espiritu Santo, and Tulagi ran between 25 and 100 percent. American troops had even brought the disease to the malaria-virgin Russell Islands. It was a disaster. But MacArthur was now fully on board with a whole new approach to malaria. He and his high-ranking entourage even attended a training session for officers on how to identify enemy mosquitoes and distinguish between *vivax* and *falciparum*. Then he established the Combined Advisory Committee on Tropical Medicine, Hygiene, and Sanitation—a joint team of American and Australian malaria experts—and gave them "broad authority to develop plans and policies to be followed by all Allied forces." Of fifteen health-related directives sent to troops from MacArthur in 1943, fourteen addressed malaria.[9] Still, he was losing eight soldiers to malaria for every one lost in battle, overall, and thirty to one in the worst places.

The navy, which took care of the medical needs of sailors and marines, fared no better. They had "an average of 5,332 men on the sick list daily because of malaria—4,148 of whom were marines."[10]

In a July 5, 1943, "secret" letter to his superiors in Washington, Major General Norman T. Kirk wrote: "The malaria situation in the South and Southwest Pacific is bad, as I originally mentioned. The attached report, which is authentic, shows that some six divisions have already been incapacitated because of malaria infection. Thus malaria handicaps the war effort in that area. The same thing will happen in North Africa if troops occupy any of the islands in the Mediterranean or southern Italy itself. . . . Malaria control is a command function and the protection of troops from mosquito bites is essential. . . . The war effort in these areas can be badly crippled, if not lost, by malaria infection alone."[11]

Nine days later, in a "secret" cable to all malarious military areas, addressed to MacArthur, Eisenhower, Stilwell, and Harmon, General Marshall wrote: "[M]alaria portends to be one of the greatest dangers to military operations in this war. . . . Regulations and directives designed to give protection have been widely distributed. The most important measures are careful screening at bases, careful supervision by antimalarial personnel and servicing of forward units with antimalarial supplies. Drugs such as atabrine are only temporary measures." He ordered them to establish the highest priority for shipment of antimalaria supplies, especially screens, and give more attention to troop discipline.[12]

The U.S. high command finally got it; they finally understood that they had to fight malaria and do it the hard way—the way Lowell and many others had been advocating for well over a year.

RUSSELL'S malaria units reached Tunisia with orders to join the jump to Sicily. They arrived to a completely wrecked Bizerte armed with microscopes, lab equipment, canisters of repellent, knapsack sprayers, sacks of Paris Green, and netting. They wore U.S. Army uniforms, but instead of combat boots they had mosquito boots, and instead of rifles they carried dipsticks. A fair number were conscientious objectors who believed in the Allied campaign but could not personally commit acts of violence. For them, mosquito work was ideal. These units of oddball naturalists and pacifists didn't fit in with foot soldiers. And commanders didn't know what to do with them, as military heroes led men with weapons, not sprayers.

General Patton's troops grew battle-ready during a brutal winter in the gravelly hills of Tunisia. "They have made the psychological transition from the normal belief that taking human life is sinful, over to a new professional outlook where killing is a craft,"[13] observed Ernie Pyle, who was embedded with them. In the barren hills they fought off "snakes, two-legged lizards, scorpions, centipedes, overgrown chiggers, and man-eating ants"[14] while throwing grenades at enemy soldiers hiding within shouting distance. They learned to move like "ghosts," unheard and unseen from three feet away, until the glow of a big gun lit the sky and illuminated a line of dark helmets and steel rifles.[15] Patton's men wanted to kill more Germans and make their way to Italy—with people more familiar than the teetotaling Arab and Berber tribes of North Africa. Patton's soldiers looked forward to drinking wine and chasing women—women who looked like the girls they grew up with, but who could be bought for a song. These brothers in arms spoke the same battlefield vernacular and sang the same X-rated versions of "Dirty Gertie from Bizerte."

To reach Italy they had to first take the triangle-shaped island of Sicily in an invasion code-named Operation Husky. Churchill and Roosevelt believed its occupation could be used to convince all Italians to desert Germany and side with them. A quick mop-up there with the right diplomatic mission in Rome, and Americans would be in Italy by the fall. Troops were pumped and ready. But they also were seeded with malaria. Some three hundred men couldn't make the jump because they were sick in bed with fevers, even though malaria season hadn't started yet.[16] Reports were sent back to Russell in Washington about the probability that troops were picking it up in Liberia while en route to North Africa.

To Patton, it didn't matter how they got sick; they were a disappointment, period. As for the antimalaria units, he wanted no part of them. They would be in the way and weren't needed. Americans would be in Sicily for so little time, why worry about the mosquitoes there? The units were left behind.[17]

LONG before dawn on July 10, some 2,590 ships and planes approached the western beaches of Sicily in forty-mile-an-hour gales to land two great armies: the British Eighth and the American Seventh. The plan had seven divisions wading ashore while paratroopers dropped from the skies to secure a defensive line inland. But the severe storm, which kicked up out of nowhere, created choppy, swelling seas that threw the boats off course. Sixty-nine British gliders loaded with paratroopers crashed into the dark seas. U.S. pilots lost their way and dropped thousands of men way off the mark.

That day, nearly a quarter million Allied soldiers scattered along the beaches and forests of Sicily to overpower a roughly equal number of Axis soldiers, mostly Italians. Only two German divisions stayed on the island. The rest had been withdrawn weeks earlier and sent to Greece and Sardinia, where the Nazis believed the next Allied attack would take place, as outlined in false papers attached to a floating corpse in a British officer's uniform.[18]

Even without the bulk of Germany's Mediterranean forces, the first few days were rough, according to an army history of events:

> During the first three days the U.S. Army and Navy moved 66,285 personnel, 17,766 deadweight tons of cargo, and 7,396 vehicles over Sicily's southern shores. By the end of the first day, the Seventh Army had established a beachhead two to four miles deep and fifty miles wide. In the process it had captured over 4,000 prisoners at the cost of 58 killed, 199 wounded, and 700 missing. But the situation was still perilous.

> Axis counterattacks had created a dangerous bulge in the center of the American line, the very point where the bulk of the 505th Parachute Regiment should have been if its drop had been accurate.[19]

On July 11, Patton sent in another two thousand paratroopers. Allied soldiers defending their positions mistook them for enemies and shot down twenty-three planes, killing nearly two hundred of their own.

Meanwhile, a struggle between General Bernard "Monty" Montgomery, who led the British Eighth Army, and Patton, leading the American Seventh, grew fiercer. Each planned to win Messina, at the northwest corner of the island, in view of mainland Italy.

To get there first, Patton made a now famous "cavalry-like raid"[20] on the defenseless city of Palermo, which, by the time of the landings, was in ruins from months of Allied bombings. Two battalions from the Seventh Army swept down slopes of crumbled houses, along a smoldering landscape that gently spilled to a quiet bay of wrecked fishing boats. Italian officers surrendered to the Allies and civilians welcomed the Americans with wine and cheese. Kesselring, with his troops caught off guard and on the wrong islands, "sourly" reported to Berlin that the "occupation of western Sicily must be considered as complete."[21]

Meanwhile, an air raid of more than five hundred bombers, with orders to avoid the Vatican, destroyed parts of Rome. The Italian king's fears were now too real; his country would be trampled by the fighting elephants. Mussolini complained to Hitler: "[S]acrifice of my country cannot have as its principal purpose that of delaying a direct attack on Germany."[22] A week later, on July 25, the Italian king forced Mussolini's resignation. The streets of Rome erupted in violent relief; citizens ransacked Fascist party headquarters, burned the furniture, and shouted, "*Viva l'Italia!*" The plan to secure Italy even before Allied troops landed there was well under way.

Patton, in his headquarters in Palermo's Royal Palace, plotted to move his Seventh Army to Messina before the British Eighth. Calling the march to Messina a "horse race, in which the prestige of the U.S. Army is at stake," he ordered his commanders to move forward quickly, even at the cost of additional men.[23] Getting to Messina after the British would disgrace the United States. He even offered the navy "a captured fleet of Volkswagens"[24] if they would help him reach this one goal.

. . .

THROUGHOUT August, malaria hit Patton's army like a sledgehammer. Huge waves of sick men rolled into the island's meager thirty-three hundred available hospital beds. Confirmed cases totaled twenty-two thousand Italian *vivax* and African *falciparum* malaria. An equal number of soldiers were diagnosed with "fever of unknown origin," at least half of which were probably malaria, the other half consisting of dengue and sandfly fever.

Ernie Pyle was among them. His decision upon landing in Africa to never take atabrine probably led to a malaria infection that exploded into fever in Sicily. Doctors said he had a blend of afflictions that churned into a hybrid disease they called "battlefield fever." But that was a guess forced by a lack of proper lab equipment to examine blood for confirmed diagnoses. The chaos around medical supplies that plagued North Africa followed troops into Sicily. With few choices, fever cases had to be evacuated, mostly by navy hospital ship. A 20 percent loss in strength resulted, which slowed Patton's advance on Messina.

In August, he visited two station hospitals to feel out the problem. He arrived at the 15th Evacuation Hospital on August 3 to an "odor of disinfectant and blood and oozing wounds. Green light filtered through the canvas, and the sound of labored breathing filled the ward as if the tent walls themselves suspired. . . . On a stool midway through the ward slouched a private from the 26th Infantry, Charles H. Kuhl," diagnosed with "psychoneurosis anxiety—moderate severe."[25] But he also had malaria and chronic diarrhea, which could have triggered the kind of trembling and fatigue that often got diagnosed as mental breakdown. Patton asked Private Kuhl to state his condition. When the boy responded, "I guess I can't take it," Patton slapped him across the face with his gloves, then dragged the young private out of the tent by his collar and kicked him with his cavalry boot. The whole time Patton called Kuhl a coward, a yellow-bellied bastard, a son of a bitch and gutless. A week later the irascible Patton let his blood boil again at the 93rd Evacuation Hospital. There he singled out a feverish and listless private named Paul Bennett from the 17th Field Artillery unit.[26] Private Bennett, like Kuhl, was unable to articulate the full range of his medical issues and simply said it was his nerves. Incensed, Patton pulled his pistol on the young trembling boy, told him he ought to be shot, and then hit him in the face, twice. When Patton left the tent, he broke into tears: "It makes my blood boil to think of a yellow bastard being babied."[27]

The pressure of war got to everyone, even Patton. But Patton's problems were self-inflicted. He wanted to reach Messina before the British and made a monster of himself in the process. Diseased soldiers, with mental and physical afflictions that would remain with them the rest of their lives, went unappreciated by medically ignorant leaders like Patton. Had he considered that disease, especially malaria, was three times as damaging as enemy fire to his goal of quickly reaching Messina, he might have planned differently and made medicine, malaria control, and frontline care of the sick a priority. Instead, fevers of all kinds went untreated because of supply and equipment problems. The sick were seen as occupying beds needed for the injured, as if taking on deadly microbes while fighting for him was less honorable than taking a bullet. The sick were dragged from the front lines and put on hospital ships headed back to North Africa.

On August 17, Patton reached the "windswept heights west of Messina where Highway 113 began a serpentine descent into the city" many hours before a British colonel from the "4th Armored Brigade arrived, with bagpipes and a Scottish broadsword in the back of his jeep." The prize was Patton's.[28]

British and American soldiers endured weeks of bloody battles. They crawled from one German-looted village to the next, over craggy, booby-trapped cliffs, through populations of devastated, starving Sicilians, all under salvos of enemy fire. None of the men on the front lines or mopping up blood in field hospitals cared who took Messina first—the bagpipes or the cavalry boots. The Seventh and the Eighth suffered equal casualties, both from disease and battle wounds. Dirty, tired, hungry, and sick, they looked forward to the respite from gunfire so they could change their socks and shave, before suiting up to move to the malaria-infested plains of southern Italy.

CHAPTER 21
Malaria and the Madmen

Publicly, the War Department had no malaria problem. Details of the shocking malaria rates for troops were in "blackout"—kept secret.[1] As such, the work of the Malaria Project received little press, and the War Department forbade publication of study results. Chief investigators received travel budgets for monthly meetings in Washington, where they were sworn to secrecy and agreed to quietly exchange information, all of which was to be held in confidence. They argued that study results should be published in medical journals to share information and inspire insights from scientists in all fields. But the War Department controlled the decision, and the answer always came back, "No."

Lowell Coggeshall understood this. But he had a hard time keeping quiet when confronted with blatant misinformation about their situation, as was the case with a *Reader's Digest* article by the famous Paul de Kruif, a biologist turned writer and author of *Microbe Hunters*, an international bestseller about the first microbiologists (still in print today, assigned by university professors as a must-read). De Kruif's *Reader's Digest* article carried the headline "Enter Atabrine—Exit Malaria," and declared that a five-day course of atabrine cured malaria.[2] *Cured* malaria!

The piece appeared on December 1, 1942, in this broadly read periodical, two weeks before Lowell arrived home from Africa. In lyrical cadence—which made de Kruif's work readable and immensely popular—the opening lines said: "American scientists have triumphantly opened a front against a sinister enemy that must be whipped before we can defeat the Japs and Nazis. . . . It was a prewar mistake of the Germans that their dye trust I.G. Farbenindustries, underestimating the ingenuity of American chemists, let us in on a hint of atabrine's secret. . . . Their Nazi masters knew that no bid for world conquest could be made without a substitute for quinine, because malaria soldiers are just too sick to fight. With the chemical cunning for which they are renowned, German test-tube wizards fashioned more than a thousand dye compounds, tested them out with appalling patience on malaria sick canaries and rice finches, till they arrived at Atabrine's yellow magic in 1932. The results were epochal."[3]

Lowell and his peers understood de Kruif's hyperbole. This was his art. He punched up the science to sell books and magazines. He used a doctorate in microbiology from the University of Michigan and experience working at the world-famous Rockefeller Institute to help the public understand science, and help other writers do the same (for example, he ghost cowrote Sinclair Lewis's Pulitzer Prize–winning novel *Arrowsmith*).[4]

But he sometimes painted too colorful a picture of his subjects (the British version of *Microbe Hunters* omitted de Kruif's depiction of Ronald Ross to avoid libel suits). He was an old-school scientist harboring naive notions of magic bullets. These tendencies gummed up his article on atabrine, especially one line that, to malariologists, must have sounded like fingernails on a chalkboard. It said that just "15 little pills of atabrine, given over five days, were all that was needed to cure the great majority of malaria victims." From there he just kept scratching that board, ruining his credibility: "It was magical the way atabrine brought the worst malaria sufferers back from the grave." This "master weapon" of "Hitler's Wehrmacht" now belonged to the Americans.[5]

Then the piece made all kinds of statements that weren't true: that atabrine could cure relapses; that the Germans had "sold" it to the Americans; that by 1939 a Winthrop chemist had figured out how to make it exactly as the Germans had; and that Winthrop and Merck & Co. were making enough of it to *cure* all American troops of malaria—at just six cents a head (a conclusion he probably derived from a Winthrop announcement in September 1942 that the company had dropped its price to $4.50 per thousand pills).[6]

De Kruif closed with a kicker of a kicker: "In the postwar tomorrow we shall be faced with a battered, maimed, starved, sick world. You cannot rebuild a world with half of its people drag-footed, shaking with malarial chills, burning with malarial fever—*now curable at six cents per capita!* And we shall meet at least part of our obligations toward reconstruction by producing and distributing billions of yellow pills of Atabrine that will eventually conquer malaria."[7]

He didn't check the facts and ended up stepping into a controversy fostered by well-known columnist Howard W. Ambruster, who had accused de Kruif of being spoon-fed propaganda from Winthrop's press agents at the firm of Baldwin, Beech and Mermey. Ambruster called several manufacturing mishaps at Winthrop espionage, including contaminated atabrine sold to the military. In his writings, published in *The Nation* magazine, he charged that these documented cases of contamination were a direct result of Winthrop's twenty-year relationship with I.G. Farben. Ambruster challenged de Kruif to disclose his sources for the *Reader's Digest* story.

This aspect of de Kruif's story didn't matter to the experts on the Malaria Project, relative to his startling misstatement of facts. The coming year—1943—would see more malaria than the world had ever known, and more than the War Department thought possible. And the atabrine problem was far from solved. But with no alternative, the army had to figure out how to use this drug in battle. A few weeks after de Kruif's article, Brigadier General Charles C. Hillman of the office of the army surgeon general told reporters that "never before had an adequate supply of drugs been so essential to victory."[8]

Immediately, Lewis H. Weed, chairman of the National Research Council's Division of Medical Sciences, wrote the editor of *Reader's Digest* that de Kruif's "colorfully" exaggerated "misstatements" were "harmful" and could jeopardize the national effort to find a new drug so desperately needed for troops.[9]

On January 7, a month and a week after de Kruif's article appeared, the *New England Journal of Medicine* ran a biting editorial, calling de Kruif's piece "erroneous" and assuring readers that treating malaria was anything but "simple."

But the misconception lived on. So Lowell wrote a rebuttal published on May 1 by the *Journal of the American Medical Association*. He sidestepped the Ambruster controversy and instead offered a four-thousand-word explanation designed to highlight the real, scientifically based fears about malaria: "With the exception of Corregidor and Bataan, where no drugs were available near the end, all of our excessive rates are occurring in men taking atabrine routinely." Of de Kruif he wrote: "Some attention has been given to a recent article entitled 'Enter Atabrine, Exit Malaria.' A more appropriate title would have been, 'Exit

Quinine, Enter Atabrine,' as atabrine is as effective prophylactically and therapeutically as the rapidly diminishing stock of quinine, but it will not cause an infection to disappear." Nor could it prevent malaria, he wrote. Like other researchers on the project, he held serious reservations about atabrine. "It is clearly evident . . . we have no drug that will prevent or reduce the large volume of malaria that is constantly increasing," he wrote. "If the damage were solely confined to the troops in the affected areas the problem would be serious enough to cause considerable apprehension. Unfortunately this is just the initial step in the chain of events that is likely to create a more serious disaster."[10]

Lowell wrote many papers, gave many speeches, and chaired many meetings in which he talked about the absolute necessity of mosquito control, especially screens, in stopping malaria's march on U.S. troops. He used this *JAMA* article to talk about potential consequences of inaction: "A major proportion of our troops overseas is in tropical areas in close contact with the huge native malaria reservoir. If we ever approach the proposed figure of eleven million men in the armed services, which means practically half of our adult male population, and if the same proportions are to serve in the tropics as now, then we can gain some appreciation of the probable consequences." As it stood, he wrote, Allied troops were acquiring "probably more cases per day than we experienced per year in the last war." He warned that if the kind of "secondary eruptions" of malaria that followed the Great War's troops back to England, Italy, and Russia were to follow U.S. troops back to the United States, American anophelines would be able to spread the disease across the country. "The stage is all set for trouble," he wrote. "We have the susceptible population and the vector, and the disease seems to be on its way. We must conclude that the potential danger of malaria during the present war is a greater worldwide menace than ever before."[11]

He issued this warning two months after his assistant at the University of Michigan pressed jars of American anopheline mosquitoes on the bare flesh of three U.S. soldiers sent home from Guadalcanal with relapsing malaria. All mosquitoes became infectious and easily transmitted the soldiers' disease to syphilis patients at Battle Creek. Lowell's fears were confirmed and he felt the public should know.

THIS was Lowell's new role—gadfly—because he was no longer qualified to lead the Malaria Project. By the time he returned from Africa, the project had grown in size and stature and was ruled by older, more famous scientists who brought depth of experience in a range of disciplines. They were highly skilled researchers, technically proficient in the molecular minutiae of medicines interacting with blood and tissue. The most important were the chemist Eli Marshall at Hopkins and the blood expert James Shannon in New York. These two giants in their fields worked in tandem to answer perplexing questions about Germany's synthetic drugs. Marshall's team examined them, adjusted them, rotated the molecules around to make derivatives they called analogues, and then tested them all for toxicity and action against bird malaria, while Shannon took the best of the drugs and used them on his state hospital patients on Welfare Island. But none panned out. They were all too toxic.

Atabrine continued to be the only drug available for troops. It didn't appear to work well, as troops turned yellow and still came down with malaria. That led investigators to ask two questions: Was Winthrop making it properly, and were the Allies using it properly? Marshall set out to answer the first, and Shannon the second. In the process, they, too, got trapped in a growing political controversy over Winthrop and its former relationship with I.G. Farben.

For the war effort, the problems centered on Winthrop's control of the German patents for these 8-aminoquinolines, atabrine and the oxygen-choking plasmochin. The U.S. Senate Patent Committee announced it would hold a hearing on the matter, but then never scheduled it. Republican congressman Bertrand Gearhart of California threatened to push the matter in the U.S. House of Representatives, but then he didn't either. This riled Ambruster, who alleged that Congress had caved to political pressure from Winthrop's parent company, Sterling Drug Inc. (successor of Sterling Products). Innuendo filled the gaps.

Sucked in was *Harper's Magazine*. In 1943 it ran an article by Yale law professor Walton Hamilton. In it, he described a Nazi invasion of America through diabolical business partnerships that trapped greedy and naive American companies in a dependency on German know-how and raw materials. His prime example was the deal between Sterling Products and I.G. Farben in 1923 to create Winthrop. Hamilton called it a front for German products: "This is the story of how Sterling Products . . . found itself finally in the toils of the Nazis. The attack upon the United States did not begin with Pearl Harbor. . . . The war behind the war was already in full swing. . . . Its objectives were to build up in Germany a huge capacity-to-produce and to sterilize the resources of its enemies." Hamilton then told a chilling tale of how I.G. Farben "dangled" the "bait" of "markets and monopolies" to capture America's top business executives in "entangling alliances." "In time, 'Berlin' imposed its will upon its corporate subjects . . . the yoke was fastened before the victimized American corporations were entirely aware of the nature of the bondage." A "neo-feudal" alliance formed and the "doomed" Sterling became "vassal to the fascist state." He continued: "A compromise of its national patriotism was never formally made. The results emerged from a yield-as-you-go sort of expediency. New York had chosen loyalty to Leverkusen; the fealty to Berlin was forced upon it."[12]

This, of course, infuriated Sterling's executives. So in March *Harper's* published a letter that gave Sterling's president, James Hill Jr., space to respond. He accused Hamilton of living in the past, as if it were the present. By the time the article ran, Sterling had severed all business dealings with Germany; sacked Winthrop's president; replaced him with the former chief of the drug division of the U.S. Food and Drug Administration, Dr. Theodore G. Klumpp; expanded Winthrop's plant to increase atabrine production by 8,000 percent; and ran all other Sterling factories at full capacity to meet war production needs. Hill asked: What more could they do?[13]

To which Hamilton responded with another description of Sterling's decades-long dependency on Germany, and how the company would very likely revert back to this same profitable but subservient setup after the war, to the detriment of American economic and technological progress. No less than American sovereignty was at stake.[14]

Winthrop had successfully been demonized.

But this sort of debate was for lawyers, commentators, and politicians, not scientists. Drs. Shannon and Marshall naturally wanted no part of it. They had their own information on Winthrop that cleared the company of espionage, but not of greed and duplicity. They gathered it the year before, in 1942, while running a study the army and navy wanted on Winthrop-made atabrine. The War Department wanted to know whether spoiled stocks—some highly toxic—reached U.S. troops "by accident, or by the intention of saboteurs."[15] So Marshall and Shannon assembled a committee of chemists. For five months, from May to September 1942, they compared samples of atabrine from four different labs: Winthrop, Abbott Laboratories, England's Imperial Chemical Industries, and Bayer's Leverkusen plant in Germany. Toxicity tests on mice, frogs, rabbits, guinea pigs, rats, dogs, and cats turned up no differences. The chemists thought maybe the variability in side effects were a result of the malaria itself, so they tested all the samples on chickens sick with *P. gallinaceum* and ducks sick with *P. lophurae*, the Coggeshall strain. Still nothing significant turned up.[16]

Their chemical analyses had found an impurity, which technicians called toxebrin. But this was no smoking gun. Marshall's chemists, in the final report, concluded that the impurity was not sufficiently toxic to trigger "the alarmingly high incidence of reactions"—the severe vomiting and uncontrolled diarrhea; the unbearable stomach cramps; and the disorienting psychosis, which included intense paranoia and depression.[17] They were stumped.

Shannon tried to tease out the variables in human studies, but this time he needed a sizable population of healthy adults who could speak reliably about the symptoms they were experiencing—not state hospital patients. He quickly got permission to use 250 Ohio State University medical students, 241 inmates at the New Jersey State Reformatory in Rahway, and 332 inmates at Sing Sing prison in New York. Researchers divided them into cohorts—groups whose side effects could be controlled and compared in parallel studies. This wasn't an efficacy study, so they weren't infected; they were just asked to articulate, as precisely as possible, side effects.

During this 1942 investigation serious concerns arose, because no pattern emerged. The Ohio State students were divided into five groups of roughly fifty, each put on atabrine from one of the four labs, plus a placebo. The placebo group had no side effects, while the other groups all experienced severe side effects in the range of 28 to 39 percent. When mild side effects were added, the rates jumped to over 50 percent, with the exception of those on the Winthrop-made pills, who averaged just 47 percent.

In the Rahway groups, average side effects ranged from 16 to 37 percent. But 14 percent of the placebo group also reported side effects, totally confounding the results. Meanwhile, at Sing Sing, those experiencing severe adverse reactions averaged only 4 percent. Committee members concluded that the differences in manufacturer could not explain the problems experienced with this drug.[18]

AT the time, Winthrop executives held tight to their patent rights on atabrine. They were willing to license other companies to make it for the war effort, but after the war they would have to enter licensing agreements. This was untenable, as no company would submit to investing in equipment and capacity, only to be tied, postwar, to a competitor company under unknown terms.

The possibility was raised of streamlining the processes so the British could make it reliably for their troops, and Winthrop's president, Klumpp, hedged again. "If the British Central Scientific Office should desire to utilize the information for actual manufacturing purposes in behalf of our United Nations war effort we shall, of course, be very happy to have them do so, but should such a contingency arise we think it would be necessary for us to arrive at a formal understanding with whatever British manufacturing interests are involved, since it would touch upon complicated patent and commercial considerations having significance after the war period."[19]

This, of course, meant the British would be licensed to make it, under Winthrop's conditions. Committee members included this quote in their final report, perhaps as a jab at Winthrop. These scientists were exasperated, as they worked hard, and not successfully, at trying to tease out problems with atabrine so the military could use it properly and stop the hemorrhaging of troop strengths in every theater of the war.

The army and navy, the Board of Economic Warfare, the War Production Board, the Department of Commerce, the Justice Department, and the Public Health Service (aka the Malaria Project) discussed the possibility of the U.S. Alien Property Custodian seizing the atabrine patents as German property. This would have given the custodian authority to open the patents to other companies, license-free. To avoid such a calamitous outcome, Winthrop licensed one company, Merck, to help meet demand.[20] Then Winthrop produced more contaminated stocks of atabrine and other drugs, which led to a very public recall of all of them. More editorials shamed Winthrop for its history with Germany. The likely problem, however, was that atabrine was extremely difficult to make and required enormous amounts of chemicals. Six intermediates were needed. To make three million pounds of the drug, Winthrop needed to draw from the American industrial suppliers twenty-five million pounds of required chemical materials—so much that suppliers couldn't meet demand. The many processing stages also exposed workers to toxic dust that got into eyes and lungs. The dust caused severe conjunctivitis and deep staining of the eyes and skin, plus other toxic manifestations, making it hard to retain employees.[21] Winthrop and the other atabrine manufacturers provided workers with protective face masks and gloves to mitigate the effects of this unpleasant work.

By the end of 1943, Winthrop had reduced toxins found in atabrine, ramped up production, and allowed other drug companies to make it license-free—but only for the war period and only for military use. The other companies complied because the war was on and atabrine was a critical war material.

Just as enough of it was available for wide distribution, numerous reports from the Pacific documented a rise in serious rashes. Medical officers called it lichen planus, and doctors were instructed to take men off atabrine at the first sign of it. Alarming reports described several Chinese soldiers who were kept on atabrine even after the rashes appeared, until their skin fell off. Autopsies showed their organs had broken down. These deaths, recorded in reports from different station hospitals, scared medical corpsmen away from using this drug.[22]

Plus, atabrine still wasn't working in some people, even after Winthrop streamlined the making of it to avoid contaminants. Of 98,050 Australian troops in New Guinea in late 1943, 3,140 were lost to battle casualties, while 47,534 were evacuated for disease, of which 28,909 were diagnosed malaria, and many more were very likely undiagnosed malaria.[23] Three divisions evacuated from New Guinea because of malaria—the 32d U.S. Division and the 6th and 7th Australian Divisions—were taken off high doses of atabrine for lower doses to see what would happen. But the drug continued to produce different results. Meanwhile, the U.S. 41st Division, which remained operational in the Buna and Gona region of New Guinea, was on atabrine to try to lower the attrition rate. But many still came down with malaria, while side effects so dramatically sapped the division's strength that it was at risk of complete collapse. General MacArthur's commanders complained that more replacement troops just meant more sick men, not a stronger fighting force.

THE one saving grace of having standardized production of atabrine was that Shannon could now run studies on dosing and absorption without worrying about defective or contaminated pills. He dosed up his state hospital patients with atabrine and then used a photofluorometer to track the drug's fluorescent dye as it saturated tissue and flooded blood cells. Some patients were given heavy up-front doses that tapered off, while others were given small, slow-building doses until it plateaued. His patients on the large up-front doses must have been vomiting and soiling themselves, and groaning from stomach cramps, as happened to troops on Hill 609 near Tunis and in the trenches of Guadalcanal and New Guinea.

Shannon's studies offered none of these details. Nor would he have known whether atabrine-induced psychosis had developed in any of the patients, because they were already psychotic. But he learned what he needed to know by looking at his patients through their photofluorometer readings, and he relied on nurses to help patients get through the unpleasant side effects.

Doing this, Shannon figured out that atabrine absorbed at different rates, based on each patient's individual body chemistry. But once soft tissue was saturated—once a patient turned yellow—atabrine performed similarly in them all. Shannon produced spreadsheets of data to see that fluorescent yellow measured by the photofluorometer correlated beautifully with reduced parasite counts in blood samples. He then led a larger study using his photofluorometer to track atabrine absorption in malaria patients at the United Fruit Company's hospital in Golfito, Costa Rica.[24]

This provided definitive proof that atabrine worked. Shannon, by now the Malaria Project's chairman of the Panel on Clinical Testing of Antimalarials, established this as an indisputable fact. The trick was to ensure that atabrine remained in the blood, which meant it first had to saturate tissue. This could be achieved in small daily doses over about two weeks, which produced more moderate side effects. They all disappeared once the body adjusted.

As Shannon grew to understand this drug, he observed that when atabrine reached a saturation plateau in the blood, it could cure *falciparum*! This is what Lowell Coggeshall had observed anecdotally in Accra and Liberia. Shannon went further by observing that patients had to stay on the stuff for several weeks after development of symptoms to ensure that all parasites had been killed off. Far from the five-day cure asserted by Paul de Kruif, atabrine took up to a month to prevent recrudescence and serve as a real cure for *falciparum*.

For *vivax*, the results of Shannon's research were disappointing. Atabrine could stop symptoms—kill the blood-stage parasites—but was of no use against the tissue stage, and therefore did nothing to stop relapses.

These were invaluable observations. But they were incomplete because they didn't begin to answer questions around using atabrine, reliably, as a prophylactic—to *prevent* symptoms from ever developing. *That* was what the military wanted. For this, much larger and healthier test populations were needed.

Neurosyphilitics, as important as they were for cultivating different strains of malaria, were, on the whole, sick and dying people. They often were weak, frail, and thin from loss of appetite. Many were incontinent, demented, irritable, distracted by incessant masturbation, prone to tremors and seizures, and unable to walk in a straight line. Their value in studies seeking to unearth the nuances of how best to use atabrine as a prophylactic was limited, because they were physically very different from soldiers in a war setting. They weren't analogues and therefore couldn't reliably be used to estimate when and how a long-term course of atabrine might effectively prevent symptoms in healthy men.

The realization that neurosyphilitics were not an ideal study group was, for Shannon, probably a relief. He had grown increasingly uncomfortable with what he was doing to patients, and advocated for a better process. He wasn't alone. Few if any of the project investigators thought the work in the state hospitals should continue beyond providing malaria therapy and, in the process, maintaining reservoirs of parasites for experiments on conscientious objectors or other volunteers. Perhaps they felt uncertain about the ethics, knowing that these patients were incapable of understanding what was being done to them. Shannon at an earlier meeting had described his ability to push to extremes the toxicity of atabrine and other drugs in his syphilis patients, but wondered, To what end? The real work in measuring the efficacy of these drugs would require much larger volumes of human *material.*[25]

They needed another source.

CHAPTER 22
The Convicts

G. Robert Coatney, famously irascible and difficult, even among the Malaria Project's edgy and opinionated chief scientists, secured a new, better supply of bodies.

During the spring and summer of 1943, he tested experimental drugs on birds infected with the chicken-killing *gallinaceum*. The tight controls on it that Lowell promised to maintain in 1942 had completely broken down. That Dr. Richards had ordered Coatney back then to destroy his samples because he wasn't authorized to have them had been forgotten. Shortly after that dustup, Coatney got his hands on more samples, probably from Lowell. Of course Coatney had to have *gallinaceum*. He was the most skilled at handling bird malarias. Every principal investigator needed it to do his work, and Coatney was a principal among principals. Plus, they all thought the government's fears were overstated, bordering on hysterical.

His bosses at the NIH had asked Coatney to use his *gallinaceum*-infected birds to test a secret compound, NIH-204 (a phenanthrene amino alcohol).[1] He found that it worked very well. His chicks were saved. So he brought it to St. Elizabeth's to test it on neurosyphilis patients. But because he wasn't a physician, he couldn't legally touch them. To do that, he hired a young doctor out of Johns Hopkins, Dr. Clark Cooper. Coatney later said he liked NIH-204 but it "had some side effects that we'll call cosmetically unacceptable." These were painful pimples filled with pus. ". . . It was a good antimalarial, you just couldn't use it."[2]

So Professor Marshall's team at Hopkins rotated the molecules around and sent new versions to Coatney and Cooper, to be tested on more syphilitics at St. Elizabeth's. But soon Coatney said he had "exhausted the supply of patients" there. While he and Cooper discussed the problem, another of Coatney's physician assistants, Dr. David Ruhe, overheard them and said to Coatney, "Chief, have you ever thought of using prisoner volunteers."[3] To which Coatney said, "My goodness gracious, I'd never thought of it, but I'm going to think about it now."[4]

COATNEY, son of a Nebraska plumber, studied journalism in the 1920s at Grand Island Baptist College. But he loved biology and went on to get a PhD from Iowa State University, with a focus on blood parasites of birds. Nebraska had few jobs for his skills, so he ended up at Peru State Teachers College training teachers for the Nebraska school system. This was just a job, not satisfying to a true biologist, and not the pedigree of the other Ivy Leaguers and Rockefeller disciples recruited into the Malaria Project.

Coatney landed with the project because he had a passion, odd as it was, for blood parasites and birds. Back in the 1930s, while still in Nebraska, he knew malariologists needed good bird malarias for research and tried to catch one in the migrating birds that flew over his home. In five years, he captured fifteen new parasite species in blood taken from owls, crows, magpies, and others. "Among . . . the Peru faculty I was an oddity," he later wrote. "I didn't play golf, I didn't hunt, and I didn't follow the fate of the athletic teams. All I did was teach, roam the woods, and work in the laboratory." In 1937 he found his ticket out of Nebraska in a mourning dove. Its blood teemed with a bird malaria called *P. relictum*.

He announced it later that winter, at the December meeting of the American Society of Parasitologists. He explained that this bird malaria lightly infected canaries and doves, but turned deadly in pigeons. Then he told everyone how he used plasmochin—the German drug for humans—to save the pigeons.[5] He knew this would excite the crowd because he was describing a bird malaria that responded to a human drug—a perfect research specimen. His peers begged for samples. But he refused; he said he wasn't finished studying it. A few months later, in the spring of 1938, he got what he was waiting for. It came in the form of a letter from the Public Health Service's chief of malaria investigations, none other than Dr. Louis L. Williams—whom Lowell had flown with in the de Havilland, cranking Paris Green over Quantico. Williams asked Coatney to share this bird parasite, to which Coatney said, "Sure, if you give me a job." Shortly thereafter he caught birds and studied their blood for the Public Health Service's Williams Malaria Research Laboratory at the South Carolina State Hospital. He craftily turned his favorite hobby into a career, earning valuable recognition for his highly specialized skills.

He joined a team of malaria experts working under the direction of Dr. Martin Young, who used *vivax* and *malariae* to treat the state hospital's syphilis patients. This work intrigued Coatney, and as he and Young became friends, Coatney learned how to work with these human malarias. Thus, Coatney became the only person he knew of within the Public Health Service with experience testing drugs using both human *and* bird malarias.[6] Because of this, he was among the first to be recruited onto the Malaria Project. He used sharp elbows to push people around until he muscled enough freedom to set up his research unit his way, which, in the end, served the project extremely well.

COATNEY'S plan to use prisoners as test subjects wasn't new. Other medical researchers for the war effort used prisoners, including James Shannon. But no one had Coatney's vision for recruiting these rapists, murderers, armed robbers, and repeat offenders into large, useful studies.

In late 1943 he asked the director of the Bureau of Prisons, James Bennett, if the Malaria Project could set up a research station at the Atlanta Federal Penitentiary. Coatney remembered him responding: "Oh my goodness. That's a tough place down there. . . . You know, that's not a minimum custody prison."[7] Coatney, of course, knew this. But he needed this prison because of its large size and proximity—a little more than a hundred miles from Young's project at the South Carolina State Hospital and just fifty miles from another Public Health Service–run malaria research station at the Central State Hospital in Milledgeville, Georgia. From these two facilities, Coatney would use the blood from syphilis patients undergoing malaria therapy to infect prisoners.

Bennett called warden Joseph Sanford, who initially refused because the bulk of his prisoners were making mattresses for the army and mailbags for the U.S. Postal Service, yielding about $2 million a year for the prison.[8] Coatney remembered with pride how he used his usual bullish demeanor to change the warden's mind: "I explained to Mr. Sanford, by Jimmy, the war was on, and what was two million dollars" compared to the "crying" need for malaria drugs to win the war?[9] The warden relented and allowed Coatney to present the idea directly to the inmates.

More than two thousand of them gathered for his talk. This is how Coatney remembered it: "The warden brought the men together in the Great Hall of the prison with its catwalk near the ceiling and guards with rifles. . . . I was not exactly calm and collected under those conditions. The warden gave a short talk and then turned the meeting over to me. I explained the need for new antimalarial drugs for the armed forces and the need for volunteers. . . . I made it doubly clear that participants would have to be of military age; in excellent physical condition; be willing to accept infection with malaria, either by mosquito or by blood inoculation; and to accept medication and routine tests as required." At the end, they would be paid $50 and have six months taken off their sentence, plus receive "an attractive certificate signed by the Surgeon General" that said participation held broad "significance to mankind."[10]

Coatney warned that some of them "would be so sick they'd wish they'd never seen, ever heard of us, much less seen us."[11] Then he pointed his finger at Dr. Ruhe, said he was a medical doctor, and promised he would take care of *all* their medical needs. "When I stepped down, these people came over just like a wave! I was really surprised," Coatney recalled. One guy asked, "Doc, I want to ask you a straightforward question. . . . Did you really mean what you said, that that man over there is going to take care of *all* our medical problems?" To which Coatney said, "That's exactly the way it's going to be, if any of you sign up to do this." With that, the prisoner shouted: "Come on, boys; let's go!" Coatney soon had consent forms from five hundred men.[12]

This was bold and ingenious. He resolved the need for human subjects in a way that held up well against the medical ethics of the time. The likelihood was strong that none of the prisoners would die from the experiments, given their overall good health. The death rate among syphilitics they used, while not measured, likely ran along the same graph as before the war—at 10 to 20 percent. For every thousand patients used, one hundred to two hundred died from the malaria because they were already so sick. But the cure rate, in all likelihood, also ran along the same graph as before the war, up to 30 percent. For every thousand syphilis patients used, three hundred got their lives back; they were saved from a killer disease. To this day, the medical ethics relating to the use of state hospital patients make malaria researchers cringe. But the prison work doesn't. Back in the 1940s, this kind of human testing was considered problematic only in that it offered time off of sentences that prosecutors had so vigorously fought for. This concession had to be kept secret for fear it would invite criticism from penal conservatives. Otherwise, by 1943, the convention of using prisoners in medical studies, with their written consent, had been well established and fully accepted by the medical community.

Coatney started his Atlanta prisoners on extreme doses of atabrine to measure the worst that could be expected in side effects. All the men were under forty-five years old, briefed on the hellish symptoms to come, and told they would need to stay in the study for up to eighteen months.[13]

In time, Coatney's template would expand to other prisons, and thousands of convicts would volunteer. Some would use it to their full advantage.

CHAPTER 23
Supply and Demand

Overseas, malaria raged. The fall of 1943 in the Pacific saw ten hospital admissions for malaria for every one battle casualty and a majority of the Allied divisions were incapable of effective military duty.[1] Combat commanders "flagrantly disregarded" the rules by going around at night shirtless and in shorts, when regulations clearly stated that long pants and sleeves were required during evening hours.[2] The problem was that the average temperature in Melanesia was a balmy eighty degrees. Men peeled their clothes off to keep from roasting.

Even the editors of *Yank* magazine failed to make the perceptual connections. This World War II army publication, with a circulation of 2.6 million readers, ran many stories on the problem of malaria and war. Yet, in late May 1943, it printed a photograph that infuriated the region's otherwise mild-mannered assistant malariologist, Major Harold Jesurun.[3] While flipping through it, he turned a page and couldn't believe his eyes. There, in a story about the campaign for New Guinea, was a photo of a GI in shorts.

Shorts!

Major Jesurun had shipped out with the 4th Malaria Survey Unit, the fourth to be trained. In March 1943, his was the first malaria unit to arrive in New Guinea. The unit was horrified to find conditions so ripe for malaria. Jesurun spent months collecting mosquitoes, examining and identifying anophelines, finding their breeding grounds, and working with the 5th Malaria Control Unit to chemically bomb them all. He hired locals to spray fields, dig drainage ditches, and clear debris from clogged rivers. He was among Russell's ideal recruits, a graduate of Columbia University with a medical degree from the University of Michigan.[4]

Jesurun penned a pointed letter to *Yank*'s editor: "I [was] very much surprised, disturbed and somewhat discouraged when turning to page 13. . . . I saw a picture of a GI in *SHORTS*." Jesurun's letter lacked the color and poetry—and expletives—of Dr. Egeberg's concerning the sheep, but it embraced a similar sentiment. He worked his tail off doing everything he could to prevent malaria, and the only publication the GIs actually read negated his efforts. It felt like sabotage. "Atabrine is given to modify the disease, make it milder so it won't be as severe as otherwise," but it didn't prevent malaria, or cure it, he wrote. To stop the disease from eroding troop strength, each man had to take the regulations seriously and protect himself from mosquito bites by wearing long pants and shirtsleeves. This was critical "to keep as many men on their feet at any one time to beat the Japs and get the war over with sooner."[5] He wanted *Yank* editors to be more careful, because the implications were dire.

But this *Yank* photograph merely reflected reality in New Guinea. Every element there—the heat and humidity, the blasted landscape, the infected Australian troops and Papuans, and missing antimosquito supplies—conspired against the malaria units. Even as supplies finally arrived in mid-1943, malaria kept on infecting the men, because commanders received the repellents, bug bombs, and netting with lackluster interest. They were hot and tired and didn't really understand what the fuss was all about. Everyone got malaria. Some got it so bad they couldn't work or fight, so they stayed in bed moaning. One commander said in a letter to the surgeon general that they were like rotten apples, spoiling the others. Send them home, he argued, and let the stronger men take their hits from malaria, get better, and return to work.

Russell's specialized units encountered such severe troop apathy that they could make no progress against their immovable object, malaria. At night, men left screened tents to relieve themselves in outdoor slit trenches and were exposed to malaria. During nightly air raids, they ran to foxholes and were exposed. For sex, they sneaked out after dark and were exposed. On night guard duty, they patrolled the perimeter and were exposed. For entertainment, they watched movies in open fields and were exposed. None of these exposures could be eliminated, because each served an essential function, was part of survival, or was crucial to morale—especially the movies, which relieved the monotony of the long, steamy days.

RUSSELL returned to Washington from the Southwest Pacific to a desk piled high with monthly reports from his malaria units. One report in from the China-Burma-India Theater described the problem of movies. The theater had seventy-two movie projectors that were moved from place to place and organization to organization, showing movies in two to three hundred locations per month, all in open-air settings, usually in cleared fields or jungles. Too many men sat in the dark for two hours, getting bitten by mosquitoes. So Russell turned this into an opportunity. He drafted a circular letter, signed by the army surgeon general, that created a new policy: Outdoor movies would have dispensers of mosquito repellent, and guards at the entrance gates were to make sure everyone smeared the stuff on exposed skin before taking a seat.[6]

The malaria problems in the CBI, however, were much bigger than movies.[7] This theater was sprawling, controversial, at the end of the Allied supply line, and the last priority of war planners. Rescuing Chiang Kai-shek's China from the Japanese and the communists came in third, after Europe and the Pacific. But Roosevelt was adamant that China not fall. If that happened, he asked, "How many divisions of Japanese troops do you think will be freed—to do what?"[8] The region remained part of the war strategy, and that meant kicking the Japanese out of Burma.

Imperial troops landed there in the spring of 1942 to seize the southern seaport of Rangoon, on the Bay of Bengal, and shut down the railway connecting the seaport to the Burma Road—which was the only Allied supply route to China. In May 1942, the British-led Burma Corps marched south to try to take back Rangoon, but failed miserably.

Leading the inglorious effort was Lieutenant General Viscount William Slim, a charming British commander with jutting chin, sensible mustache, and lanky build—consistent with his surname. He led thirty-five thousand men: four thousand British officers and NCOs commanding thirty-one thousand Indian and Burmese troops. Against eighteen thousand invading Japanese troops, Slim should have prevailed.[9] But Japan's skilled jungle fighters won one victory after another, driving the Allies up and out of the country, aided heavily by Burmese people taking revenge on their British rulers by revealing their positions, sniping at them, and sabotaging the railroads.[10] In a month, Japan held Rangoon and shut down supplies to China.

Even though Slim led the defeat of Allied troops, historians blame the American general in charge, Joseph Stilwell—described by some as "sulfurous" and made of vinegar. He was sent to "the end of America's thinnest supply line" to win over China with diplomacy.[11] But he undermined American diplomatic efforts already in place—referring to the American ambassador to China as "pickle prick"—and then mishandled his military duties.[12]

Stilwell served ostensibly as Chiang Kai-shek's chief of staff, and as such, Chiang put him in charge of the Chinese Fifth and Sixth Armies. Chiang had ordered Stilwell to keep his armies concentrated in the north to hold enough of northern Burma for the construction of the Ledo Road to run supplies from India into China. Chiang wanted his two armies together to keep them strong. He believed sending them to Rangoon would thin them out, weaken their strength, and open up southern China to an invasion.

Stilwell, who called the Chinese leader "the Peanut," had no intention of taking orders from Chiang. When the time came, Stilwell marched Chinese troops southward to defend Slim's retreating forces. Then the Japanese did as Chiang predicted: They used the dense jungle to annihilate his thinly spread armies. In the end, Stilwell and many of Chiang's men were cut off from China and had to retreat into India. Chiang's two armies were shredded and no longer near China to defend their people from advancing Japanese troops. His trust in the Americans disintegrated.

As for Slim's Burma Corps, it also disintegrated. The Burmese soldiers went home and everyone else marched toward Assam. Along the way they destroyed towns and villages in a scorched-earth retreat designed to deny the Japanese any resources, including buildings, oil depots, crops, supply houses, and clean water tanks. The resulting devastation left the population destitute. Troops even blew up bridges before all units had passed. When the monsoon rains started in May 1942, Slim's forces were still two weeks from crossing into the safe haven of Assam. "From then onward, the retreat was sheer misery," he remembered years later. "Ploughing their way up slopes, over a track inches deep in slippery mud, soaked to the skin, rotten with fever, ill fed and shivering as the air grew cooler, the troops went on, hour after hour, day after day. Their only rest at night was to lie on the sodden ground under dripping trees, without even a blanket to cover them."[13] The rains, however, also stopped the Japanese so that the retreat wore on free of air attacks and ground ambushes.

When the Burma Corps marched across the border into Assam, nine hundred miles from Rangoon, they were spent. They came along different paths, some narrow, precipitous, and rough, others along streambeds and rough-flowing rivers. The most unfortunate trekked a grisly route from Kalewa through "Death Valley" along ninety miles of malarious terrain, hauling equipment on bony backs covered in shreds of filthy uniforms. "All of them, British, Indian, and Gurkha, were gaunt and ragged as scarecrows."[14]

Slim lost more than thirteen thousand men. His command also lost all their tanks, more than 130 big guns (returning with just twenty-eight), and most of their four-wheel-drive "lorries" and jeeps. Once in Imphal, Slim was told to bivouac his men on a hillside of dripping trees and mud. They had only the soiled clothes they stood in. "Imphal was a thousand miles from Calcutta at the extreme end of a most rickety line of communication, stretched to breaking-point. India itself was deficient of everything, and it was impossible to get forward over that distance at short notice what a destitute corps required."[15] Eighty percent were sick with malaria and dysentery. Without drugs, many died right there where they were ordered to stay. The rest waited for orders to take back Burma.

In March 1943, those orders came. There was no way around it. Airlifted cargo into China over the eastern Himalayas, nicknamed "the Hump," brought in only so much. Pilots from the Pan Am Africa route joined the Air Transport Command and now made hair-raising cargo lifts over sixteen-thousand-foot peaks, using no radio service, land charts, or weather information. More than six hundred planes carrying more than a thousand crew members were lost as ice formed on the wings, buckled the steel, and flipped planes on their backs, never to be heard from again. Survivors landed several hundred thousand tons of cargo in China. But this represented only a fraction of the need. The Ledo Road had to be finished; the Burma Road had to be reopened; and thus Allied troops had to repel Japanese forces back into Thailand and Indochina.

CBI was a twenty-four-thousand-mile stretch of malaria-ripe country. This theater had more than fifty different anopheles mosquitoes, but only twelve of them triggered the vast majority of malaria outbreaks.[16] The most important one in India, Ceylon, and southern China, *A. culicifacies*, bred across the plains in fresh or brackish water, fed on humans or cows, and digested blood meals while resting in huts or cowsheds or dung cakes or chaff. Normally this mosquito wasn't a major malaria vector, because she preferred cows. But whenever heavy rains increased mosquito breeding, her palate expanded to include humans. She became an excellent malaria vector because she could carry heavy parasite loads.

Climatic variations often prevented her from breeding. But then other anophelines filled in, some breeding in shaded waters clogged with vegetation, others in sunny, clean waters, and still others under thick jungle canopies. Some were highly anthropophilic—preferring only humans—while others preferred mammals, except when thinly populated areas suddenly teemed with armies of men who consumed all the livestock (as happened during the war). Then these mosquitoes also settled for human blood, and spread malaria.

As in the North Africa, West Africa, South Pacific, and Mediterranean areas, the combination of good mosquito carriers and movement of hundreds of thousands of troops guaranteed fantastic success for CBI's mosquitoes, and therefore the region's many strains of *vivax* and *falciparum*. Mapmakers casting shadows over the Pacific and Mediterranean to show the spread of malaria and its severity were now busy shading in areas across India, Burma, and into China. Malaria's success had never been so intense or widespread.

In western Assam, troops slept in "*bashas*" made of bamboo covered in burlap that mosquitoes could fly right through. But many had bed nets, and some even used them. So malaria rates were only about four hundred per thousand per annum. By far the worst malaria was in the advance sections—eastern Assam through Burma and into southern China. Transmission peaked in June 1943, then again in August, and again in October, each time exceeding a thousand cases per thousand men per annum. These rates reflected hospital stays only, and left out so-called "fevers of unknown origin," most of which were malaria.

This theater was so short of men and supplies that U.S. malaria units were put to work doing jobs other than surveying and controlling malaria. Commanders also feared atabrine side effects would further erode their troop strength, so they didn't use it.[17] Bed nets sent to the forward areas were of extremely poor quality, made of thick burlap that restricted airflow and developed mold. "They were hot and smelly," according to reports, so no one used them.[18] The mosquito repellent was also terrible, made of petroleum that stung sunburned skin, was "sticky and oily," "made the user feel hot," and even "dissolved plastic." Soldiers refused to use it, except when they went to see Betty Grable and Rita Hayworth, as Russell had made it "impossible to go to GI movies unless one submitted to a liberal application of repellent."[19]

COLONEL Earle Rice, the theater malariologist, arrived in February 1943. He had worked on malaria at the Pasteur Institute in India, as Russell had. In 1938, he studied malaria around a $40 million dam project called the Santee Cooper, in South Carolina, doing work similar to that done by Lowell Coggeshall around the Lake Murray dam project. Dr. Rice was commissioned in the army in 1941 and started out teaching malaria courses to recruits at the Army Medical School. He had impeccable training for the CBI assignment.

For more than a year he wrote to Russell about "the crying need for more intensive indoctrination of troops." His own efforts to teach troops had been unsuccessful "due to the apathy of unit commanders." He wanted malaria hygiene to be as vigorously enforced as "sex hygiene," and malaria prevention to be placed above VD prevention, because the malaria rates were far worse. He made a suggestion that stuck: that mandatory reports be made of every malaria infection to shame the individual and pressure unit commanders to impose better standards, or risk disciplinary action.[20] He had only ten malaria survey and control units for more than fifty widely dispersed combat units. "Grave concern is felt toward the future unless larger units in greater numbers can be obtained. . . . The outlook is hardly one to be looked upon with complacency."[21]

Rice had been worn down and needed to leave. A secret cable arrived in Surgeon General Kirk's office, urgently requesting that Russell "or a malariologist equally experienced" be sent immediately.[22] Russell, however, had already received new orders. On August 28, 1943, he was promoted to full colonel and assigned to Africa. Before he left, he tried twice to recruit Lowell Coggeshall to replace Earle Rice.[23] Lowell turned down Russell's offers, ostensibly because the National Research Council wanted him in the United States. Which was true: the NRC resented the army using the draft to rob the Malaria Project of scientists.

Lowell's true reason for staying out of the army, however, went a little deeper. The army wouldn't give him a high enough rank. His skills should have started him at lieutenant colonel or possibly colonel, but the army wanted to start him as a major, so he refused. Navy officials caught wind of this and acted quickly.

Lowell was in the middle of helping one of the army's newly acquired military hospitals set up standards for running clinical trials on malaria-wrecked soldiers sent home for study, when he received a call from Rear Admiral—soon to be Vice Admiral—Ross T. McIntire. The admiral was the navy's surgeon general and President Roosevelt's personal physician. He asked Lowell to catch the next flight to Washington.

Admiral McIntire told Lowell that the marines from Guadalcanal were "near mutiny," and not just because of malaria. The navy's medical corps had discovered, to their horror, that thousands of marines had contracted filariasis, the mosquito-borne worm that caused elephantiasis. If these boys from small-town America started seeing their brothers in battle balloon up like circus creatures—especially in the scrotum—the damage done to their collective psyche would likely ruin whole companies for battle.

McIntire told Lowell that many thousands should have been sent home. But because of red tape they were stuck in convalescent camps, where they slept, moped around, and ate the same chow week after week, month after month. The monotony sank in. They suffered depression, anxiety, and psychosis. Their bodies grew soft and limp from inactivity. They were slipping away, as if they'd never been trained to fight.

"They're sick. They're worn out. They've had terrible casualties. Would you come to the Navy?" the admiral asked Lowell. He wanted Lowell to design and run a new five-thousand-bed facility in the cool, dry mountains of Klamath Falls, Oregon, McIntire's home state. Lowell would have a lab to study filariasis *and* run clinical trials on the South Pacific's terrible malaria. Troops would live in barracks with the goal of being rehabilitated for return to battle. Then McIntire told Lowell what he needed to hear: that he would enter the navy with the rank of a lieutenant commander, be immediately made a commander, and soon be promoted to captain, one rank below a one-star rear admiral.[24] This was an offer Lowell couldn't refuse.

McIntire on the spot called to an assistant: "Get my car!" Within minutes they were at navy headquarters near the White House. The admiral walked Lowell to the head of the recruitment line. When an underling offered to weigh Lowell, McIntire said, "I've already weighed Dr. Coggeshall. He satisfied the weight requirement. Go ahead and complete the paperwork and commission him." Years later Lowell recalled that he had weighed significantly less than the navy requirement. He hadn't changed much during the twenty-six years since he first tried to join the navy, when he was only sixteen and sent home to gain weight. Throughout the decades he remained skeletal, a naturally skinny man. But this didn't matter. He wasn't going to battle; he was to undo what battle had done, at least for the lucky thousands put under his care.

By the end of the day, Lowell was a reserve navy commander in a position that fit well for everyone. The NRC's Malaria Project would have one of its most skilled and knowledgeable lead investigators working directly with an inexhaustible supply of live infections from overseas. Every experimental drug of any value would be given to him, to be tested against these men's seemingly intractable fevers. This special mission was perhaps the best of all the jobs offered to scientists coming out of the project.

MEANWHILE, Paul Russell prepared for his assignment as theater malariologist for North and West Africa to replace the likable and well-respected Dr. Louis L. Williams. Williams had just turned fifty-four and was perhaps too old for the stress, long hours, unclean conditions, and difficult travel in this theater. He suffered a heart attack that sent him to the U.S. Army's 12th General Hospital, a sprawling tent city in Oran for the sick and injured.

If this new Africa assignment worried Russell, he didn't say so in his diary. He had just reached his fiftieth birthday, and had a family history of heart trouble. He had just endured more than three months in the South Pacific's insectary-like climate, sleeping in the mud, bouncing over terrible roads, taking rough flights under the constant threat of enemy fire, and eating almost no fresh food, only the army's canned rations. Now he was off to another Allied region that offered even less comfortable travel and accommodation.

He left on September 14, 1943, without penning a worry. He flew the old Pan Am route down to the elbow of Brazil, then over the Atlantic to the hump of Africa. He arrived first in Dakar, a former German outpost, before the Vichy French capitulated to the Allies. Now British forces ran the Rufisque airport. They put him up in "filthy" officers' quarters with no screens to keep out the malarial mosquitoes. He later noted of this place: "*complete lack of any interest in malaria control by officers commanding transient barracks*, whether US or British, or French." The next morning he took off for Marrakesh and then Algiers. On September 21 he was officially made malariologist for the North African Theater of Operations, U.S. Army.

He had plenty of work to do in Algeria, Morocco, and Tunisia. But General

Eisenhower wanted Russell to investigate a distressing problem. Hundreds of pilots of the Eighth Air Force had arrived in England to fly bombers over Germany, but they couldn't because they were sick with malaria. They had picked it up during an overnight stop in West Africa. Russell flew to Accra, where he met three officers sent from Washington to help him investigate.

ONLY a year had passed since Lowell left these airfields completely sanitized. Accra, Roberts Field, and Robertsport at Fisherman's Lake were under control. He had been so thorough in Liberia that Pan Am's hospital emptied and his handpicked nurses went home, because they didn't have enough work to distract themselves.

By now, however, "malaria hygiene" had completely broken down. But venereal disease rates actually exceeded that of malaria, reaching almost 100 percent, according to army sergeant George "Doc" Abraham.[25] So, in Liberia, sex control took precedence over malaria control.

Sergeant Abraham landed there in late 1942 with more than two thousand black troops and seventy-five white commanders. He was white.[26] They had two jobs: protect the Firestone rubber plantation and expand the runway at the Roberts Field Army Post (today's Roberts International Airport). Black units included the 41st Engineers General Service Regiment, the 812th Engineer Aviation Battalion, the 802nd Coast Artillery Battery, and the 25th Station Hospital. Female nurses were sent, too, but had to be replaced with men, in part because of what went on there and in part because of severe menstrual bleeding doctors blamed on atabrine. One woman bled for eighteen days, then six days later began another ten days of bleeding. Anemia and debilitating cramps affected 80 percent of the nursing staff and all were sent elsewhere.[27]

The army blamed the rampant VD on "camp followers." These were several hundred Liberian women and girls whose husbands, fathers, and brothers were conscripted into the Liberian army or hired to work for the U.S. Army. The women migrated from the interior and soon learned they could exchange sex for money. By day, these U.S. troops built the longest airstrip in Africa to service heavy bombers en route to England for air raids on German industries (including Bayer's labs in Leverkusen and Elberfeld). By night, they visited with "camp followers." In no time nearly all the men and women were "peppered" with venereal diseases.[28] No report focused on the origin of these sexually transmitted diseases, though it's likely they arrived with troops, or maybe Pan Am. The army just needed to get them under control. Of 222 women examined for VD, all but fourteen were infected. Most had gonorrhea.[29]

Camps were set up to contain and control the sex trade. This was never officially approved by commanders, because no one was supposed to talk about it. Little is known about what happened there, beyond the descriptions provided by Americans. Whether any girl was abused is unknown. In a passing reference in Sergeant Abraham's memoir, he mentions a girl found dead in a hut the soldiers had set up for her to service them, but offers no details. He writes about these girls narcissistically, from the standpoint of the soldiers' needs—how the men were shipped to this strange place at their sexual peak in need of "comfort" that these girls, as young as eight years old, provided.

He writes little of the political situation in Liberia. How President Barclay and his True Whigs party comprised only about 5 percent of the population. They were descendants of returned slaves from the United States who, in 1821, as part of the American Colonization Society, boarded schooners for the dangerous voyage to the hump of Africa. They colonized Liberia's coastline and, in 1847, created the independent Republic of Liberia. By World War II, twelve thousand Americo-Liberians and another sixty thousand elite coastal Liberians ruled an estimated two million "aborigine" farmers, many of whom left their fields to work on plantations harvesting and processing export products that included coffee, cacao, ivory, and, of course, rubber, the country's most strategically important raw material.[30] The people living in the interior had no rights. The Americo-Liberians called them "natives" and "tribesman" and "aborigines" and, even more derogatorily, "savages" and "bush niggers."[31]

These disrespected and misunderstood masses had lived here for millennia, largely in thatch-roofed villages. They wore almost nothing. Some carved and tattooed their flesh; some genitally mutilated their daughters. They spoke many different languages and had their own tribal political structures, all of which were subordinated to the Whigs, for whom they could not vote. Attempts to push back on this unjust system had long since ceased, after brutal slaughters by armed Americo-Liberian forces, with support from the United States.

President Barclay allowed the United States military to operate in Liberia in exchange for military training of his Liberian troops. They learned to use bigger, better weapons—planting seeds for a bloody, tumultuous future.

In many ways, rank-and-file American soldiers sympathized with the "natives," who had left their villages and farms to do hard labor for the U.S. Army. They lacked sufficient food and often scrounged for it. "We gave them leftovers," remembered Abraham. But that violated the rules. "The Army had an inconsistent view of such matters. It didn't mind having soldiers shacking up with syphilitic prostitutes, but it wailed when soldiers gave natives scraps of food." Once, to discourage garbage picking, pails were laced with lye: "[A] cry pierced the jungle—the wildest cry I ever heard. A Kpessi boy stole garbage from a pail . . . and nearly died from it." The lye had turned "his mouth black and disintegrated his teeth to stumps before we could administer a vinegar emetic, which barely saved his life," wrote Abraham.[32]

THE 14th Malaria Survey Unit arrived in the summer of 1943 to find a state of soldiers and sex workers. Major William F. Diller wrote about his first assignment here in a report published after the war.[33] He was asked to examine a number of the prostitutes, called "belles," who had developed obscenely swollen and deformed breasts, and village men who had scrota the size of "large grapefruits."[34] Medical corpsmen feared this was elephantiasis and that troops would be next. Instead of working on malaria, Diller studied filariasis, the cause of elephantiasis.

By the time he arrived, camp commanders had rewritten rules for rationing sex, which they called "jig-jig." They first tried burning down native villages, including Village E, which they called Talking Charlie, which was really an independent brothel set up by the village women.[35] Then they forbade men from fraternizing and had MPs arrest violators. But the crowded stockades couldn't hold them all. One medical officer complained: "It is humanly impossible to watch every man at all times. In fact, the greatest offenders are those who should enforce law and order, the MP personnel."[36] So a "policy of regulated toleration" was established.[37] The men cleared two areas on the military reservation, fenced them in, and, in each, built three hundred small mud huts with thatched roofs. One army-inspected prostitute was assigned to each hut. These six hundred or so women and girls were strictly forbidden to have sex with other "natives." They were there to service black American troops only. Each girl wore an inspection tag. If her weekly gynecological exam revealed she was infected, her tag was taken and she was quarantined to a third village and held under guard until her infection cleared. The two active prostitution villages were called Shangri-la and Paradise. The inactive village was creatively called Idlewylde.[38]

White officers had a different setup; they rented girls on a semipermanent basis. Sergeant Abraham said these "wives" gathered during the day to fuss over their appearance. And their "husbands" gave them soap, toothpaste, toothbrushes, and other toiletries so that over time they looked, smelled, and acted in ways more familiar and pleasing to the men.[39] They also remained VD-free.

Major Diller used the women of Shangri-la, Paradise, and Idlewylde for his filariasis study. He called the brothels "Health Center Villages" and described the women and girls as being young and "of average or superior health and vigor."[40] From them he needed a drop of blood, which he took by force, if necessary. Abraham remembered one "native belle" so terrified she "resorted to biting, hair pulling, eye-gouging and even kicking below the belt."

Diller managed to take two rounds of blood from hundreds of the girls. He found about 10 percent of them infected with filariasis. Then he tested 431 of the Americans, and none—not one—carried the worms.

Filariasis infected the locals but not U.S. soldiers, he discovered. This disease was therefore not a threat at all. Doing the study robbed him of time to focus on malaria infections, which the theater medical inspector, Lieutenant Colonel Thomas G. Ward, called "a disgrace to the U.S. Army."[41] To him, all of West Africa was a disgrace. Liberia's maleria infection rate was eight hundred per thousand per annum; Dakar's was two thousand per thousand per annum; and Accra's was five hundred and eighty per thousand. These were service posts where troops did the grunt work so others could carry on with the gruesome business of war. The men lived in relatively permanent installations that *could* be sanitized and screened, but weren't. Even hospital staff failed to protect themselves. In Dakar, for example, half of the enlisted staff of the 93d Station Hospital came down with malaria during the height of the 1943 malaria season.

JOYCE Abramson, a twenty-two-year-old army nurse, sailed from the United States to West Africa and reached the Gold Coast's main shipping port of Takoradi on October 27, 1943. From this dusty, seedy seaport, she and the ten nurses she arrived with joined a large convoy for the half-day trip eastward along the coast to Accra. The nurses' quarters there were in the old Pan Am barracks, which she described as "very nice" and "enclosed with screens all over" and "mosquito nets when we went to bed." That evening she and the other nurses, all of whom were replacements for those sent home because of malaria, walked across the compound for a welcome party thrown by the British. Wearing a long-sleeved shirt and her army-issued mosquito boots, she took her first atabrine pill, as ordered. But about ten days later she came down with malaria—*falciparum*, most likely.[42]

Joyce was sure she got it that first night going to this dinner. She spent two weeks in the very hospital she was supposed to be working in—the 250-bed 67th Station Hospital, built in 1941 by Pan Am. In early 1942, Lowell Coggeshall had arrived to wrap the entire place in screens. His screens made such an impression that Joyce could still remember them seven decades later. "Every passageway was enclosed with a screen," as were the vestibules between two swinging screened doors, she said. She arrived a full year after Lowell's departure, yet his screens had held up. The antimosquito work, however, had not. No crews were taken from the runways to clear channels and fill wetlands. No one cranked Paris Green into ponds or sprayed barracks. No one slicked waters with diesel. And no one patrolled the base for the kinds of empty cans and trash that caught water—and mosquito larvae. A year later, when Nurse Abramson arrived, *gambiae* was everywhere again.

On November 12, 1943, she suffered through the fires of fever and ice-cold chills, unaware that Paul Russell had just landed there. She would be in West Africa for another year. He would be there for just another week. But his stay would make hers possible. Army historians later called Russell's trip there the "turning point."[43]

The work really began on November 15, when his partners arrived from Washington, via Natal, Brazil. They were Colonel William A. Hardenbergh, Lieutenant Colonel Karl R. Lundeberg, and Major Elliston Farrell. Medical historians called them the "malaria control commission," but they had no formal title. They were just there by order of General Eisenhower. They stayed in the barracks, ate the food, took notes on behaviors, and even attended a show outside at night, where they observed little use of bug spray, few mosquito boots, and soldiers in rolled-up pants and sleeves.

Rumor was that Russell was there to roll some heads. And that was what happened. The malariologist, Lieutenant Colonel Daniel Wright, was replaced by Major Farrell, who used the same standards and policies Lowell put in place a year earlier. Eisenhower agreed to make malaria control a command function, and to give Farrell the needed authority to impose drastic changes.

By the time Joyce Abramson got out of the hospital, Russell was long gone. But she remembered the changes like they were yesterday, especially the MPs walking up and down the aisles every ten minutes during outdoor movies, making sure everyone wore mosquito boots and proper clothing, and smelled of insect repellent. She also remembered the changes in Dakar, where she moved in early 1944 to work at the 150-bed 93rd Station Hospital. She lived inside the walled "Native Center," which functioned more like a "green zone" within which all essential buildings and troops were isolated from the people and culture living around them. This sanitized oasis provided much rougher accommodations than in Accra. But by the time she arrived, the hospital and barracks had screens, the grounds were being sprayed regularly, patrols hunted down anopheline breeding sites, and everyone took daily atabrine. Even the protected paths down to the MP-guarded beaches were dusted weekly with Paris Green.

Here, commanders also controlled "jig-jig." One brothel stood right outside the compound, for everyone to see. "When the GIs would go out, they had to stop right inside the gate, and had to go in to get their protections, their condoms and whatever, before they left," Joyce said. "It was a no-no thing to talk about. We knew it was there. And we had a few cases of VD, but not a lot."[44]

FROM the end of 1943 to the beginning of 1944, malaria infection rates in West Africa dropped substantially.[45] Each location had a malaria survey unit and at least one malaria control unit, with ample supplies. Screens were flown in and nailed to every building. Pay was docked for anyone caught without mosquito boots and proper clothing after dark. The brothels were all sprayed and dusted. Russell and the rest of the "commission" set in motion a process that made malaria control as important as VD control. In Accra in 1944, the malaria rate fell to about forty per thousand per year. And it hovered around seventy per thousand in Dakar and Liberia—not perfect, but a huge improvement.

CHAPTER 24
Whoville

On July 8, 1943, General MacArthur's chief surgeon, Colonel Percy J. Carroll, sent a memo to Washington. He said twenty thousand experienced troops, acclimated to the energy-sapping heat and humidity of the Southwest Pacific, were better than forty thousand inexperienced replacements. Malaria was a preventable disease. But it required education and strict discipline "from the commanding general down to the buck private in the rear rank." He said unit commanders thought they were conserving troop strength by teaching men how to survive in battle, "yet the same commanders stand to lose from 40 to 70 percent of their men from malaria infection and still do not make any effort to master the techniques of preventing this disease. The only thing we can and should do is to make the word '*malaria*' so familiar to each individual that it will always be before them as a serious disease to be avoided at all costs."[1]

He called for "a well organized educational campaign, using all available media," running nonstop programs so the message stayed "constantly in the soldier's mind," from the rear to the advance bases. Colonel Carroll then recommended that Special Services work on the problem. "It certainly will pay us a million times over."[2]

This memo articulated an educational program similar to the one Lowell Coggeshall used on Pan Am crews, and Russell tried to convince the War Department to adopt. But Colonel Carroll offered a twist: Don't just train the men; indoctrinate them. Turn on the War Department's propaganda machine, the Special Services section. Assign personnel there to work on nothing but malaria. Include an editorial unit to write articles, a motion picture unit to make educational shorts, a radio unit to provide daily messages, and a research unit to study the men's attitudes and come up with ways to change their behavior. "Dramatize the problem so that it will reach the soldier emotionally," Carroll wrote. Proper discipline would be the equivalent of having two new divisions in the Southwest Pacific Area alone.[3]

By now, malaria was officially the number one medical problem of the war and treated that way by top commanders. Carroll's call for creativity instead of brute force for the rank and file was convincing. He helped kick off one the most effective media campaigns in U.S. military history. The War Department came to life with clever artwork designed to coerce, manipulate, guilt, and entertain the men into malaria discipline.

In 1943, cartoonist Theodor S. Giesel, also known as Dr. Seuss, was a captain in the army's animation department and was assigned to the malaria work. He created a character named Ann, a female *ANN*opheles mosquito that was a cross between an old advertisement he had drawn for the insecticide Flit, and characters he hadn't yet created for his future book *Horton Hears a Who.* Ann had long, skinny arms and legs, a wide smile, big creased cheeks, round innocent eyes, and a head with just two hairs floating up in gentle curls. "This is Ann: She's dying to *meet you*," read the opening caption of one of the first pamphlets produced for the Special Services malaria campaign. Ann drank blood out of a wineglass and peeped through a keyhole, looking for unprotected GIs. One caption read: "Her trade is dishing out MALARIA. . . . Ann moves around at night (a real party gal) and she's got a thirst. No whiskey, gin, beer or rum coke for Ann . . . she drinks G.I. blood. She jabs that beak of hers in like a drill and sucks up the juice . . . then the poor G.I. is going to feel awful in about eight or fourteen days . . . because he is going to have *malaria*." On the next page, a naked, cross-eyed GI lies limp on a cot, as this Whoville mosquito flies off. "NEVER give her a break. She can make you feel like a combination of a forest fire, a January blizzard, and an old dish mop. She will leave you with about as much pep as a sack of wet sand and now and then she can knock you flat for keeps. . . ."

The pamphlet offered more. "Head nets, rolled-down sleeves, leggings and gloves may seem like sissy stuff and not so comfortable—BUT, a guy out cold from MALARIA is just as stiff as the one who stopped a hunk of steel." Turn the page and two Whoville-like men walk past a sign that reads: "Mosquito-ville 2 Miles," with the following caption: "NOW IF you really are looking for trouble and you don't want to miss—just drop down to the nearest native village some evening. The places are lousy with fat little Anns sitting around waiting for you with their bellies full of germs. They stock up on MALARIA bugs from the home-town boys and gals and when they find a nice new sucker they give him the works." Turn the page and there's a classic Dr. Seuss animal, a horse of sorts, with another message about horse sense and using repellent. And a final point: "Never give Ann a break. She'll bat you down and it won't be funny."

Geisel's pamphlet went to every theater and was followed by a rapid succession of posters and pinups, all stamped: "By order of the Secretary of War." They had such an impact that troops started calling mosquitoes "Ann."

Special Services also sponsored three episodes dedicated to malaria in a popular Warner Bros. cartoon series about a private named Snafu—for Situation Normal, All Fouled (sometimes "Fucked") Up. This animated, dunderheaded private was created by Dr. Seuss, put to Looney Tunes music written by Carl Stalling, and voiced by actor Mel Blanc, best known for his characters Bugs Bunny, Daffy Duck, Porky Pig, and many more. "Private Snafu v. Malaria Mike" featured Snafu swimming buck-naked in a pond. He tosses his repellent, which inspires a *male* mosquito to fill his bayonetlike nose with malaria and dive-bomb into Snafu's flesh. "Target Snafu" featured a battalion of anopheline mosquitoes—all *male*—marching in step, sharpening their noses, flying in formation, doing reconnaissance looking for holes in bed nets, and finding one, Snafu's.

Finally, in "It's Murder, She Says," the anophelines are females (as only females bite for blood—males drink fruit nectar). The main character is in a low-cut cocktail dress from the "good ole days" when she was "really some stuff." But now she's overweight and downing martinis. In a raspy voice, "Annie" tells two younger, prettier anophelines a sorry tale set to classic Carl Stalling scores and dramatic orchestral sound effects. "To look at me now you'd hardly believe it. . . . Why, I used to be the toast of hot spots, *QUEEN* of the swamps," she tells the two young Anns. "The world was my playground—the South Pacific, India, North Africa, Italy. . . ." With each mention of a new location, the music changes to suit the culture, an animated map of the region flashes across the screen, and Annie floats by in culturally appropriate dress with heaving cleavage and blood cocktail. "Yup, all over the world I knocked them on their 'eels. . . . My percentages *was going up*—when the big shots turned on the heat. It was the same old vice squad that made it hot for me down in Panama," she says, referring to Dr. Gorgas's famous defeat of malaria during the canal project. She continues: "Parasitologists! Entomologists! *MALARIologists*! Every damned *-ologist* in the country got on my trail, and I do mean trail. They started a cleanup. . . ." Across the screen flashes a newspaper with banner headline: "Get Annie!" And she continues: "First they wised the boys up. . . . And, then, oh brother, did they move in on us. They smashed up our hangouts. They *really* poured it on, *with oil*! They dusted us off with *Paris Green*. They busted in all the joints from Burma to *BizertEE*."

She pauses in horror, then offers a glimmer of hope: "The percentages ain't what it used to be. But thanks to Snafu a smart operator can sneak in for a one-night stand!" She shows the two young Anns pictures of Private Snafu, shirtless, sleeping under a tattered bed net, pushing aside repellent, and tossing pills. "Repellent, huh, he never touches the stuff," Annie huffs. "And that goes for the *AtabrEEne*, too." The young Anns giggle and laugh, then go after him and whistle a catcall as they fly straight for his rear end—to the sound of falling bombs. Raspy Annie gulps back another martini, blows a strand of hair out of her face, and looks at the audience with a final word: "Well, dearie, as long as *that* guy is around, a little gal can still make an honest living."

Another fifty malaria-related cartoon trailers were produced to open every movie shown overseas. Radio broadcasts brought a new weekly message on malaria prevention, while malaria units put on stage performances to dramatize how the disease spread.

Widely distributed bumper stickers read: "Prevent Malaria—Shorten the War" and "Stay on Authorized Roads—Malaria Breeds in Road Ruts—Do Your Part to Stop It." These messages were also printed on ration boxes and matchbooks sold at army post exchanges. Dozens of posters depicted men in embarrassing positions designed to shame troops into protecting themselves from mosquitoes. One had a goofy-faced, bare-bottomed GI diving into water, his rear end glowing in the moonlight as a giant mosquito bites it. In another of this series, called "Fight the Peril Behind the Lines," a GI in only boxer shorts and boots hangs from scaffolding with three angry male mosquitoes pointing bayonet noses at his chest.[4]

Other ideas included sexy "Margie mosquito" working at night, with a caption that read: "One night's indiscretion may cause you to pay and pay and pay." Another showed philandering female mosquitoes pointing at a GI in shorts, with the caption: "Oh look! There is Sloppy Joe. We all eat tonight." Others showed a GI glistening with repellent and the caption: "Stay away from him, dearie. He'll turn your stomach with that nasty smelling stuff he has on his skin," and, "My dear, that man's blood is positively revolting. It tastes so strongly of Atabrine."

Many worked from racial stereotypes, depicting mosquitoes as bucktoothed Japanese with slanted eyes, some wearing thick round glasses. And more still had women in provocative poses. They had nothing to do with the message, but got the men's attention.

DISNEY made a film for troops about screens called *Winged Scourge. Time* magazine wrote about it in July 1943 and described Lowell Coggeshall as its inspiration. Lowell's four-thousand-word essay in the *Journal of the American Medical Association*, the one rebuffing de Kruif's naive notions about atabrine, had made Lowell a public health celebrity. Reporters went to him for insights and gave him due credit. "The best way to fight malaria is not by drugs but by fighting mosquitoes. That is what Dr. Lowell Coggeshall of the University of Michigan concludes after helping Pan American Airways throw an airline across Africa in 60 days for the U.S. Army." *Time* called Lowell's work a "striking achievement" and gave him credit for convincing Walt Disney to get involved.[5]

The British Directorate of Army Kinematography also produced propaganda films using morality and duty to guilt men into protecting themselves from mosquitoes.[6]

Back in the United States, Wally Boren's syndicated homespun column "Wally's Wagon," with a Sunday circulation of six million, made a special plea to women: "Gals, Use Your Influence!" He wrote: "Me an' the U.S. Army has cooked up a conspiracy. . . . Seems that *nothin'* the Army can do will persuade all the boys to take care that they don't get bit. You can tell 'em how dangerous it is, but they just laugh an' say, 'Who's afraid of a lousy mosquito?' Some won't even take their malaria pills. . . . Plumb discouragin', ain't it? . . . I believe that you gals ought to write your husbands or your boy friends or your sons or grandsons--whoever it is you got out there. Write 'em one of them front-hall 'promise me' talks you're so good at. I'll bet you'd sure get some action!" Soon letters from home begged the men to protect themselves from malaria.

These messages kept hitting the men every day, where they lived, until they started seeing stagnant water as mosquito breeding grounds; torn bed nets as a way in for biting anophelines; and sex in the village as exposure to malaria, as well as VD. In this way, malaria and VD prevention were married, and opportunities to teach men how to avoid both, in one message, were fully exploited.

IN July 1943 the U.S. Army Services of Supply instructed all "officers, all grades" to prepare for "malaria consciousness" classes. SOS was the same organization Dr. Egeberg dressed down with his sheep comment. But now, just three months later, this office was fully on board with a plan that had medical officers training unit officers, who would then be held responsible for educating all enlistees. The tone of the directive said: Listen up, *unit officers*: Take this seriously, or else. It was signed by command of Major General Richard Marshall, the commanding general of SOS in the Southwest Pacific Area, who was MacArthur's former deputy chief of staff, soon to be his chief of staff.[7]

Then, in September 1943, the War Department created a mandatory four-hour course in malaria control for everyone—enlisted men and officers. Medical officers had to take leave from their posts for a four-week course with Mark Boyd at the Army Medical School in Panama. These army doctors sat before giant posters depicting malaria's life cycle, from the sticks inserted by mosquitoes, to the spheres that invaded red cells, to the sexual stage that formed egg sacs soon after hitting the mosquito's gut.

A pamphlet from the army surgeon general's office, written by Paul Russell, warned medical officers to take their training seriously. "Each medical officer will be held responsible for the adequate instruction of the entire personnel under his medical supervision."[8]

This all happened at lightning speed for the military. By the end of 1943, everyone was learning about malaria. To make the courses more interesting for troops, and to help with attendance, billboards of buxom, hippy women were used. One in New Guinea had a curvy strawberry blonde with ruby red lips lying on her side naked. The caption read: "Remember This, Take . . . Atabrine!"[9] For

Middle Eastern, Indian, and North African audiences, a similar message accompanied a billboard-size bare-breasted woman with jet-black hair and translucent parachute pants.

All troops, from the lowest-ranked privates to sergeant majors, and all commissioned officers up to the generals, attended seminars on malaria. Soon malaria rates began to slow. The accumulated effort of all that had come to pass forced malaria, this immovable object, to budge.

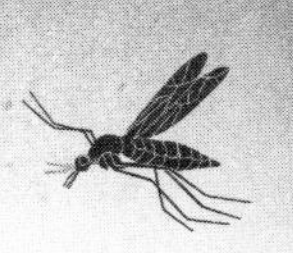

CHAPTER 25
The Breakthrough

On September 3, 1943, Canadian and British forces poured into Italy from Messina. Five days later the Italians switched sides to join the Allies. At dawn the day after, some 450 ships reached the beaches of Salerno and Paestum to begin landing the fifty-five thousand troops of the U.S. Fifth Army, officially launching Operation Avalanche.

The first days were a "death struggle" around Salerno.[1] The Germans kept trying to "fling" Allied troops back into the sea, and the Americans and Brits kept driving forward, making their way inland. By day four, the outcome remained a toss-up. The Americans might have retreated, if a plan existed for it. However, they had no way to evacuate fifty thousand men from shore to ship; the ships would have to move too close, exposing them to German panzers and big guns. So more warships and air cover arrived. "More than a thousand 'heavy' sorties had been blown at Salerno," wrote Rick Atkinson. The "heavies" dropped "760 tons of high explosives per square mile, annihilating intersections, rail yards, and villages." The barrage forced the Germans northward, and as they retreated they destroyed everything in sight. "The scorching and salting of the earth had begun. Horses and mules were stolen or shot. . . . An estimated 92 percent of all sheep and cattle in southern Italy, and 86 percent of all poultry, were taken or slaughtered. 'Rail rooters'—huge iron hooks pulled behind locomotives—snapped railroad ties like matchsticks. The echo of demolition rolled from the mountains, and oily smoke smudged the northern skyline."[2]

They also destroyed generations of ditching and canals, and the pumps used to drain the malaria-ridden bogs south of Rome. German troops turned the clocks back a hundred years as they retreated northward, and floodwaters drowned the countryside.

American commanders were doing their job: forcing Hitler to maintain a large fighting force in Italy to try to prevent the Allies from advancing northward. This robbed him of needed manpower in Russia and to finish his Atlantic Wall—of bunkered big guns and booby-trapped beaches along the northern coast of Europe. Allied strategists viewed the destroyed pumps and resulting floods as merely a nuisance, at first. But then American public health officials investigated and found that the Germans had cleverly left a few pumps working, only in reverse. The pumps sucked in seawater, which killed other species of mosquitoes, but for Italy's main malaria carrier, a subspecies of *A. labranchiae.* This species bred well in salt water.

The Nazis were attempting germ warfare—using malaria to cripple Allied troops. Historians later accused Germany's top malariologist, Erich Martini, of using his professional training for "evil."[3] German military leaders, on their own, could not have known the breeding habits of this one mosquito subspecies. However, Martini knew it well because he had studied *labranchiae* with American and Italian malariologists in the 1930s. Together they discovered that it was the primary transmitter of malaria in Italy. Martini must have told military leaders that a few pumps operating in reverse would wipe out other competing mosquito species and create a vacuum for *labranchiae.*

Allied troops were on course to move through the area in late spring or early summer of 1944, at the beginning of the malaria season.

MEANWHILE, in the United States, Malaria Project researchers observed with surprise that atabrine was actually a much better drug than quinine. Now that dosing was understood and side effects were reduced, researchers could see that both worked similarly against infections—they could quell fevers and prevent death. But quinine required six times the volume of drug to establish the same results. And cinchonine (low-grade quinine from poor-quality South American cinchona bark) required seven and a half times the amount. In terms of getting the drugs made from available chemicals, loaded onto cargo vessels, and added to medical kits and ration kits, atabrine was a much more attractive option. Even if Japan had never captured Java, and the Dutch made ample quinine at affordable cost, it was six times heavier to transport. In a war without enough cargo vessels and air transport capacity, atabrine was now seen as the far better option. (That is, at least it didn't take up as much space.) It would do until researchers found something to actually cure, or at least prevent, malaria.

On that front, the failures kept piling up. SN-113, SN-232, SN-112, SN-213—drugs made from a range of compounds thought to have antifebrile properties—all produced some level of treatment, but at ten to thirty times the dose of quinine (which was used as the standard for comparison). SN-374 and SN-377 also showed action against parasites, but at doses eighty times that of quinine; and at ten to twenty times the dose of quinine, SN-2922, SN-4271, SN-4270, and SN-4130 did absolutely nothing. Anything above that was too toxic to explore.[4]

The sulfa drugs suffered a similar fate. Back in 1942, chief investigators were certain that one called sulfamerazine would replace atabrine and end the quagmires associated with it (from Winthrop's attempts at controlling production to the toxicity crippling troops). Lowell had used a close analogue of this sulfa drug when he was in Africa fighting against *falciparum* infections, and reported great results; other project scientists used sulfamerazine on their state hospital patients and found it worked well, and wasn't toxic. They also thought they saw evidence that it killed the sporozoites carried in mosquito saliva. If this could be supported by larger studies, the War Department would have a cheap, easy-to-make drug able to prevent mosquitoes from infecting soldiers. Troop strengths would return to normal and the malaria problem would be solved in a matter of weeks. Civilian and military scientists wanted it tested in the field against naturally occurring malaria. Dozens of letters flew back and forth between the National Research Council and the War Department. Medical corps doctors were sent to James Shannon for proper training in running field studies, while Lieutenant Colonel Roger G. Prentiss and Brigadier General Larry McAfee convinced Secretary of War Stimson to grant authority for top-secret trials in overseas theaters.[5]

By early 1943, military investigators were running tests on several hundred soldiers intentionally exposed to malaria in Assam, New Guinea, the South Pacific, and the Caribbean Defense Command. Atabrine was replaced with sulfamerazine, code-named "Mary." Everyone working on it had to sign a secrecy agreement.[6] New troops—referred to by scientists as malaria virgins, having never been exposed to the disease—were shipped into highly malarious areas to do construction jobs, some of which were contrived specifically for the study. All those on Mary were lied to.

In Assam, for example, the study began in late spring 1943. It included a hundred Americans from the 151st Medical Battalion, Company C, and a hundred from the 478th Quartermaster Regiment, Company H, plus two hundred Chinese soldiers from the 1st and 2nd Reserve Company, 6th Motor Regiment, 38th Division. They were split into groups of fifty and put on Mary, atabrine, or a placebo. Officers told the men they were in a vitamin study. But the "vitamin" was really a sugar pill, and the "activator" pill was really Mary.[7] Results showed Mary to be no less toxic than atabrine, no more effective, and to achieve the same level of treatment, it required five times the dose.[8] Russell lamented that Mary was not the "magic wand" everyone hoped for, and that malaria control remained "a matter of hard work, good discipline and smoothly flowing supplies."[9]

A vaccine study run by Michael Heidelberger, considered the father of modern immunology, also turned up very little. For him to do his studies, the army tried to deliver what he needed most, which was highly infected blood. Medical corpsmen took cups of it from malaria-felled soldiers in the Southwest Pacific—as much as half a liter at a time—then airlifted it over the Pacific and continental United States to Heidelberger's lab on West 168th Street in New York. The bleeding was sometimes so heavy that the sick soldiers required transfusions. Heidelberger needed these large volumes of blood to concentrate parasites for vaccine preparations.[10] But the blood kept arriving denatured and unusable. And Heidelberger was getting no further than Lowell had in figuring out a way to work effectively with these tricky parasites.

MEANWHILE, on October 30, 1943, the army issued Circular Letter No. 111, describing the dire consequences of failed malaria discipline in all theaters. It said the equivalent of six divisions and counting had been incapacitated in one theater. In another, the equivalent of a battalion a month was being evacuated with malaria. All told, the anopheles mosquito had produced more casualties than enemy aircraft and land forces combined, it said.

THE project's deflated chemists met frequently to discuss the growing list of failed drugs. Each had worked in birds and passed toxicity tests in mice, rats, dogs, and cats. Thousands of animals suffered terribly and had to be euthanized for the cause—to have their organs studied for toxicity. Breeders were recruited to provide more dogs.

By the end of 1943, the chemists had made nearly seven thousand drugs from fifty different structural types, 15 percent of which showed action in birds, and of those, a few dozen passed the toxicity tests. Still, not one ended up being nontoxic and effective against human malarias. In written comments, the chemists openly questioned their own study design. What if the reverse were true? What if drugs passed up because they didn't work in birds, or were too toxic in mice or dogs, ended up being active and safe in man?

Pharmaceutical companies asked project scientists to take their compounds straight to human studies, as the animal studies were obviously unreliable. But that was out of the question for this group. No way would they start trying random compounds without solid data on toxicity. So Marshall at Johns Hopkins tried a new approach. Instead of injecting experimental compounds into his birds, he put the drugs in the feed to mimic oral administration at different doses. His team of chemists used hundreds of birds for each cohort and plotted the data as cumulative frequency curves. They did this for quinine, and when they calculated through the curves they saw that the ducks' parasite levels dropped quickly if they were on a low dose that incrementally went up, versus those started on a high dose.[11]

Through these challenge exercises, researchers saw that their dosing charts were too simplistic. A different amount of drug was needed to treat the different types of bird parasites, and to treat the different strains of these malarias.[12] The more project scientists learned, the more they felt in the dark.

THEN a light switched on. It was just a flicker at first, started at the September 2, 1943, meeting of a newly organized entity: the Subcommittee on the Coordination of Malaria Studies (which was really just a restructuring of the project's hierarchy).[13] The most important lead investigators were there, including Lowell Coggeshall. It started off in the usual way: Members heard updates on previous problems. Someone complained that atabrine continued to arrive overseas with a shellac coating, while another said, No, that wasn't right; the drug hardened with age, causing delayed absorption. The Brits asked whether pilots should be given atabrine, as it appeared to cause vertigo, to which the Americans said, No, they had just done a study; it caused no vertigo. Another group said they detected more cancer in rats on long-term doses of atabrine, to which someone said, Well, that may be, but Winthrop has had employees on it now for six months and they're fine. Someone else brought up potential liver damage caused by atabrine to announce that the Medical Corps had started biopsying the livers of casualties known to have been on long-term atabrine and that no confirmation of liver necroses had turned up yet. (By 1945, however, several autopsies showed atabrine had caused toxic shock to the liver.[14])

Other atabrine-related matters were discussed, including planned U.S. production of three billion tablets for the coming year. This coincided with strict rules that forbade downplaying its value, or publicly discussing concerns about toxic effects.[15] Atabrine propaganda in the field was "dogmatic and forceful" so that the drug's good features were emphasized and its shortcomings were minimized. Medical officers questioned the intellectual honesty and scientific accuracy of the propaganda, because they saw reports linking atabrine to skin disease, gastric distress, and psychosis coming in from the field.[16] But those reports were suppressed. Atabrine had by now reached every theater. "Education" of troops (cartoons, posters, bumper stickers, billboards, signage, stage performances, and scantily clad women next to "Take Atabrine" signs) had softened troop resistance to taking the drug.

HALFWAY through the day, somewhere between breakfast and evening cocktails, Professor W. Mansfield Clark, the chief chemist at Johns Hopkins, cleared his throat and announced that a new substance "captured from the enemy in North Africa" might be of interest. The room buzzed with questions but Clark had very little information on the drug, which he called sontochin.

No one knew whether it was German or Italian or what. But Dr. Jacobs at the Rockefeller Institute in New York had been studying its chemical composition, Clark said. He wasn't ready to report definitively but believed it might be an old German drug. This crowd, thirsty for a new lead, insisted that the bird studies start at once, knowing they could take as long as three months to complete (which was a bone of contention, as some felt the process was overkill, and others, like Marshall, absolutely insisted on the necessity of using at least a thousand birds per drug to allow for quantitative analysis of the data with curves, logarithms, and probability charts). So they all agreed that Rockefeller would work with Johns Hopkins to make sure studies began as soon as possible.

Then they returned to regular business. The military brought up its desire to have a tracer molecule put into atabrine to help catch slackers not taking their daily pill. No one liked this idea, so they agreed to the concept but assigned no one to do the work.[17]

TWO weeks later, on September 16, project scientists met again. They opened with more disappointing reports on NIH-204. James Shannon tried it on twelve of his insane patients, and ten had pretty bad toxic reactions, which included pus-filled pimples, reduced heart rate, and blood in the urine. The chemists sat around arguing over what went wrong. One doctor said the drug made his chickens "exceedingly dopey," as if "narcotized." But he wasn't looking for toxic effects; he was looking for activity against bird malaria, and this drug for that looked very good. The NIH team found it not too toxic in lower animals, but project chemists complained about their work because they had no controls, and the dogs threw up much of the drug because the dosing was too heavy. Then another doctor pressed G. Robert Coatney to discuss the toxicity of NIH-204 in patients at St. Elizabeth's. This seemed to irritate the irascible Coatney, as he replied flippantly that he saw "no instances of priapism"—prolonged erection.

Army and navy representatives, hearing the debate, canceled their plans to test the drug on troops and told project scientists that they could keep the 6 kg of NIH-204 the military had ordered. As the meeting spiraled down into another bellyaching session, Eli Marshall of Hopkins read aloud a complaint from Wiselogle that said chief scientists weren't filling out paperwork properly. To which one doctor questioned the value of all the paperwork, to which Marshall said, yes, yes, it's not perfect but it's all we've got. Then he pleaded with everyone to please adhere to poor Wiselogle's requests, as he was the only one keeping track of all the compounds coming in and out of the project.

The dour mood changed when Marshall reported on the mysterious drug sontochin. He passed around copies of the one report on the drug, written by Jean Schneider of France, and told the group how it was captured and how Brigadier General F. A. Blesse, the surgeon for the North African Theater of Operation, had flown Schneider from Tunis to Algiers to help translate his work into English. On July 8 everything—Schneider's pills, powders, reports, and spreadsheets—were packed up and shipped to the Office of the Army Surgeon General.[18] From there it all went to the Rockefeller Institute in New York to be studied.

Meanwhile, more captured documents on sontochin turned up. During the surrender of Tunis, Captain Theodore Woodward of the medical corps reported to headquarters in Algiers that a French doctor at the city's Pasteur Institute, Paul Durand, had kept statistics on 13,564 cases of typhus. To manage a crush of panicked civilians, Durand combined a "killed mouse lung" vaccine with sontochin to try to reduce mortality. He told his interrogators that the chemical structure of the drug was unknown, at least to him. All he knew was that the Germans used it for malaria. So Captain Woodward packed up Durand's supply and shipped it to Lieutenant Colonel Perrin H. Long, an infectious disease expert at Johns Hopkins before joining the Army Medical Corps.[19] And Long shipped the sontochin samples back to the army surgeon general. He included a memo from Captain Woodward that said Afrika Korps POWs had carried only atabrine for malaria. This meant the Germans hadn't yet started using this secret drug sontochin for troops. The Americans were at least not behind and might even surge ahead, now that they had their own samples of the drug.

Attendees at this September 16 meeting perked up. They all found Jean Schneider's report intriguing, especially where he said he had been "induced" to run a clinical study using sontochin, and that he and Durand had brought the drug to Tunisia from Nazi-occupied Paris two months before the Allied invasion. Schneider noted that he couldn't provide the drug's chemical structure; all he knew was that the Germans called it Methamsalz or, more generically, "methylene-dioscy-naphtoique"—which caused the chemists to wrinkle their brows, because clearly this was wrong.

Schneider said in his report that he had surrendered thirty thousand tablets and four boxes of intravenous solution to Colonel Loren D. Moore, the theater's American malariologist, but that he kept samples as a "safeguard" for France, in the event it proved to be truly valuable. He said he felt "morally responsible" for giving Moore "the results obtained" from his studies, and assured that "the greatest secret regarding the results and the nature of the product employed has been guarded till now." He added: "Since the liberation of Tunisia by the Allied forces, I have believed it my duty to submit to competent authorities, all that I have assembled . . . realizing that the supreme intent of the actual struggle, and that of France, forces me to make known the results obtained."[20]

Schneider's sentimentality toward this drug and what it might mean for postwar France was ignored. The room focused instead on his actual data. Sontochin, it appeared, worked better than atabrine, had no side effects, and might even *cure* malaria.[21]

This crowd had heard these types of claims many times before. First came sinine, made from the root bark of the sinine tree, also called Pai-chi'ang-kan, of southern Yunnan. Reports from medical corpsmen in China said it was thought to be "more potent than and just as satisfactory as quinine for the treatment of all kinds of malaria."[22] Similar reports came in on drugs like benzedrine, neoarsphenamine, thorium dioxide, thiobismol, Forbisen, pulverized leaves of *Laurus nobilis* (laurel from Spain), and something called fraxine, which ended up coming from the same bark as sinine. One team of American medical corpsmen even swore that therapy with oxygen-nicotinic acid prevented cerebral malaria. Another team argued that they'd seen lower infection rates among men with high cholesterol, so they tried drugs to elevate plasma cholesterol and said it worked. Also tried were nylon intermediates, rubber accelerators, mud from the Nile River, Lugol's solution, and eggshells suspended in choice whiskey.

"The Board's scientists tested them all," one member explained. "It was easier . . . to make the tests and send a scientific report than to write a letter explaining convincingly why the lower leaves of the cotton plant, for example, were not likely to prove a cure for malaria."[23]

But Schneider's report appeared more authentic than most. Plus, sontochin came from a German lab. *That* meant something. And he had run a full clinical trial, complete with control groups. Still, everyone remained guarded—fooled too many times to get excited. That is, until Mansfield Clark made a final point on the matter. He said the drug's chemical structure had been deciphered. As it turned out, it had already been tested in the United States by none other than Lowell Coggeshall and his colleague John Maier, way back in 1941 when they worked for Rockefeller. They found it to be "quite active" against bird malaria in ducks. This was the drug Lowell had wondered about as he left Rockefeller for the University of Michigan!

Then Clark delivered the kicker: Sontochin was made by Bayer but its U.S. patents belonged to Winthrop Chemical Company, whose chemists had apparently ignored Rockefeller's conclusion that it was "quite active" and shelved it instead. The room exploded with questions. Clark didn't have anything else to say, except that he had enough samples to distribute for a wider investigation and that more details would be available soon.

IN the following weeks, Clark dug in to find the truth. Without revealing too much, he asked Winthrop to give him samples of sontochin. He reported at the October 7 meeting that the samples had "apparently been lost in the mail" but that he was tracing them down.[24] Everyone at the meeting wanted to know why Winthrop did nothing with this drug back in 1941.

As this inner circle of scientists mulled over the facts, Marshall tried to explain. He said this drug was one of the first tested by the Malaria Project, back when the project ran on a shoestring and the number of labs involved could be counted on one hand. The drug had been assigned a survey number, retroactively, by the reliable and hardworking Wiselogle, who by now had four secretaries helping him organize and catalog reports and data charts coming in from thirty universities and forty laboratories.[25] He didn't start until 1942. At that time he assigned this shelved Winthrop drug as SN-183. That was when K. Blanchard, the young ace chemist hired by Marshall to help him at Hopkins, looked at SN-183, saw that it was like atabrine, only different, with its nitrogen side chain in the fourth position—a 4-aminoquinoline—and asked whether he could work on it. But his bosses said no. They were tied up with drugs that, at the time, showed great promise: the sulfa compounds, quininelike drugs from low-yielding cinchona bark, and atabrine.

Still, in 1942, Blanchard persisted. He guessed that this 4-aminoquinoline could be played with; the side chains could be moved around, maybe the methyl group removed and replaced with a chloro group. "No one was interested!" wrote Coatney, who sat in on the meetings.[26]

Now, millions of dollars and eighteen months later, the drug floated back to the surface because it was captured in battle! "This disclosure . . . that the two compounds (SN-183 and sontochin) were identical created havoc bordering on hysteria," Coatney said eight years later, almost to the day, during a speech at the annual meeting of the American Society of Tropical Medicine and Hygiene in Atlanta. "We had 'dropped the ball' and in so doing had lost valuable time."

No one at that meeting recorded the hysteria. Instead, the minutes merely say that Marshall "gave an interesting review" of sontochin; that it had been sitting in the "open files"—not held as confidential or top-secret, just there, for everyone to see—all this time. The damning number, SN-183—this reminder that sontochin was among the earliest drugs tested—was "declared dead."[27] Sontochin was issued a new number, SN-6911, signifying it was the 6,911th drug tested by the project. Then everything about it was declared a secret under the Espionage Act, forcing the group to put down their pens and just listen—no note taking.

Marshall went on. Using samples Schneider sent, his chemists launched studies. Results were in from one using it on ducklings infected with Lowell's bird malaria, *P. lophurae*. Sontochin, it turned out, was eight times better than quinine at clearing the parasites. He said Schneider's human studies were "fairly extensive" and showed less toxicity than atabrine in man. But given their bad experience with NIH-204, he felt it necessary to run sontochin through a complete "pharmacological work-up"—in mice, rats, dogs, cats, etc.—before releasing it for project-sponsored human trials. A collective groan could be heard. But these were the standards. Everyone on the clinical testing panel would just have to wait.

The group voted to compel Winthrop to hand over all its files on this drug and elected Dr. Richards, chair of the White House's Committee on Medical Research, to make the request directly to Winthrop President Klumpp.[28] They also agreed that after a fast workup of the toxicity in lower animals, "[T]he drug should be promptly exploited in man."[29]

As everyone else left the room, key representatives from the army, navy, National Research Council, NIH, and the Committee on Medical Research huddled in executive session. They needed to tighten the administrative grip on this growing and unwieldy project, *and* stiffen reporting rules with checks and balances to avoid another ball dropping. Thus they created a new overarching organization called the Board for Coordination of Malaria Studies, or the board, for short. All the other groups had to submit proposals for investigations through this group of independent overseers. They were there to assist investigators with problems and integrate the planning of projects—and keep an eye on all work paid for by the OSRD. George Carden was handpicked to oversee everything and report directly to Richards. Sontochin would require care, especially around the delicate matter of international diplomacy—as Jean Schneider made a claim on it for his beloved France, which was now an ally. Then there was the thorny problem of Winthrop Chemical Company, which held the patents. Somehow the U.S. government had to seize ownership of this drug.

CHAPTER 26
Stateville

Sontochin changed everything. This 4-aminoquinoline looked like the 8-aminoquinolines, only different. Not exactly novel but different enough, presenting many, many possibilities. Project chemists had a new set of molecules to twist and turn about. After two years of dead ends, and nearly seven thousand new compounds that went nowhere, they finally had something. The OSRD wanted SN-6911 fully vetted.

Drs. Marshall and Shannon, however, reported that project facilities for clinical testing were wholly inadequate. No one knew how many analogues would clear the hurdles for human testing, but they guessed that at least four or five might. Shannon had only fifteen to twenty patients per month available for use at his psychiatric hospitals, and they were already being used in studies of quinine alkaloids from South American cinchona bark. Coatney had his prisoners, but they were being used in atabrine studies.

With so little *clinical material*—Shannon's sanitized name for human test subjects—it took Shannon nearly three months to run preliminary studies on NIH-204.[1] For SN-6911 and its analogues, the project would need more people willing to be used in experiments. The army had a lot of this so-called *clinical material* at several hospitals, including Harmon in Longview, Texas; Kennedy in Memphis; Bushnell in Brigham City, Utah; and Lowell's Percy Jones in Michigan. Each had two hundred beds filled with soldiers sent home with relapsing malaria.[2] But this *material* was already tapped. Harmon was using the toxic drug plasmochin to try to stop relapses; Letterman was testing a chest respirator to try to clear organs of the "latent" parasites that caused relapses; and Percy Jones was trying different methods to trigger relapses, to understand the mechanisms. Overseas, convalescent camps in Australia and Fiji were also tapped, because they were used for running atabrine tests. Field hospitals in India and New Guinea tapped for secret tests of experimental drugs were still measuring outcomes using sulfamerazine (Mary) and sulfapyrazine. Mary alone was given to more than 850 intentionally exposed soldiers to tease out any benefit to this drug.[3] And totaquine (one of the alkaloids made from poor-quality cinchona trees) was occupying other hospitals with the capacity to run studies.

GEORGE Carden, the secretary of the board, was elected to find more bodies. On December 6, 1943, he wrote a fruitful letter to Dr. Clifton T. Perkins, Massachusetts's chief mental health official. Carden asked whether the project could open a research station at Boston Psychopathic. The hospital would get new staff paid for by the government, and $2,000 in spending money. In exchange, the project would administer malaria therapy to end-stage syphilitics, who would also be used in drug studies. He also asked permission to give malaria to patients with "other neurological and psychiatric disorders." He said the literature suggested those patients might benefit. "A decision on the extent to which this is true is impossible on the basis of information now available," he wrote. He made no mention of how Jauregg had shown that these other groups did not benefit, at all, from malaria therapy. But that was beside the point. The project needed test subjects to help find a cure for malaria, and the U.S. government would make it worth Perkins's while in new staff and money.[4]

Perkins agreed, and one of Massachusetts's most prominent physicians, Allan M. Butler, was chosen as lead investigator. The Harvard medical professor and pediatrics specialist at Massachusetts General Hospital had already been conscripted into war research. Beginning in 1942 and until he joined the Malaria Project, he explored several projects connected to shipwrecked servicemen. In one, he exposed two conscientious objectors to intense sunshine, and gave them limited food and almost no fresh water. For ten days he measured the effects on their bodies. In his conclusion he said he was "skeptical" of reports that suggested drinking seawater could help save shipwrecked men, as it would "increase fluid removed by the bowel so that dehydration will result, if vomiting does not precede it." He ran the study with a budget of $1,700.[5]

On January 21, 1944, the board approved Butler as a lead investigator and gave him a budget of $116,290 ($1.55 million in 2014 dollars).[6] The board sent him the rulebook for running malaria studies, and a letter explaining the project's "restricted" classification: "This means that information concerning the nature, progress or results of the work should not be disclosed to persons other than those immediately interested in it, unless special authorization has been given by the Committee on Medical Research. Aliens are to be entirely excluded from contact with this work."[7]

Similar letters and similar grants with instruction manuals went to new investigators at Gailor Memorial Psychiatric Hospital in Memphis, and two more mental health facilities in New York: Manhattan State Hospital and Bellevue. All began running small studies on drugs already tried on birds and dogs—ramping up capacity for human trials while sontochin underwent intense investigations in birds and lower animals.

Shannon in New York and Coatney in Atlanta geared up to take control of the whole thing. But Lowell hadn't left for the navy yet and wanted to ensure these titans shared the work—and potential success stories—should sontochin pan out. So he and Dr. Taliaferro at the University of Chicago formulated a plan to capture some of the big work for their region. They knew each other well. Taliaferro and Coggeshall served together on the first malaria committee created by the NRC back in 1941. As the project grew and diversified, they ended up on the same subcommittees and were on the executive panel that had access to all confidential and restricted information. They also collaborated on projects, including an attempt to infect lower animals with human malarias—a painstaking effort that ultimately showed it was not possible at that time.

The Germans, with help from Claus Schilling, came to the same conclusion years earlier. But those studies were in German. Here, studies in English made the point Schilling had made to Herr Himmler: human-based experiments early in the drug-development process were unavoidable. People had to be used as guinea pigs. But just as Germany and the United States were on very different political and social paths, the two countries' scientific leaders used "clinical material" with very different standards.

THE University of Chicago, already tightly affiliated with the project through Taliaferro, had top-rated labs and researchers. Less than an hour's drive south of campus sat Manteno State Hospital for the Insane, and Stateville Prison, with a combined population of ten thousand—literally tons of potential "material." This combination was perfect for a major expansion of the work. This research triangle would, ultimately, produce the most important malaria-related drug studies of the war, and for another two decades after the war.

It was approved in early 1944. For a lead a investigator, Drs. Shannon and Marshall wanted the likable and friendly University of Chicago medical professor Alf Sven Alving. He was among the most competent researchers in the region. Alving, a plump Swede with a round, friendly face and thick sandy hair combed back in finger curls, was an expert in diseases of the blood and kidneys. He honed his skills from the masters at the Rockefeller Institute in New York. This was not the International Health Division made up of public health-related researchers, like Lowell. This was the *institute*, where the country's best minds developed cutting-edge technologies and techniques. Alving landed at the University of Chicago as an assistant professor because of breakthrough work he did for Rockefeller.

People adored the jovial Alf Alving. A heavy Swedish accent made his playful jabs all the more entertaining. He lit up faculty meetings and course lectures with what one colleague called "a loving laughter." The same colleague said he brought to his three sons and invalid wife a "fantastic devotion." Her condition, multiple sclerosis, broke his heart, but he "never liked to give any sign of the emotion that overpowered him."[8] Instead, his colleagues and students became dear friends who gave him healing warmth.

He took his first trip down to Manteno in April 1944 to tour the thousand-acre "dreary-neat" institution, set up like a village.[9] The state of Illinois's largest mental hospital had been built in 1937 on a flat plain of limestone, forty-eight miles from Chicago. A redbrick administration building with a bell tower and stately white pillars sat next to a hospital. Behind them was a grid of about twenty-five blocks with squat dormitories lined up like barracks. The roads dead-ended into soybean and corn fields that stretched for twenty miles, walling in Manteno's fifty-eight hundred "residents." The campus had an electrical grid, sewage plant, dedicated well, cemetery, laundry facility, and working farmland where higher-functioning patients grew much of what they ate. Its population exceeded that of the village of Manteno. Locals called it Crazytown and spread unsubstantiated rumors about sadistic staff raping the girls and torturing the boys, and how those who tried to escape turned up dead in soybean fields—which did happen once.

Alving stood out among project investigators: The state hospital work concerned him and he wasn't sure he wanted to do it. As he was an expert in measuring drug levels in blood plasma—a key to understanding how and when drugs work—his skills were needed. Taliaferro, Shannon, and Marshall had to virtually draft him. It helped that Alf's mentor at Rockefeller, the great Donald Van Slyke, had joined the project. Founder of clinical chemistry, Van Slyke had identified and explained the role of electrolytes in blood. He pioneered techniques for measuring blood chemistry as a diagnostic tool for diseases like diabetes. And he did the same with urine excretion and metabolic processes to diagnose diseases of the kidney. (There isn't a hematologist today who doesn't know his name.) The Malaria Project had collected the most important thinkers, especially in chemistry and clinical medicine. How could Alf refuse to join?

So he said yes, then made his studies *the* most valuable being done. His project code number, OEMcmr-450, would soon absorb as much money as Shannon's. ("OEM" for Office of Emergency Management, the account to which Congress appropriated federal funds for secret projects not subject to public scrutiny, and "cmr" for Committee on Medical Research.)[10]

ALF'S instincts toward Manteno proved spot-on. The place suffered from severe overcrowding and equally severe staff shortages. It was sprawling, dirty, and unkempt. Food-borne diseases broke out with some regularity.[11] A few years earlier, substandard sanitation brought in typhoid that hit nearly four hundred patients and a hundred staff; sixty people died—all within a month and a half, from early July to late August 1939.[12] "The disease spread to a dozen, a score, a hundred," wrote *Time* magazine. "Patients lay moaning in bed. Others, whipped by mad fear, beat against the screened windows. . . . Some of the attendants fell ill. All were panicky. Every night kitchen boys and orderlies disappeared. Over 45 ran away in all."[13] Typhus at Manteno made headlines for another two years.

Republicans and Democrats blamed one another in "an attempt at 'playing politics with human misery.'"[14] Authorities believed cracks in the sewer lines allowed fecal matter to contaminate drinking water. Public Welfare Director A. L. Bowen was indicted on charges of neglect (but later acquitted by the state supreme court).[15]

This was a troubled place, and Alf had a hard time keeping nurses. He used "patriotic duty" to press three discharged army nurses into working for him there. One lasted only a few days. The head of army nurses for the corps area, Colonel Pearl C. Fisher, had facilitated getting him those nurses, and when he asked her to please send more, she requested an official confirmation that his project was sanctioned as appropriate and essential to the war effort. No doubt these health practitioners felt conflicted about giving malaria to this unfortunate lot. Manteno, to those on the outside, felt a little creepy—a village of crazy people in the middle of soybean fields.

On April 27, 1944, George Carden, on behalf of the board, asked the Committee on Medical Research to write the needed letter. Two days later it was in the mail, begging Colonel Pearl to help Alving, as his work was too important to languish for lack of staff.[16] She sent more nurses, but still they kept quitting.

So Alf got creative. Other projects recruited "clinical material" from the Civilian Public Service, otherwise known as the corps of conscientious objectors. About twelve thousand pacifists enlisted for "alternative service" under the CPS. They were doled out to 150 projects to farm, fight fires, serve as orderlies, and do other needed work. Many volunteered for medical experiments, like Butler's saltwater study. James Shannon used the program heavily; he gave malaria to at least thirty conscientious objectors—or COs.[17] Other scientists gave them hepatitis or starved them to near death for studies that shed needed light on how best to revive malnourished soldiers and civilians.

The Civilian Public Service labeled Alf's Project "Unit 115-23," and designated it a high priority, which allowed him to take COs from other units. In all, he had twenty assigned to him.[18] But he had no intention of using them in experiments. His COs worked as nurses, lab assistants, and gofers. In time, he would train them to raise mosquitoes, read blood slides, and recognize when malaria grew life-threatening. They made routine trips through twenty-seven miles of cornfields and soybeans, between Manteno's insane syphilitics and Stateville's famous murderers. While Coatney, Shannon, and the others complained about a lack of staff and accused the army of stealing their best technicians, Alf made no such fuss. He had solved that problem.

ON June 3, Alf Alving took on another of his concerns at Manteno. His instructions were to use schizophrenics and other mentally ill patients, not just late-stage syphilitics suffering from paresis. The argument, articulated by George Carden, asserted that malaria therapy had *not* been proven to *not* help these other mental disorders. This was a play on logic that fit the project's needs. Clearly Alf was being asked to experiment on people, not just provide malaria therapy in a mutually beneficial way.

Alf felt wary about the lines being crossed here.[19] He wrote to Robert Loeb at the NRC for permission to hire David Slight, professor of psychiatry at the University of Chicago, as a consultant. "As none of the Army men have had any post-graduate psychiatric training whatever . . . Dr. Slight's opinion in the choice of non-paretic [nonsyphilitic] patients for fever therapy would be most important should any legal question arise. For instance, for justification of fever therapy, we would have a much stronger case if a man of Dr. Slight's standing in the profession had approved its use."[20] His request was approved.

ALF next looked twenty-seven miles west, to the rapists, murderers, and thieves of Stateville.

To say Alf felt nervous every time the steel gates closed behind him would be an understatement, and a little off the mark. The problem was in his stomach, which tightened every time the guards hollered, "Gate!" To reach the society of criminals within Stateville's massive walls, he had to pass through adjacent vestibules, each separated by an independent set of steel bars. Like locks in a canal, no one could pass to the next chamber until the gate behind him locked shut. There was no side door for visitors. Everyone experienced the same network of security. And everyone was imprisoned until released, gate by gate, under armed guards.

When Alf stepped into that first vestibule, and heard, "Gate!" followed by a loud crank and crash of steel bars locking behind him, his stomach tightened. Then a clattering of pulleys opened a rail of steel, releasing him into a holding area, and again, "Gate!" and the loud crank and crash of steel locking behind him, and the tightening of his stomach, followed by another few steps deeper into the interlocking gates of Stateville. Along the way, he'd hand over his ID, be frisked for contraband, sign his name on a ledger, and wait at the next steel bars and for the next officer to yell, "Gate!" Then more clatter of steel, and tightening in his stomach.

Once inside, Alf felt ambient hostility, confirmed by the gun ports, shooting boxes, and guard towers. The men walked in lockstep and slammed against the wall, paramilitary style, when Warden Joseph Ragen strode though, then shifted their eyes to size up this outsider. Alf assumed any one of them could reach out and slit his throat, at any time, as they marched by, faces pointed straight ahead but with crook eyes on him.

Had this been ten years earlier, the odds of something like that happening would have been high, but not now, with Ragen in charge. Impulses toward violence were mitigated by a full day of work, mandatory schooling, rules and regulations that reduced gangster hierarchies, and books, lots of books. The men were encouraged to read.

When Alf first met with Stateville's storied warden, he liked him immediately. This legend in the penal community had inherited a prison gone wild. Eighty-three shacks made of tarpaper, tin, and lumber filled "the yard." These "throne rooms" were ruled by gangs who pimped out young offenders as prostitutes, grew marijuana, gambled, and made alcohol in stills using potatoes and sugar stolen from the kitchen.[21] The guards got drunk and allowed inmates to roam freely inside the thirty-two-foot-high concrete walls.[22]

Two famous killers were among the prisoner elite: Nathan Leopold and Richard Loeb. In 1924, the two Chicago teens had committed the "crime of the century" when they walked out of their handsome stone mansions to hunt down a boy, any boy, preferably one heading home from the neighborhood's Harvard School for Boys.

They spied Loeb's fourteen-year-old cousin Bobby Franks and decided on him. Then they convinced Bobby to get in their car, drove a chisel into his skull, stuffed rags down his throat, stripped off his clothes, poured acid on his face and genitals, stuffed him in a culvert pipe near railroad tracks in the Hegewisch swamps, south of Chicago, and sent a ransom note to Bobby's father for $10,000. The note was part of their game, to commit the perfect crime and get away with it. Loeb was the criminal mastermind while Leopold assisted, on the promise that he would rule the sex that followed their evil deed.

Leopold and Loeb dressed well, drove fast cars, and were just nineteen and eighteen years old, respectively. Already in graduate school at the University of Chicago, they were sons to be proud of—until they killed Bobby, just to see if they could get away with it. Newspapers throughout the United States and Europe followed the criminal investigation as police discovered Leopold's glasses near the culvert and seized the Underwood he used to type the ransom note. Reporters surrounded the courthouse during this generation's most sensational criminal hearings.

By all accounts, Leopold and Loeb were brilliant, cocky, superior-minded, and deluded. They didn't think the rules of law—and decency—applied to them. They were disciples of Friedrich Nietzsche and interpreted his teachings on "supermen" to mean that men of superior intellect could and should test societal norms. But they weren't as smart as they thought and were quickly caught. Loeb cried like a baby, while Leopold, once cornered, remained businesslike and explained in detail the murder plot.

The world hated them for killing Bobby, and they would have surely hanged had their parents not hired the famous Clarence Darrow to head their defense. In a move considered to this day to be among the most brilliant in criminal law, he surprised the prosecution by having the boys plead guilty. This bypassed a jury and put the decision of capital punishment in the hands of one liberal-leaning judge, John Caverly. Then Darrow mounted a campaign using psychiatrists—called alienists back then—to paint the boys as mentally unstable and, therefore, not appropriate candidates for hanging. It worked. Leopold and Loeb were sentenced to life plus ninety-nine years. The public was furious, and Caverly feared for his life.

Newspapers followed Leopold and Loeb into Stateville, and reported on how their families' wealth allowed them to set up fiefdoms in this corrupt prison system; how they used their family's political connections and wealth to rule guards hired under the patronage system. New wardens came and went, while riots, escapes, and scandals occurred monthly. In 1931, for example, two thousand prisoners lit an inferno that burned down the kitchen, laundry, and furniture- and shoe-making factories. The state police and National Guard stormed the grounds to get the men back in their cells. The warden at the time, who previously had been a car salesman, left in disgrace. On his last day, a long-term inmate stole his trousers and walked out the front gate, giving the papers another escape to howl about.

Then Ragen took charge. He had been a sheriff, like his father. Law and order were his life. So when he took the appointment he aimed to clean the place up, not leave defeated, like all the others. His job might have been harder had it not been for the most sensational event ever to hit Stateville, just three months after he took over. On January 28, 1936, Richard Loeb met a young convict in a small private shower inside a building near the mess hall. He locked the door behind him, removed his clothes, and ordered his young companion to do the same, while pointing a straight razor at him. But the short, stocky con, named James Day, wanted no part of it. He kicked Loeb hard in the groin, grabbed the razor, and cut him more than fifty times while Loeb tried to get the weapon back. Loeb stumbled and collapsed to the floor of the shower, then got to his feet, unlocked the door, and staggered into the open corridor, croaking for help. Day followed and handed the razor to guards, who were pouring into the passageway.[23]

Doctors tried a blood transfusion to save Loeb, but life bled out of him too quickly. Day had punctured his windpipe, severed his jugular vein, driven deep into his neck and torso, and sliced a foot-long gash in Loeb's abdomen.

Chicago readers awoke the next day to a front-page, banner headline: "Kills Loeb; Prison Scandal: Fellow Convict Slays Slayer of Franks Boy."[24] Newspapers used the opportunity to retell the troubled history of this prison—riots that ended in a half million dollars in damage, guards murdered by prison gang leaders, and vice versa. They also retold the story of Bobby Franks's murder, and repeated many rumors about Leopold and Loeb's pampered prison life—that they were served specially prepared meals in a separate cell they used as a lounge, and that Loeb had a private shower. At one point, the main article in the *Chicago Daily Tribune* repeated a rumor that "Leopold and Loeb were to be freed through a plot in which they were to be reported as dead," which conjured images of Loeb skulking off to the Caribbean while someone else's body was buried in his place.

In reality, Loeb made an enemy of Day by hounding him with improper advances. Day wanted Loeb to stop hitting on him and asked to talk with him about it. Loeb insisted that the conversation take place in the shower, where he pulled the blade and tried to rape Day.[25] A jury found Day's testimony believable and acquitted him of murder charges.

Clarence Darrow said of Loeb's passing: "Death is the easier sentence compared with a life behind the walls of a prison. He is better off than Leopold, better off dead."[26] Darrow assumed Leopold would never leave prison. He had no way of knowing that malaria, not death, would set Leopold free.

. . .

WARDEN Ragen used Loeb's death to impose state-sanctioned authoritarian rule. He tolerated challenges from no one and demanded submission and loyalty from everyone. "He was feared and respected, beloved and despised. . . . Stateville was his sole interest," wrote James B. Jacobs, a penal expert and author of a book about Stateville prison.[27]

Ragen razed all makeshift shelters on the prison's sixty-four-acre property, instituted a "complete eradication of clubhouse comforts," and broke up "cliques among favored prisoners." He reduced the population to a comfortable 3,033; everyone had just one cell mate in the smaller cells, instead of three bunks stacked to the ceiling; everyone was put to work; and daily activities occurred under "strict military discipline."[28] No more fancy meals and accommodations for the "big shots." Everyone walked in lockstep to "feedings"—the officers' word for meals, as if leading animals to the trough. Everyone used the same tin cups and plates, and sat "in the company with the lesser array of robbers, killers and thieves."[29]

Ragen took prison employees off the political patronage system, and instituted standardized pay and merit-based promotions from within the ranks of current employees, thereby ending the revolving door of political hires.[30] To curb violence, Ragen ran mattresses through metal detectors to confiscate hundreds of hidden weapons—knives, scissors, saws, clubs, and even guns. To stop escapees he created a rapid-response team made up of prison guards and state police, trained to quickly—within minutes—lock down roads and buildings in a fifty-mile radius.[31] To bring employees into *his* favor, Ragen held goodwill picnics and rewarded loyalty with promotions. For prisoners, he helped reform the parole process to make it more flexible—to focus on the man and his prison record, not the crime he committed.

Inside the prison's four drum-shaped cell houses, convicts were under constant watch, even when they slept, used the toilet, or masturbated. In a courtyard at the center of each round building, which they called roundhouses, stood a central tower that prisoners called the birdcage because it had panoramic views into all 248 cells on all four floors, which they called "galleries." Panoptes's hundred eyes gazed on everyone in these panopticon structures. Prisoners couldn't see well into the birdcage, but guards inside, at the gun ports, could take reasonably good aim at everyone. To even look out of line was to invite "meditation," Ragen's name for dungeonlike isolation cells, which convicts called the Hole.

Here, Leopold was just another murderer among many.

In less than two years, Ragen had worked a miracle. "The job of making a prison out of the Illinois penitentiary at Stateville is complete," announced the *Chicago Daily Tribune* in March 1937.[32]

NATHAN Leopold served as an X-ray technician in the prison hospital when Ragen agreed to the malaria experiments. Alf knew nurses would never sign up for the daily indignities of those clanking steel gates, gun ports, guard towers, and line after line of leering convicts. So he planned to use prisoners not just as guinea pigs. He would teach microscopy and drug administration to the smartest men there. In time, they would learn to dissect mosquitoes and read blood slides. He needed men with a knack for the tedium—men who wouldn't tire of it, but be jazzed by the kingdom of animacules dancing on a glass slide.

Setting up the lab went smoothly, because Ragen liked the affable Alf Alving and wanted the prison to do as much as possible for the war effort. His convicts were donating blood, buying war bonds, and making uniforms for the army and navy. Now they would also find a cure for malaria.

Unlike Warden Sanford at Atlanta, who brought the whole body of prisoners down to hear Coatney's pitch, Ragen felt a less democratic process would work at Stateville. This was a very different kind of prison, for repeaters and long-termers. These were hardened criminals under heightened security who had had their freedoms obliterated. Getting volunteers would require political finesse from within the inmate hierarchy. He set up a meeting for Alf with key convicts so he could ask them to help gauge whether their fellow cons would sign up. As Nathan Leopold continued to be among the most influential prisoners there, and probably the smartest, he sat in on this meeting with six others—which took place in September 1944. Alf explained how malaria was taking more soldiers than enemy fire, that the project had discovered some promising drugs and needed volunteers willing to take them in heavy doses to measure toxicity and activity against malaria. Alf said he would need about two hundred volunteers.

Leopold spoke up first. He said he was "confident there would be no difficulty in getting twice or three times that number."[33] The following day signup sheets were posted at the four cell houses, and 487 men volunteered.

ALF took over the entire third floor of the prison hospital. By October 13, 1944, he had eight prison assistants helping him prep forty prisoner volunteers—all of whom were between twenty and forty years old, in good health, sentenced to serve at least eighteen months (the expected length of the study), and "malaria virgins"—that is, not from a malarious region. They were all white, as Alf started out working with the McCoy strain of *vivax*. Black prisoners couldn't be used because of the mysterious immunity most people of African descent had to *vivax* malaria. (It would be decades before the discovery of Duffy negativity as the reason for the resistance.) The McCoy strain was taken from one of Coatney's neurosyphilis patients in Milledgeville, Georgia, and flown to Alf for his experiments.

After Alf's prisoners suffered through the fires and ice of full-blown malaria, each group took a different drug, all amino-methyl concoctions with varying side chains and different survey numbers: SN-5241, SN-8867, SN-9160, and SN-8617. These drugs were derived as analogues from a series that had shown promise, then disappointment in the months before the sontochin epiphany. The project needed to confirm their relative uselessness before shelving them in the "open files"—the purgatory for failed compounds. The studies also gave Alf a needed test run to get him acclimated to the protocols of malaria, and give his team of conscientious objectors and prisoner technicians experience transmitting malaria by way of infected blood—using the very techniques developed by Jauregg. These rookies also needed experience tracking rectal temperatures, taking routine blood samples, and counting parasites under a microscope.

Some of Alf's convicts turned yellow; some vomited; some had lowered heart rates; several broke out in a rash—a "bright red popular" eruption. None were cured. All were in the hospital for at least two weeks.[34] Doting on them was Alf's team of army doctors and prisoner technicians. Nathan Leopold, among them, proved to be sharp, efficient, and technically reliable.

But Alf did not like him.

He needed Leopold because of his power over other inmates. Life at Stateville had its own rhythms, its own societal structure, and Leopold had the whole thing wired to his advantage. He was brilliant at controlling people. He even used the malaria work to manipulate sex out of some of the guys; Alf was certain of it. Alf didn't like it but what could he do? Sexual favors were traded like a commodity, a symptom of prison life.[35]

Alf's project freed Leopold from the eyes in the birdcage and marching in step, and the constant threat of getting tossed in the Hole. Alf Alving, in Leopold's rubric, would also potentially help him get out of jail—by writing flattering letters to his parole board. So he sweet-talked Alf and tried to manipulate him. He studied Swedish, and mastered it enough to talk with Alf. He even wrote Alf a letter in Swedish. Alf knew what Leopold was up to. So while he recoiled when this vile man used his cherished language to try to manipulate him, he didn't show it. Alf was smart, too. He played along—an interloper in this society of convict citizens. His goal was to "harness" Leopold in a way that served the project's needs, which were really the needs of the war.[36]

ALF ended up liking many of his inmate assistants, especially Kenneth Rucker and Charles Ickes. Both were terrible men—by standards on the *outside*. Rucker murdered a black man and wondered, under his breath, in hushed conversations, why the jury convicted him. "He was just a nigger," Rucker told Alf's young son, Carl, one day. Carl didn't know what to say. Years later, as a medical doctor and research scientist for Walter Reed Army Institute of Research, Carl could still see the venom Rucker spit when he said the N-word.

Warden Ragen and the parole board felt inmates who worked for Alf became better men, including Rucker, who was an excellent lab technician—so good that Alf listed him as an author on his papers. Charles Ickes was also an excellent technician and named in Alf's papers. He didn't kill anyone but had done time over and over, first for auto theft and then grand larceny. In 1939, at age thirty-one, he was arrested again for robbery, labeled a "habitual criminal," and given a life sentence.[37] Despite his third-grade education, Ickes caught on to lab work fast. He cheerfully endeared himself to Alf and became his friend. He was the opposite of Leopold.

Also working for Alf were other hardened criminals who found purpose under Alf's supervision. John Dempsey murdered a policeman in 1928, received a fifty-year sentence, and was sent to the old stone prison in nearby Joliet. But while testifying against his accomplices, he escaped. Three years later he was caught and sent to Stateville. Hurlon "Jack" Burlison at age twenty-one was part of a Chicago street gang that shot and killed a police officer while robbing a café in December 1931. He was sentenced to fifty years. Joseph Russano was a popular boxer who, at age twenty-three, got a life sentence for slaying a police officer in 1939 while attempting to rob a couple in Chicago's Humboldt Park.[38] Harold Rose at age twenty-five beat a man to death during a quarrel and, in 1941, was sentenced to life.[39] Charles P. Stein at age twenty-two killed a man while robbing a restaurant and, in 1933, was sentenced to ninety-nine years. The judge said to him: "You are the most cold blooded, unemotional defendant I have ever seen and should be locked up so that you will never bother society again," perhaps because Stein, after hearing his sentence of ninety-nine years, asked the judge whether he could have the $40 he stole during the robberies so he wouldn't "go to the pen a pauper."[40]

In white orderlies' uniforms these convicts manned a hospital room that served as Alf's laboratory. They gave malaria and experimental drugs to other convicts, including Joseph Di Chiari, sentenced to ninety-nine years in 1935, and George Ross Wagner, sentenced to ninety-nine years in 1940—both for murder by arson; Ralph Denning, sentenced to ninety-nine years in 1934 for murder and robbery; Frank Bialek, sentenced to ninety-nine years in 1933 for the murder of a policeman; David Allen, sentenced to 199 years in 1931 for murder; Leopold Jones, sentenced to life in 1937 for habitual crimes; Richard Kamrowski, sentenced to one year to life for robbery and sixty-five years for three charges of rape; Peter (Piccolo Pete) Pace, sentenced to 199 years in 1933 for the murder of a policeman; and Clarence James, sentenced in 1932 to one year to life for robbery and forty years for rape.[41]

The project grew and grew, until Alf had four army doctors supported by twenty-three conscientious objectors and a dozen inmate assistants. Some worked the malaria ward, where prisoner volunteers undergoing "treatment" and "relapse" experiments lay on iron cots with fresh white linens and pillows, burning through the furnace blasts and ice storms of their malaria infections. Under large windows with streaming sunlight they ate freshly made food from the hospital kitchen—which they devoured after their temperatures finally turned down and they could eat again. In another room on the third floor, lying on iron-grille cots atop crisp white linens, were healthy men waiting for their "date with Ann"—to be bitten by anopheline mosquitoes.

Leopold, Ickes, and the other prison technicians—which, in addition to Dempsey, Burlison, Rose, Russano, and Stein, also included H. Harrison, M. Klein, J. F. McGuire, S. R. Onesti, R. Perry, and C. Winchester—did the grunt work. On "biting days" they pressed small jars holding female anopheline mosquitoes onto the bare skin of fellow convicts. The jars had featherlight screens through which the insects landed, tilted their butts in the air, and began to drink. But before their bellies filled with blood, Leopold and the others removed the jars, moved to the next man lying on a cot, and pressed each jar's half-fed mosquito against his flesh. This went on and on and until each jar with a mosquito bit ten men and each man had been bitten ten times.

This is how Alf controlled exposure. Each test subject had been bitten exactly the same number of times by the same mosquitoes, so their infections would be of similar intensity and the measurement of drug activity reliably comparable.

In early 1945, disaster struck. Alf started a healthy, hearty inmate on a drug cataloged as SN-8233. A few days later he died.[42] "Nolan" Milam had queued up to volunteer with no visible issues.[43] As he sweated through symptoms and complained of chest pains, the staff quickly took him off the drug, which was an analogue of the dangerous blood-choking German drug plasmochin. Milam, who was toward the upper end of the age limit at thirty-nine, had an undetectable heart condition and couldn't handle the stress. Mortified, Alf shut down his work to await the autopsy report, which confirmed that Milam died of a heart attack.

Alf was told to say nothing. The death had to be contained. But Milam had a wife, now a widow, and he was part of the prison community, with friends wanting to know the truth. That he was incarcerated for murder was beside the point. Alf didn't differentiate based on the crime committed. He appreciated the men's dedication to the cause, which Milam showed just by being there. The other volunteers needed an explanation of what happened. As word spread quickly through the prison, more than three thousand inmates filled the yard to ask questions and demand answers. Morale around the project plummeted. And the guys stopped showing up for their scheduled experiments.

Alf pleaded with Washington to act. Finally, in July, a month before A-bombs incinerated Hiroshima and Nagasaki, Milam's widow received an explanation, written by Dr. Richards, chair of the White House's Committee on Medical Research. The letter explained the circumstances of her husband's heart attack, and offered the government's sincerest sympathies and adulations for the patriotism her husband showed in volunteering for this important war-related effort.

As soon as Alf received a copy in the mail, he brought it to Warden Ragen, who posted it on the bulletin board of the prison hospital. At the same time, 150 of Alf's prisoner volunteers and technicians entered the yard as "missionaries," proselytizing the good works of Alf Alving and his team. "We heard through the grapevine that there was considerable argument for a day or two. The end result, however, was most astonishing. We have had quite a number of new volunteers who were converted to the worth-whileness of our experimental work. . . . We feel there has been slight, if any, perceptible diminution of enthusiasm in the prison," Alf wrote in a letter to Richards, personally thanking him for defusing what had grown into a smoldering, soon-to-be-explosive situation in this volatile prison population.[44]

ALF grew increasingly comfortable with the prison culture. Soon he would bring his spry, witty son Carl with him regularly. Carl was his father's sidekick, precocious, bright, and brave—enough to want to hang out in this scary prison. Carl wanted to do everything with Alf. At age six he saw his life ahead of him—a medical doctor and researcher, just like his dad. While Alf disappeared into his malaria work, Carl received red-carpet treatment from the warden. He walked with Ragen on rounds to the giant drum-shaped cell houses and cafeteria. He marveled when the men stood at attention and all but saluted as Ragen walked past, Carl skipping alongside, wide-eyed and impressed. Then, once they passed, the cons returned to their "hang-dog look," to march back to tiny cells and the watchful eyes of the birdcage.[45]

Ragen showboated for Carl by identifying the men's heinous crimes—the elevator operator to the warden's apartment, a murderer; the warden's personal barber, who shaved his neck every morning with a straight razor, another murderer. When they walked past lines of prisoners, Ragen would point to this one, a rapist, and that one, a bank robber, and the next one, a notorious gangster. He told Carl funny stories, too. One convict had woven years of squirreled-away dental floss into a ladder long enough to scale the prison's wall. Guards shook down his cell and found it a day before his planned escape. Then there was the clutch of conspirators on kitchen duty who brought coffee to the gun towers, as was the routine. This time they spiked it with a sedative stolen from the hospital. "One after another the guards passed out. But one guy didn't drink his coffee and foiled the escape," Ragen boasted. He told these stories with glee, and humor, and pride. No one left his prison except through parole. And those who tried to leave without papers ended up in the creepy, dungeonlike Hole, shackled to steel bars.

Carl eyed Stateville's criminals with trepidation. But none conjured more fear in him than Nathan Leopold. Leopold looked right through him and left a cold chill, like something dark had just taken inventory of his mind and soul. He couldn't open his mouth when Leopold was around, or even look into Leopold's eyes. He was just a boy, younger than Bobby Franks, and appropriately spooked.

CHAPTER 27
Taking Command

On September 27, 1944, the *Daily Mail*, Paris edition, reported: "The most evil secret weapon yet conceived has been put into operation in Italy—and it has been conquered. Hitler's plan . . . was this: Deliberately to propagate the malaria mosquito in the hope that troops would be as helpless against the scourge as those tropical populations who used to die in the millions." The article said the Nazis made "sour bogs" with the help of German entomologists, referring to Erich Martini, and that the stage was set to poison Allied troops with malaria. But "Colonel Paul Russell . . . chief malariologist to the Allies" had completely neutralized the threat.[1]

RUSSELL had become a public health war hero.

He made his first trip to Italy in January 1944 to assess the Nazi sabotage of the bonifications and resulting floods. He steamed in from Bizerte on a rickety old army ship, the *Shamrock*, which had already capsized once, served as a transport vessel in the Oran landings, and was just four years from being turned into scrap metal. Now, reoutfitted for hospital duty in the Mediterranean, it was painted white with a red cross. Unstable, uncomfortable, and small, it anchored in the harbor under Vesuvius, next to a sunken steamer turned into a pier. During this short trip, Russell watched as Fred Soper dusted thousands of civilians with DDT to stop a terrible typhus outbreak; spent a night in a cave that had been turned into an air-raid shelter; got shaken from bed by an early-morning German raid on Naples and the harbor; and toured the bonifications, which he called "crippled."[2] He then returned to Bizerte to set up the Allied Forces Malaria Control School, for training new survey and control units. By mid-March he was back in Naples to reverse the Nazis' horrific attempt at germ warfare.

While Russell discussed plans with Soper during his first night back, Nazi bombs fell all over the town and harbor, killing many civilians. One shell whizzed past his face and blew a hole in the street below. Covered in plaster and glass, he ducked with everyone else into an air-raid shelter. Then Vesuvius came "alive" with the "greatest lava flow in seventy years." It exploded on March 22, raining cinders three inches deep on Salerno. At nine a.m. the skies were "dark as night."[3] He drove around taking inventory of the wrecked countryside while ash fell like snow and "thousands of peasants" searched for shelter and food. On an icy mountain road he almost careened over a cliff. But his old army command car had outmoded running boards that grabbed and held the retaining wall. While the front wheels dangled over a twenty-five-foot drop, he pulled the driver into the backseat and with "knees knocking" they "hastily" exited the car through the rear door. They hitchhiked while Vesuvius erupted in a "cauliflower of billowing smoke and cinders, changing into an enormous toadstool, then a spectacular fountain."

He met with malaria experts in different locations through the next few days, dodging flak and falling ash, and getting rousted from bed by air raids and bombs. Four days after escaping the teetering car, while on his way to a malaria conference in Volturno, his driver intentionally drove into a ditch to dodge a lone German pilot flying straight at them, firing his machine guns. Again Russell escaped from a wrecked car into chaos, Vesuvius rumbling and spewing black ash, Russell navigating around one disaster after another in a place that appeared caught in some kind of biblical end.

By May, however, things quieted. Vesuvius calmed down. The air raids turned up dry, with fewer and fewer actual bombings. Summer uniforms arrived, marking the start of mosquito season, and Russell's reason for being there.

The German campaign to flood the countryside had turned ninety-eight thousand acres of fertile, agriculturally productive land into soggy bogs.[4] Thousands of homes stood half-submerged. Villagers walked through waist-high, sewage-filled waters trying to rescue belongings. The Germans applied surprising specificity to their destruction. They stole the pumps and dynamited the channels; blew to bits flat-bottom boats that cleared the lakes of vegetation to prevent mosquito breeding; and stole nine metric tons of quinine from the Department of Health building in Rome and hid it all in Volterra.[5]

MALARIA, now a biological weapon, lay embedded in an environment that Allied troops would soon march through to pursue the northward-advancing Germans. And *that* presented the perfect conditions for a definitive test of the new wonder insecticide DDT. The mosquitoes were coming, the population was unprotected, and medical supplies ran low. An old factory with industrial mixing capacity blended thousands of tons of Paris Green, which was sprayed from P-40s. Russell and his colleagues also had about seven thousand gallons of oil mixed with different proportions of DDT.[6]

Florida labs had demonstrated this chemical's miraculous action against mosquitoes. Once the stuff was absorbed into their cells, the poison worked slowly and eventually they all died. Controlled lab tests showed that the killing effect of indoor residual spraying—saturating walls, floorboards, and ceiling shafts with DDT and oil—lasted months. When a mosquito landed to digest a blood meal, DDT poisoned her. No one knew yet how well aerial spraying would work against larvae growing in the huge mosquito-breeding bogs created by the German sabotage. But Russell would soon find out.

He worked though channels to convince General Mark W. Clark to hold off on moving his troops northward, to give malaria control and survey units time to sanitize the region, warning that the Nazis had created an environment that would expose Clark's entire army to malaria.[7]

If this were the first serious malaria threat of the war, Clark might have dismissed Russell as overly cautious, as commanders had done in the Pacific. In Italy in 1944, however, the dramas that played out on the other side of the world—especially Bataan and Guadalcanal—were remembered. General Clark got it; he understood what malaria could do. Of course he would wait. Russell said he needed a month.

Then he moved to Rome to get started. His first act was to find Italy's chief malariologist, Alberto Missiroli. He wasn't on the blacklist, which meant his "skirts were clean"—no Fascist taint.[8] Russell took him on as a consultant. They "went to work at once on malaria control in the Roman Campagna."[9] As fast as the rising armies of *A. labranchiae* molted off the flooded bogs, Russell's units ran air raids to shoot them down—spewing DDT out the back of P-40s. They also restored the damaged hydraulic systems,[10] and taught local laborers to do door-to-door residual spraying. DDT proved miraculous. This crash program sanitized the entire Campagna and Lido di Roma—in just three weeks. Russell's team made the area safe for General Clark's massive army a full week ahead of schedule.[11]

ON June 6, 1944, while General Clark's men drank wine, visited the brothels, and enjoyed R & R in the sanitized Lido di Roma rest center, more than seven thousand vessels crossed the English Channel to land 1.3 million Allied troops on the French shore. Many were massacred on the booby-trapped beaches of Normandy, but most survived. D-Day had commenced and a new portal into Europe had opened. They had breached Hitler's Atlantic Wall; his fortress was now crumbling. Allied headquarters in Africa and the Mediterranean shrank as the Germans retreated. Everyone eventually packed up and moved north. The malaria problem had thus been solved, for the military.

They left an Italian countryside flung back into the nineteenth century. That summer in southwestern Italy, cases of malaria exceeded a hundred thousand in a population of 245,000.[12] But it could have been worse. Within a year, using the know-how and tools brought to Italy by Russell's Malaria Control Branch in partnership with the Rockefeller Foundation, the Italians would soon reverse the devastation. The successes of the American experiments with DDT convinced Missiroli to launch a large-scale experiment in Latina and a subsequent five-year plan of action announced in 1946 to eradicate malaria from Italy.[13] Then Missiroli and the Rockefeller Foundation, with Russell back in his position there, applied these new techniques in an "internationally observed experiment" that eliminated malaria from Sardinia.[14] This would become the template for a worldwide eradication effort in the decades to come.

IN the Southwest Pacific, Russell's malaria survey and control units attached to commands in every location. General MacArthur ordered as many units as he could get. By D-Day he had sixty of them working with his battle commanders. The War Department's propaganda machine had worked magic. Even as soldiers tired of the relentless messages and "ridiculed" and "booed" the silly cartoons of dim-witted Snafu getting his rear end bitten by Ann, they changed their behavior.[15]

Atabrine discipline also improved. By sheer force of will, Allied troops turned yellow, and called their daily lineup for a pill "atabrine parades."

Sent to work in New Guinea were fiercely dedicated and competent American and Australian malaria control and survey units. Major Jesurun's 4th Malaria Survey Unit was joined by another thirty like it, matched by an equal number of control units. Denton Crocker, who had spied the nearly naked dancer in the French Quarter during training in New Orleans, finally shipped out to New Guinea in January 1944 as part of the 31st Malaria Survey Unit. He traveled by transport, passing through the locks of the Panama Canal, spying hundreds of porpoises and sharks, and thousands of man-o'-war jellies. His transport passed the toothy Marquesa Islands and stopped at exotic Bora Bora, where the only cloud in the sky hung over the cone of its extinct volcano. Captain Cook, Paul Gauguin, Joseph Conrad, and Somerset Maugham stirred as Crocker devoured books about the South Pacific.

His unit arrived in Milne Bay, at the eastern tip of New Guinea's tail, on February 25, 1944. The bay hummed with cargo ships, destroyers, and submarines, plus natives in dugout canoes with sails. Bombers and fighters patrolled the skies over this established, sanitized camp. Crocker's unit included Bill Brown, a Harvard entomology student who was an expert in ants. "His first act ashore was to reach into a shirt pocket, pull out forceps and a vial of alcohol, and begin collecting ants."[16] These men still had just undergraduate science degrees. But they'd go on to earn PhDs, write books, and become professors at prestigious universities. Unlike the units that came before them, they arrived with their equipment—nine truckloads of it, plus a twenty-four-foot long equipment tent for the 30th, 31st, and 32nd Malaria Survey Units. They chased cuttlefish, zebra fish, squid, moray eels, sea urchins, sea slugs, brittle stars, serpent stars, sea gar, giant sea cucumbers, leopard sharks, and, on land, two-foot lizards, huge biting grasshoppers, giant fruit bats, the marsupial bandicoots, wallabies, and twenty-foot snakes.

But mostly they did their work.

Crocker grew a beard that made him look like Pierre Curie. His twelve-man unit, meanwhile, grew tight. They slept in pyramid tents on cots with mosquito bars and nets. They patrolled for larvae and committed their murderous spilling of Paris Green to wipe the insects out. When DDT arrived, their work went faster and easier. They brought knapsack sprayers to the villages and trained locals on how to use them. Then they moved to Finschhafen, along the northern coastline. There they set up tents on a bluff overlooking the sea and synced up with the 38th, 39th, and 59th Malaria Control Units, which included men from Cornell's biology department. Together they did their mopping up of malaria and moved onto Hollandia, just after the bloody battles to take the area from the Japanese. By September 1944, New Guinea was a staging area for moving Allied troops and sailors into the East Indies.

Bug bombs were fashioned inside netting over slit trenches so troops were protected from mosquitoes while relieving themselves at night. They kept repellent in their shoes while sleeping, so they could grab it—and use it—as they fled for foxholes during air raids. No one saw a movie without a thorough dousing of repellent. Every soldier sick with malaria was sent to a medical officer, who filled out a card with the soldier's name, rank, serial number, and type of malaria. Theater surgeon staff compiled the cards and wrote reports on whose men were getting sick, then sent warning letters to any commanding officer whose unit exceeded fifty infections per thousand per annum—which amounted to no more than a few infections per company.[17]

These units became part of the propaganda. In circular letters to other theaters, New Guinea was listed as having a "secret weapon" against malaria. But really it was just these hypervigilant survey and control units using "rigorous application of all known malaria control measures."[18]

Medical officers were trained to see the signs of atabrine-induced exfoliative dermatitis—the new name for the skin rash this drug sometimes caused—with orders to switch these soldiers to quinine. (The early hoarding of quinine in 1942 provided enough of it for this limited use overseas.)

From May 1944 to the end of the war, malaria reports from the Southwest Pacific read like checklists that itemized the surveyed mosquitoes, mitigation methods used, number of "natives" employed to do the ditching and clearing, type of anopheline mosquito responsible for transmission, number of gallons of Paris Green and DDT used to kill larvae, number of villages sprayed with pyrethrum and DDT to kill mosquitoes, and the malaria rates—which were always under fifty per thousand per annum.

The army surgeon general ordered a theater-wide study to determine atabrine's role in reducing malaria rates here. The report said atabrine was working, but the mosquito populations had been so decimated by the malaria units that the conclusion was flimsy, and read more like propaganda than a serious study of atabrine. In reality, a combination of efforts created an unstoppable force that was given superpowers by this new agent, DDT.

CHINA and Burma would never get a handle on their malaria nightmare. They had nearly thirty malaria control and survey units. But these units were scattered across uncharted jungles, trying to convince three hundred thousand mostly Chinese troops to protect themselves from mosquito bites with wholly substandard supplies.[19] The mesh netting for screens was so loosely woven that a large spider could crawl through. Theater malariologists sent samples of the mesh to the army surgeon general's office, begging for better supplies. They had no pyrethrum, and no hope of getting DDT for residual spraying anytime soon. Nearly 100 percent of "noneffectives" were made that way by malaria. And the infection rate for Chinese troops stood between three and ten times that of the Americans, depending on location. Pamphlets and posters in Chinese didn't get the message out, because most of the Chinese troops were illiterate. Many could not be convinced that mosquitoes carried malaria; they believed the fevers came from "wet feet, uncooked pork, etc.," wrote one frustrated malaria survey officer.[20]

This theater had just too many hurdles. But Japanese forces there were also crippled by malaria; Allied troops were at no greater disadvantage. They just fought through the fevers. Evacuation of men because of malaria ended; each combat commander simply sent feverish troops to medical tents for atabrine, then returned them to the front until their relapses broke through and they needed another round of treatment. More quinine was shipped to this region for the unusually large number of skin reactions blamed on atabrine, which investigators began to think was somehow connected to "jungle trees."[21]

D-DAY changed the focus of the Malaria Project. Investigators still huddled in monthly meetings, arguing over the value of bird studies, and the proper procedures for toxicity tests in mammals, and whether the 8-aminoquinolines were worth anything. But D-Day marked the beginning of the end for Germany and Japan. No one talked of finding a drug better than atabrine for troops. The awful yellow pills would get the Allies through the rest of the war.

Malaria Project investigators turned their focus to the war's aftermath.

Atabrine was a terrible peacetime drug. The side effects had to be downplayed for the war's sake. But they were pretty harsh, especially the psychosis and scary skin eruptions. A much better drug was needed to extinguish the parasites coming home in the men's blood. Everyone hoped that the captured sontochin, or one of its analogues, would be *it*.

CHAPTER 28
Klamath Falls

Lowell Coggeshall, by now, had grown matter-of-fact about malaria; he didn't need studies to understand it. He had communed with the disease for decades, which distanced him from the frenetic men running the Malaria Project. These other scientists didn't follow their instincts, because they had none with respect to malaria. They were data-driven and grew rigid, and they had a new partner, Brigadier General Neil Hamilton Fairley, Australian Army Director of Medicine. He proved that the best study results came from healthy men put under combat conditions, using wild malaria taken from troops fresh off the battlefields. He ran a huge camp in Cairns where his staff raised seventy-five thousand mosquitoes and gave malaria to nearly nine hundred Australian soldiers who volunteered for drug trials.[1] He produced results using sulfa drugs that disproved Lowell's theory that they might be useful as a prophylactic against malaria. Fairley's atabrine studies also led Shannon to understand key factors in absorption. His studies were superior to the Americans' because he had such a huge number of healthy soldiers to experiment on, which in 1944 inspired the U.S. Army to set up similar-sized drug studies, including one at Fort Knox. Fairley's work won him a spot on the Malaria Project, and everything he did and said mattered. He was another titan at the meetings.

George Carden managed diplomacy between the elite researchers of the project and the U.S. Army and Navy investigators, whose studies weren't always helpful. Meetings often turned blustery with all the big personalities. Lowell went to verbal blows with Coatney at least once, which was very much in Coatney's personality, but not Lowell's. He liked to laugh, bring people into his inner circle, and develop lasting connections. The ice storms at these meetings grew increasingly frigid. Lowell's considered opinion, while still listened to, mattered less and less.

So perhaps he felt relief when, in January 1944, he left to start his commission as a naval reserve commander and take control of Admiral McIntire's experimental rehabilitation center. Lowell packed up his family and drove from Michigan to the rugged lumber town of Klamath Falls, in the cool, dry mountains of southern Oregon. This was cowboy country south of Crater Lake, and within view of the snowcapped peak of Mount Shasta.

Lowell oversaw construction of eighty buildings on an eight-hundred-acre compound carved into a hillside a thousand feet above downtown Klamath Falls—four miles up along an old wagon trail called Old Fort Road. This facility was created for the sole purpose of treating malaria and filariasis picked up by marines and sailors in the Pacific. Its existence was a testament to the nightmare these diseases created for the military.

The town was on a railroad line for easy access to supplies, but was otherwise isolated—a day's drive from a major city. The people were insular but patriotic. After a well-honed effort to endear political and social leaders to the idea of a marine barracks in their hills, the townspeople opened their big, strong rancher arms and embraced everything about Lowell's camp. And Lowell embraced them back by joining the Rotary Club and huntsmen, with whom he shot geese, duck, and deer that his family ate for dinner. At home here, away from the self-important project managers and all-day meetings in DC, Lowell socialized and made friends. He even tried to convince the navy to build the barracks' competition-size swimming pool in Klamath Falls proper so the community would have a nice natatorium after the war, but the navy wouldn't allow it.[2] The effort, however, set the right tone for a solid, hearty relationship between this town of twelve thousand residents and the five thousand incoming marines.[3]

The average temperature was sixty degrees Fahrenheit, the average annual rainfall was twenty inches, and the average humidity was 30 percent. That made Klamath Falls the antithesis of the wet, hot hell of Guadalcanal, New Guinea, Bougainville, Tarawa, Samoa, Eniwetok, and Saipan. The crisp mountain air gave the men energy and appetites, so they ate more and regained strength lost during long stays in convalescent camps in the Pacific.

This was medicine, by Lowell's standards. The navy created Klamath Falls to make sure the men were *not* pampered. This was not a hospital. It was a camp to harden the bodies and minds of softening marines—so they could return to war.

Lowell was there to study and treat their filariasis and relapsing malaria, and try new compounds coming out of the Malaria Project to see if the drugs could stop the relapses. But he also educated the men about their diseases. He told his five thousand malaria victims and twenty-five hundred filariasis victims that they harbored self-limiting germs. He showed them images of their invaders and explained. These parasites couldn't live in their bodies forever. Eventually each man's immune system would expel every last microscopic monster. The more physically and psychologically fit they became, the faster these nasty germs would be flushed from their bodies as waste.

He likened the barracks to "a training camp for a football team." He said: "We don't want the men to overtrain nor do we want them to break training . . . they must be kept busy enough to eat well, and tired enough to sleep well. The stronger they are, the better they can combat any ailments."[4]

The navy had more than a dozen other special facilities in the United States for managing war casualties. They ran on very different philosophies, centered on the idea that the men would recover by way of mental and physical relaxation. But many just ended up deeply troubled and in special wards for psychotic tendencies. These rest hospitals were all over the country, from Monterey Bay, California, to Springfield, Massachusetts. By the end of 1945, they held eighteen thousand marines and sailors, most in need of psychiatrists.[5]

Admiral McIntire wanted to try a different format for malaria and filariasis, and came up with the plan to create Klamath Falls. Congress first approved $1.5 million for a twenty-five-hundred-capacity camp, but the head count of relentlessly relapsing malaria cases was tallied, and then filariasis cases from Samoa were added, and before the barracks were finished, McIntire realized he needed to double its size. Appropriations soon increased to $4 million.[6]

The first marines arrived on May 27, 1944. Half of them came from the convalescent camps in the South Pacific, where they spent too much time in bed, replaying the horrors of battle. The other half had come from the carnage on Tarawa.[7] As soon as they arrived, they were paid and given a thirty-day furlough to go home, rejuvenate their faith, hug their mothers, and find romance.[8] The married ones were encouraged to bring their wives back to live in off-base apartments, open to all ranks. The ones with sweethearts were told to get married and bring their brides back to Klamath Falls. This was Lowell's strong belief: Happiness was half the battle to recovery.

During the day they drilled and held parades. Lowell's son, Richard, watched with amusement as medical corpsmen, who also had to march and drill, turned in the wrong direction "like a high school drill team banging into each other."[9] The men also did "shit details," like clearing rocks from roads, shoveling dirt, and building fences. When they felt ill, and their commanding officers accused them of "goldbricking," Lowell intervened to explain that filariasis and malaria appeared in cycles, and when symptoms broke through the men needed rest. The marines grew to like and respect Lowell, whom they called Doc Cogg.[10] The men's strength and morale grew by the day. "A more active, healthier outfit of Marines couldn't be found this side of combat theater," wrote one observer.[11]

Many of the men complained because they wanted to go home for good, but Lowell needed to keep them there, to test new drugs against their relapsing malaria. So he worked with the socialites of Klamath Falls to make the place as inviting as possible. And soon the men were using liberal leave—two nights a week—to attend dances, dinners, and picnics organized by a proper, educated group of young women who called themselves the Klamath Commandos. They had a center at 815 Main Street and wore uniforms of navy blue, with a cap, white deck shoes, and bobby socks. They ran drills, had ranks, and maintained a duty roster. They even had a song set to the "Notre Dame Victory March."[12] They weren't party girls; they were socially appropriate, with a duty to make servicemen feel welcome.[13]

But prostitutes also flocked to Klamath Falls to work the saloons and entice the marines into their brothels, called the Palm Hotel, the Iron Door, Irene's, and Myrtle's.[14] Historian Richard Matthews said the marines he interviewed years later recalled "with amazing details the names, rules, and quality of the town's four most prominent houses of prostitution," Irene's being the best, because the women came up from San Francisco.[15]

Townspeople supported other servicemen from a nearby army camp and navy air station, "but the marines were combat veterans, and therefore garnered the most respect and attention."[16] This created a remarkably gentle transition back from war.

. . .

ABOVE town, inside the marine compound, Lowell led a staff of 150 and ran the hospital "like a bomber plant," taking care of patients with malaria and filariasis.[17]

The filariasis was easy; Lowell just had to treat the men's fear. They were terrified of the slim worms living in their lymph glands. Most, if not all twenty-five hundred of them, had either seen or heard of the effects of "mumu"—the dreaded elephantiasis. They had marveled at the deformed women with pumpkin-size breasts, and legs so swollen they looked like tree stumps. But most frightening were the Samoan men with clublike penises that hung to their knees, and the famous wheelbarrow-size scrota.

Lowell calmed their fears. He told them that biting mosquitoes spit microfilariae worms into them, but not enough to cause elephantiasis. If a mosquito spit in five microworms, the infected person only ever had five adult worms. The horrendous disfigurement of elephantiasis seen on Samoa—and by men frequenting Shangri-la and Paradise in Liberia—was the result of huge swarms of worms that had been delivered by infected mosquitoes over many years, one handful at a time, until the numbers reached such overwhelming proportions that they clogged the lymphatic system and caused the grotesque swelling. Only years of exposure and accumulation of worms can cause elephantiasis, which is why the deformities develop mostly in older people.

Lowell told the men that the swelling they experienced was probably an allergic reaction to the worms invading their lymph nodes, not actual elephantiasis. In some cases their sex organs grew to twice the normal size. Lowell applied an arsenic-based rub to test a theory—suggested by someone on the Malaria Project—that the poison might absorb through the skin to kill the worms.[18] It didn't seem to work. So Lowell pulled from his medical kit standard hot packs, which he'd first used as an intern in Indianapolis on the grocer's tarantula bite. The packs he pressed against the marines' swollen parts. As with the tarantula bite, they seemed to help, or at least didn't hurt, while offering great psychological relief.

Their infections were so mild that Lowell never saw microfilaria in blood smears, which meant they weren't infectious. No American mosquito would pick up the disease and spread it to others. He did, however, find the adult worms in biopsied lymph nodes. Some were calcifying, which meant they were slowly dying. He concluded, correctly, that the worms would be gone soon.[19] He shared this information with the men, which helped. But it didn't defuse a persistent rumor that filariasis caused sterility and impotence. Lowell couldn't prove the rumor wrong without a study. So he ran one; he told the married men to get started on baby making for the sake of science. Anyone who hadn't moved his spouse to Klamath Falls was encouraged to do so. Of the 249 married men with filariasis, forty-five within six months reported their wives pregnant.[20] "Mumu men have fathered twice as many babies as wormless veterans," he told a group of tropical disease experts, which was reported in *Time* magazine.[21]

The navy had diagnosed ninety-five hundred cases of filariasis from 1943 to 1944; eighty-two hundred of them were marines; the rest were Seabees and medical corpsmen. Not a single case developed elephantiasis. Of those sent to Lowell, most went back to the Pacific to finish out the war.

His malaria cases were an entirely different matter. They had contracted *falciparum* and *vivax* while overseas, but the *falciparum* burned out. The *vivax* persisted. Some of the marines had already suffered more than thirty-five relapses by the time they reached him.[22] And the attacks continued. These men were perfect "clinical material" for measuring a drug's ability to stop relapsing *vivax*.

THE Malaria Project by 1944 employed more than four hundred investigators working in three areas—synthesis of drugs, animal testing, and human studies. Lowell flew to DC for bimonthly meetings and found he quite liked this new addition to the clinical testing panel, Alf Alving from Chicago. They grew to be friends and enjoyed conversations, while Bob Coatney, a compact man of pent-up energy, rattled off precise meta-analyses of reports on drugs, and identified those not worth the paper they came in on. James Shannon offered yet more brilliant insights derived from close studies he made of the different strains and their reactions to the different classes of drugs in state hospital patients, while Australian researcher Fairley brought data on his big studies at Cairns that the Americans learned from and duplicated. British officials always sat hat in hand, needing chemicals to make their own drugs and run their own studies, as their labs in London struggled through shortages of everything. The meetings were crowded and loud, and filled with big personalities.

Lowell grew to trust Alf's work. When Alf's prisoners complained of blurred vision while on one drug, Lowell agreed to give it to his marines, and then take them out for target practice at Klamath Falls' new shooting range—and send the results to Alf. When Alf finished his work on one of the newer concoctions that seemed less toxic than the rest, Lowell asked to have that drug sent to him. When Alf took an interest in the plasmochin analogues—the 8-aminoquinolines—Lowell watched and waited to follow up on Alf's good work. Whatever the clever and curious Alf explored, Lowell wanted to follow.

THE problem all investigators shared was that time was running out and they still had nothing novel. Everyone involved needed this captured sontochin to *be something.*

But Winthrop put a hold on progress again. For two years Winthrop had delayed and withheld stocks of atabrine and plasmochin requested by investigators for their studies. Meetings routinely had agenda items with such headings as "Problems Related to Winthrop." At one meeting, a very frustrated investigator complained that after a long wait for five hundred grams of plasmochin, it arrived too impure to use.[23] The problem was mainly that Winthrop was having a hard time making these drugs in part because they didn't fully understand the processes and in part because of supply shortages. So project chemists in the university labs made needed samples for the studies.

On sontochin, Winthrop executives grew bold. They insisted that the company's patents be recognized for sontochin *and* analogues made from its base formula.[24] In exchange, the company would give out royalty-free licenses for the duration of the war (the same agreement it forced on the project during the struggles over atabrine). Winthrop refused to provide the samples of sontochin that Clark and Marshall at Hopkins had repeatedly requested. When George Carden inquired, Winthrop president Klumpp said the holdup was Carden's fault; the contracts he sent failed to adequately recognize Winthrop's ownership of this entire class of drugs (sontochin and all other 4-aminoquinolines).

Winthrop also held up progress on the 8-aminoquinolines (plasmochin and its analogues) on the same claim—that it owned the patents and the project needed to make clear that these drugs belonged to Winthrop. But Bayer in Germany had written the patents for Winthrop to file.[25] This was shaky ground on which to make demands of ownership, during a time of war.

By early 1945, Carden's patience had run out. Back during the atabrine muster, he was fairly new to the project, and the army didn't want him to do anything that would offend Winthrop and slow production of atabrine. Now, however, he was executive secretary of the board and a trusted confidant of Dr. Richards (head of the Committee on Medical Resarch). Vannevar Bush, head of OSRD and the boss of everyone working on science projects for the government, from malaria to the atomic bomb, wanted the situation resolved: "I am disturbed about the whole antimalarial affair, and I believe that it needs rigorous and wise handling," Bush wrote to Richards.[26] So Richards turned the legal struggle over to Carden, who handled it so well it won him the coveted position of chief of the Malaria Division of the Committee on Medical Research.

He canvassed the Malaria Project's pharmaceutical consultants at Sharp & Dohme, American Cyanamid, Dow Chemical, Eli Lilly, Abbott Laboratories, Merck, and Parke-Davis and Company.[27] He asked them to comment on a plan being explored with members of Congress that would rewrite patent laws so that certain protections applied only if the patent owner developed the product to the point of establishing its usefulness.[28] The consultants, of course, worried that a new law to get around this one problem might "open a wedge to government control of industry" and "might not be entirely justified . . . since there are no instances to date in which companies other than Winthrop have materially benefited by the work done by OSRD contractors."[29] Carden agreed, but in the process got the message across that the government was ready to play hardball with any company staking a claim to a drug they didn't know how to make.

Mansfield Clark at Hopkins, who set up the agreement with drug companies back in 1942, promising to protect their intellectual property, knew what Carden was up to and felt violated. According to historian Leo Slater, Clark believed in such old-fashioned notions as a gentlemen's handshake, and he was crushed by the breach of an agreement he established. He saw chemists as artists, making new molecules out of scraps of synthetics and dyes. In this case, their art took aim at a disease, to save people. The commercial and nonprofit chemists deserved to own their work, even if public funding was used in the process. Clark also believed the drug companies would act honorably and sell the drugs made under these circumstances at reasonable prices. The project had to honor the companies' rights. Or, Clark said, he would resign.

But Bush and Richards came at the problem from the perspective of the Committee on Medical Research, which was an arm of the White House. Their jobs were as political as they were scientific. They were writing checks to fund research that was taking Winthrop-patented drugs through needed trials, from bird studies to toxicity tests in dogs to human trials. No way could they allow Winthrop to own and profit from all this federally funded work—even if that was the original agreement. So Richards "passive-aggressively" failed to respond to Clark's threat, and this heartbroken, principled, aging chemist resigned from one of the most important posts of his career. Clark's friends on the committee pleaded with him to reconsider. The whole thing blew up into an emotional mess. Robert Loeb, one of the leading voices on the project, but a softer personality, resigned as chairman of the board in a show of support for Clark, whom Loeb called "about the dearest, finest and most lovable soul that ever lived." But pleas from just about everyone convinced Loeb to stay.[30]

Vannevar Bush finally stepped in and eventually convinced Clark to stay on as one of the project's lead chemists, even if, on principle, he could no longer act as chair.

In the middle of this drama, OSRD lawyers found a possible solution. Project scientists at the University of Illinois developed a specialized method for making a specific side chain for an analogue of sontochin, SN-7618. (This was sontochin with a methyl group removed and a chloro group added—the one the Hopkins chemist Kenneth Blanchard wanted to study way back in 1942 when he asked to explore the shelved drug SN-183, which, at the September 1943 meeting, everyone realized *was* sontochin.) If the methods to make SN-7618 proved to be "patently novel," they could be used to protect the government's interest in the manufacture of this drug. Winthrop would have to come up with its own way of making it, because the government would have the right to block the company's use of these highly technical processes. This became the kink in Winthrop's patent armor.

Winthrop, reluctantly and somewhat bitterly, buckled to allow license-free manufacture of their German drugs. Company executives had no choice, as the Justice Department had stepped in again, asking questions. And the U.S. Alien Property Custodian discussed seizing the patents as alien property, as was threatened two years earlier with the atabrine patents. That all quieted down once Winthrop capitulated. The 4-aminoquinolines would belong to the project with project managers knowing the public would benefit from the research.

. . .

JEAN Schneider's original sontochin wound its way through testing on birds and lower animals, and came out looking nontoxic and four to eight times more effective than quinine. But this "patently novel" drug SN-7618, with the chloro group added, looked just as promising. If it panned out, sontochin could be given back to the French. Jean Schneider was now an official friend of the project and corresponded regularly with board members about the project's progress on sontochin. At one point he wrote from Tunis to ask for his sontochin samples back, and to request additional stores of it so he could continue his research. He wanted Specia, France's premier drug maker, to have a product ready to bring to market when the war ended.

Project chemists had made some two hundred sontochin analogues stable enough to be tested on bird malaria.[31] Ten advanced to human studies and five showed the most promise (in addition to sontochin): SN-7373, SN-7618, SN-8137, SN-9584, and SN-10,751. Other non-4-aminoquinolines were still in the hopper: metanilamide (SN-11,147) and a few quinoline methanols, as well as the naphthoquinone and pantothenone groups.[32] But everyone clung to the hope that sontochin or one of its offspring would prove to be *the* drug.

All of the drugs had to be vetted. So all the major researchers got samples—Shannon in New York, Coatney in Atlanta, Butler in Boston, Fairley in Cairns, and Alf in Chicago, among others. They also went to the army's four malaria testing hospitals (Letterman, Harmon, Percy Jones, and Hammond). Samples also went to the naval hospital in Bethesda. And they went to Lowell in Klamath Falls, as a final stop to see whether they could cure the naturally acquired, relapsing *vivax* still plaguing his marines.

Alf Alving's team at Stateville was the most productive. His four handpicked army doctors, twenty conscientious objectors, and dozen inmate technicians could do much of the preliminary clinical work, and their skills just kept improving.

Nathan Leopold not only performed the tedious duties well, he worked around the clock. As much as Alf distrusted him, he admired Leopold's industry. Leopold helped maintain and handle a female rhesus monkey brought there for experiments, to see whether she could be infected with different strains of human *vivax*. If the experiments were successful, the need for human studies would dramatically decrease, if not disappear—as all the studies could be done on these petite, twelve-pound monkeys that looked like minibaboons, only with pretty gray eyes. In early 1945, Alf infected a prisoner named Wallie with a *vivax* strain from New Guinea, then let his blood boil with fever until it was teeming with blood-destroying merozoites—seventy-six thousand per cubic millimeter, "the highest count" Stateville "ever recorded."[33] A doctor from the University of Chicago injected two centimeters of this parasite-saturated blood into their small Old World monkey. Her species held the distinction of being anatomically close to humans, making her kind extremely popular with scientists, who always tortured them with drugs or vaccines or diseases, and eventually even sent one into space.

Alf's female rhesus became immediately ill from Wallie's blood. Using thick leather gloves to protect against her bites and scratches, Leopold helped strap her to a table and then sliced her ear for blood that he studied under a microscope. It was filled with *vivax* parasites, but all were still inside Wallie's blood cells. All night Leopold sliced the poor creature's ear for more blood, and took her temperature rectally. He recalled years later that every time he entered the room wearing those gloves, she panicked and stared, as if she were making a thousand silent pleas to leave her alone. By morning, Leopold could see under a microscope that Wallie's blood was gone, washed from the monkey's system. Her fever also disappeared. The *vivax* parasites never crossed into the monkey's blood cells; the rhesus could not replace man.[34]

Leopold performed other jobs, too. He separated salivary glands from a mosquito's chopped-off head, did the doctors' typing, collected blood samples, operated the portable electrocardiogram machine, and recorded data. Upon Alf's suggestion, Leopold set up a microscope in the ward on bite day so he could show the men who had just been bitten by Ann what the germs in her saliva looked like.[35] He created a statistical model to infer concentrations of parasites on a blood slide, instead of doing the actual counting. Leopold said in his autobiography that the doctor he worked with was so impressed that he showed it to Alf, and Alf had Leopold write it up for the projects's *Malaria Bulletin*, authored by "Staff of OEMcmr-450."[36]

Alf's prisoner assistants learned marketable skills. For their hard work, they enjoyed many privileges. They played cards and dominoes, lived in the hospital away from the birdcage, wore hospital ward whites and, at night, comfy pajamas and robes rather than prison jumpsuits. They had fruit juice, took long baths, and worked shoulder-to-shoulder with scientists and prison staff, instead of being segregated by iron bars.

But that wasn't enough for Leopold. He wanted to be included in a study, not just serve as a technician. He wanted his name on the special list of those who agreed to "take" malaria, so that when the parole board and the governor later reviewed the list—which everyone knew would happen, once the war ended—his name would be on it. But Alf said no, he needed Leopold in the lab. So Leopold threatened to quit and Alf relented, knowing exactly why Leopold wanted to be infected, and powerless to prevent him from using the project in that way.

Leopold ended up in one of the more gruesome experiments there, perhaps not entirely by accident. He was generally disliked and had forced himself on Alf. In any case, on June 19, 1945, one of Alf's doctor assistants drew into a syringe the salivary glands taken from mosquitoes that carried a hypervirulent strain of *vivax*, called the Chesson strain (obtained from a young soldier named Chesson who picked it up in New Guinea and landed at Harmon Army Hospital in Texas in 1944, where researchers isolated it, then cultivated it in the insane to be used in project studies). The doctor then drove the needle, full of this infected saliva, into both of Leopold's thighs on two consecutive days. He said it caused "severe local itching" as though he had "bitten by a hundred mosquitoes."[37] A day later the doctor sliced from each bite site a silver dollar–size chunk of Leopold's flesh, and surgically implanted it in the thighs of two other inmates. Then everyone was sewn back up. Alf wanted to know how long these stick-shaped sporozoites in saliva stayed at the site of the bite. If Leopold's flesh transmitted the disease to the others, Alf would know the germs hung around the site for at least a day. But neither got sick, which told Alf that the sporozoites disappeared from the bite site quickly.

Leopold, who had an extremely heavy load of the germs injected directly into his body, was sure to get sick. Thirteen days later he shook his bed with "chattering teeth," a 104-degree temperature, and "a headache out of this world."[38] He now carried the same malaria strain that decimated MacArthur's troops in 1943, and that Lowell saw causing relapse after relapse in Klamath Falls—without hope for a cure.

TO treat Leopold, Alf used an experimental combination of the sontochin analogue SN-7618 and the old German drug plasmochin (cataloged as SN-971). Alf wanted to see whether low, safe doses of plasmochin could stop or at least slow relapses while the sontochin analogue treated the acute attack. SN-7618 did its job; Leopold was better within twenty-four hours. But a few weeks later, he relapsed. This time for treatment he was queued up for a different experimental drug—a special one that Alf had been working with, SN-13,276. This was a plasmochin analogue, one of many made and tested. Most caused the same scary blood disorders. But this one worked in small doses, which appeared to be safe.

In the meantime, Alf made *the* most important breakthrough of the Malaria Project, one that would produce an excellent political outcome for them all.

This involved the sontochin analogue SN-7618, with the "patently novel" process for synthesis. He noticed like everyone else that inmates on 0.3 grams *a day* became terribly ill. They had headaches, vomited, and developed itchy hives. A few of the guys' hair bleached out.[39] This had been seen in rats undergoing lethal doses of the drug. So Alf cut the dose dramatically, to 0.5 grams once a week, just to see what would happen. The side effects were much milder. To his surprise, he saw no reduction in action against malaria. So, to experiment, he cut the dose again, all the way to 0.3 grams *once a week*. This time side effects disappeared. And, again, he saw no reduction in its action against malaria.[40] This was truly amazing.

So he tried similarly small doses of sontochin and the other 4-aminoquinolines, and recorded the same quick action against malaria, with a nice even level of the drug sustained in blood plasma. These drugs, it turned out, had been greatly underestimated. They worked even in the smallest amounts, and they stayed in the bloodstream long enough to be taken just once a week. This, if confirmed by larger studies, meant the project had its miracle. Or at least a drug the team could call a miracle, compared to atabrine.

LOWELL also tested sontochin and its analogues on his marines at Klamath Falls. But he had to use truly large doses (over 5 grams in a week, more than ten times what Alf started out with), because he wasn't just curing an acute attack; he was trying to cure the tissue stage that caused relapses. He had to stop the drug in his first five cases because the subjects developed severe symptoms: double vision, anxiety, irritability, and itchy hives so intense they had to be given a sedative. He cut the dose by half and started another twenty relapsing patients on the drug. Still the men developed this terrible itching, of which they "complained so bitterly" he had to stop the drug.[41] He reduced it by half again, and tried that on another thirty-four patients. Only two developed extreme itchiness and hives, but none were cured.

He ran similar studies with the other 4-aminoquinolines and reported to the board that "dose for dose,"[42] sontochin (SN-6911) was the better drug. He had put 106 marines on atabrine, eighty-six on sontochin, and a hundred on SN-7618. The results showed that all three were equally inactive in stopping relapses, but that only sontochin had no toxic effects.[43]

At the same time, investigators at the U.S. Naval Hospital in Bethesda put forty-four students of epidemiology, all volunteers, on high doses of SN-7618, and a quarter of them developed diarrhea, headaches, blurred vision, loss of appetite, and depression—one student refused to take more because he "felt like dying."[44] Dr. Butler in Boston also tried high doses of SN-7618 and said his conscientious objectors suffered nightmares, headaches, diarrhea, and severe problems sleeping.[45] Reports of similar doses using sontochin indicated little, if any, toxicity.

It appeared sontochin was the better drug.

But there were other considerations. SN-7618 was *patently novel.* Alf had shown that in smaller doses, its side effects were relatively mild to nonexistent. Carden asked Winthrop to make SN-7618, without giving the company access to the patented Price-Roberts EMME synthesis, and their chemist couldn't do it.[46] Even when the project supplied the company with the starting materials—more than two thousand pounds of oxal propionic ester—Winthrop could not deliver. If patent lawyers went to litigation over this drug after the war, Winthrop's case for proprietary ownership would be weak at best.

SN-7618 made a great antimalaria drug not just because it worked as well as sontochin and could be used safely in low doses, but because the project—and just about everyone working for the project, from the chemists to investigators running clinical trials—could stake a claim to its success (because all the lead investigators had run trials with it). *And* no one company would capitalize on it after the war.

At the May 1945 meeting of the board, Alf was asked to keep his Stateville prisoners on SN-7618 for a full year, at 0.5 grams a week, to see whether any toxic effects appeared over time. In the meantime, National Analine and Chemical Company expanded its plant production capacity to make it, using the Price-Roberts process.[47] Thus production began. George Carden ordered as much of it as he could get and ramped up the testing. He sent five hundred pounds of SN-7618 to Fairley in Cairns, Australia, for one of his big studies on hundreds of volunteers. The ball started rolling faster for SN-7618 than for any of the others.

When investigators asked for more samples of sontochin, board members wrote that they had "no control over the supply of SN-6911; that material is patented by the Winthrop Chemical Company; and that with respect to the clinical status of the drug the Board is not extending its exploration of the compound at the present time as there are other members of the same series of more interest."[48]

This neutralized Winthrop's hold on the drug, because the French also had a claim to sontochin, established by Specia and Jean Schneider. As such, the Malaria Project never gave the drug an "in confidence" designation.[49] This allowed administrators to share the formula and all technical instructions for making it with Schneider, by now a respected partner of the Malaria Project in his capacity as research scientist at the Pasteur Institute in Tunis. Schneider then shared the intelligence with Specia so the French chemical company could make the drug after the war.

International markets for sontochin would not belong exclusively to Winthrop, and the United States government was developing SN-7618 for license-free production. Winthrop had been boxed out.

AS Allied troops liberated Western Europe and crossed into Germany, George Carden worked to slow the project down. No more compounds were needed. He had taken the expedient and not too risky move of settling on SN-7618. At least four of the 4-aminoquinolines were comparable in activity and toxicity, but the project needed just one. He wanted SN-7618 ready for the Allied victory parades so it could be used to treat returning troops seeded with malaria.

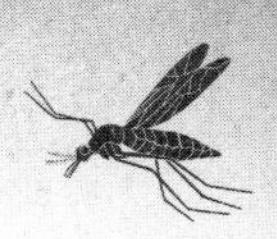

CHAPTER 29
Victory of Sorts

On April 29 members of the U.S. Army's 42nd Infantry Division arrived at the Dachau concentration camp. A Red Cross official waved a white rag and invited the Americans in. The SS had abandoned the place only days before. Outside the gates, a line of about thirty boxcars sat still on railroad tracks, each filled with tangled, emaciated bodies, many showing signs of cannibalism. Inside the camp, U.S. soldiers heard voices coming from long wooden barracks. The echoing pleas asked the same question in different languages: "Are you Americans?" Then they wailed with joy. The healthy ones stormed the men of the 47th, while the weak limped and crawled and cried. It was unbelievable. First dozens, then hundreds, then thousands of them.

U.S. soldiers described these "first encounters" similarly—at thousands of prison camps. The smell hit them first—decomposing bodies, human waste, and the acrid stench of burned hair and flesh. At Dachau in the last weeks before the Nazis fled, the crematoriums' coal had run out, so the bodies piled up next to the brick furnaces. Another pile of some two thousand bodies filled a ditch outside the fence. Many of the Americans needed psychiatric care afterward, as the horrors, and especially the smell, stayed with them for years.

Liberated prisoners kept dying from epidemic typhus carried in body lice. The U.S. Army flew in stores of DDT and used methods Fred Soper devised in Morocco then perfected in Italy to spray for lice. Long lines of ex-prisoners, still dressed in their filthy striped uniforms, stood still while U.S. soldiers used hoses to shoot blasts of DDT dust into their shirts and trousers. The mark of cleanliness was white residue at the collar and in the ears and hair. Within days, typhus-related deaths slowed. DDT alone in this one camp saved thousands of lives.

Dachau had beds for six thousand prisoners but held more than thirty thousand when the Americans arrived—five people per bed in bunks with bags of straw for mattresses. The barracks, or blocks, were divided into four sections, each with a living room and about one hundred bunks, in which five hundred people slept. They shared two bathrooms with a total of six toilet seats and six urinals, plus a shower, which allowed a weekly washing in the early years but only once a month by 1945. When the pipes froze the toilets stopped flushing and abominable sanitation grew even worse. When typhoid fever and dysentery broke out, anyone who soiled himself was dragged outside, hosed down, even in the cold, and moved to "shit beds"—that had no mattresses so they could be easily hosed. All the barracks lacked toilet paper, so a man would defecate and just pull up his pants. One prisoner said the "sanitary conditions were absolutely indecent."

THIS is where Claus Schilling went to work every day. He arrived in early 1942 with a mission not unlike the American Malaria Project's mission—to find a cure for malaria. But *this* was Nazi Germany's solution to the need for human subjects in medical research. When Schilling arrived at Dachau, the number of prisoners stood between ten and twelve thousand—crowded, but nowhere near the nightmare it became.

"When I had to take over duties and work here in the SS, in the Dachau Camp, I knew the fame which Dachau had—and I said to myself, 'the less I will see and observe things passing, the better it is,'" he told American interrogators during preliminary interviews for the coming war crimes tribunals. Speaking in English with a thick German accent, he told the Americans that he used to run a lab at the Koch Institute, which was like the Rockefeller Institute, doing purely scientific research. His questioner asked him when he stopped practicing medicine. He answered by making the distinction between his work as a medical researcher and that of a doctor giving medical treatment: "You see, 'practice' is not the right word." He was at Dachau to test his theory, to see if he could build a natural immunity in man by repeatedly giving patients malaria and controlling the infections with drugs—allowing the fevers to break through then suppressing them so the body's immune system could become acquainted with the invading germs and build a strong defense against them. If he could do this against the sporozoite stage—the one carried in mosquito saliva—he could prevent infections.

The drugs he used came from Bayer labs and I.G. Farben factories in Leverkusen and Elberfeld, all of which fell silent when the Allies occupied the Rhineland. The windows in the cavernous brick factories, which had turned dyes and coal tars into the world's first synthetic malaria drugs, were blown out. Smokestacks lay crumbled in piles of broken brick. I.G.'s headquarters in Frankfurt were occupied by U.S. forces and turned into American headquarters for denazification and dismantling of the Third Reich. The industrial phoenix that had arisen from the Treaty of Versailles to dominate the world of industrial chemicals had fallen.

Schilling knew none of this. All he knew was that the war had ended and that he needed to write up his research results from his three years infecting over a thousand Dachau prisoners with malaria. So as the SS fled the camp in the days before its liberation, Schilling stayed put. He sat in his dingy office waiting for the Americans to blow through the gates.[1] His malaria colleagues from the Allied countries respected and praised him before the war, and even funded his projects. Maybe he thought they would defend him. In any case, he was sure they would want to see his data. Given his reputation and senior place among the world's malariologists, surely they would stand up for him.

But they did not. As the horrors of the concentration and death camps emerged, his former colleagues asked: How could he work in so vile a place? Malariologists of the next generation would view him more sympathetically.[2] But no one could in 1945—for good reason.

AMERICAN investigators interviewed everyone at Dachua in the weeks and months following liberation.

Under oath, Schilling swore that in fall of 1941 he had been summoned out of retirement by Reichsgesundheitsführer Conti, the Third Reich's health minister, and then called to see Reichsführer Himmler that December. He retold the story of how Conti sent a car to take him to the train station for all-night transport to Himmler's headquarters near Insterburg, Prussia, and how Himmler asked him to explain his experiments.

"I came from Italy, you see," Schilling said. "I exposed my idea to him, to be done by inoculation of vaccination against malaria, and . . . he asked me to continue these investigations in Dachau."

His examiner asked: "Did you have an opportunity to refuse the request that Herr Himmler made?"

"No," said Schilling. "It was not possible. Under the conditions of Nazism it was not possible. . . . No choice in the matter."

"Would it be fair, then, to say that Herr Himmler ordered you to go to Dachau and to perform malaria experiments upon prisoners who were confined there?"

"Yes," said Schilling. Then he gave his history, how he had been forced to retire from Koch in 1936, so he became a "consulting specialist in malaria" at an asylum in Wittenau, where he learned how to use syphilitics in malaria experiments. "And after that I asked my superior, [the] Minister of the Interior in Berlin, if I could perhaps continue in these investigations in malaria in Italy. The reason was because, in Italy, you have an opportunity to study malaria in persons who had malaria before, you see. It makes a difference if a man has had spontaneous malaria and is inoculated afterwards."

Schilling knew before the war what American investigators had figured out during the war—that previous exposure mattered when assessing the value of a drug or possible vaccine, because that exposure influenced the course of the disease and how it reacts to agents.

He had successfully inoculated a few patients in Italy when Himmler summoned him to his mountain headquarters. By February 1942, Schilling had his research station set up at Dachau, and there he stayed until March 13, 1945. He had to leave for an operation on his bladder and prostate. He was in the hospital until April 12 and returned to Dachau three days later, and just waited. When asked what he waited for, he said, "Yes, yes, I had nothing to [do] any more with the camp, and with the hospital. I remained because I thought it would be much, much better for me, and the healing, and for my work, especially to remain here and to wait until the thing was aired out."

By "thing," he meant the U.S. Army's hunt for war criminals. He wasn't one, he felt, and couldn't wait to be cleared so he could share his results with malariologist friends outside Germany, get the results published, and present them at an international conference.

He told interrogators that Dachau's *lagerkommandant* had received orders from Berlin to give Schilling only healthy men. "I did not choose any of the men," he said. "The choice was made by the medical officer of the camp." The list of names went to Berlin for approval, and then the men were sent to him. He thought he experimented on nine hundred patients, total.

"Doctor, didn't you tell Major Larson the other day that the figure was 2,000?" asked his interrogator.

"No, no, that is not true," he said.

"As many as 1,000?" he was asked.

"Well, I would say 900 to 1,000. I can't say."

"Would it be fair to say 1,000?"

"That might be," he said.

When asked what kind of malaria he gave them, he answered: "I had some, about seven or eight strains in the beginning, but soon, I saw that it was not necessary to maintain all these things, and at last I had three different strains—a Russian strain from Ilmensee, a strain from Crete, and a strain from Madagascar." All were *vivax*, the cause of benign tertian, he said.

"Did you ever use any other species?"

"No, no."

"Did you use *falciparum*?"

"No, this is the malignant tertian, and it is too dangerous to use."

"Did you use *malariae*, the quartan strain?"

"No."

"*Ovale*?"

"No."

His interrogator asked whether he allowed his patients to burn through several rounds of fever before treating them with medication.

"That was in my work—to wait first for the attacks, and then to cure them. . . . I kept them in my ward. I had two wards, you see, under my care. Of course I kept them, and they were examined very carefully every three hours. . . . I think I couldn't have been more careful than I was." Patients were as "healthy as possible." The SS called him "Professor" and gave him ten prisoner assistants to help with the roughly four hundred patients on his roster at any given time.

When the Americans arrived, the hospital was staffed by 196 prisoners who cared for 3,162 patients under treatment for various illnesses. On a separate roster were listed "323 masculine Jews and four Jewesses." U.S. interrogators were given evidence of SS doctors allowing German medical students to practice surgical skills on healthy people. More than five hundred prisoners had their stomachs, gallbladders, spleens, or throats cut open and sewn back up, for practice. Many didn't survive.

But Schilling wasn't SS. The Americans were trying to figure out what he was.

"Doctor, do I understand that whether a prisoner was inoculated with the malaria serum or was bitten by a malaria-carrying mosquito, that you would wait until he had the fever and chills before you would commence treating him?"

"Of course, yes," he said, then corrected himself: "Wait a minute—one of my methods was to inoculate a man, and then not to wait until he got his first chills or his first attack—to treat him before then. The idea was to use this time of incubation, what we call the incubation of blood, and the beginning of the fever, to use this for immunization of the man, and to treat him and to repeat the inoculation, make the same again, and so on, until he developed immunity. That was my first line, which I followed."

He said he gave only quinine, neosalvarsan, pyrimadon, and atabrine as a treatment. Once he was asked "by Berlin" to try a drug called No. 2560, a synthetic compound similar to atabrine. He found it to be about the same quality as atabrine. That could have been sontochin. Gerhard Rose, Schilling's successor at Koch who had advised Conti against funding Schilling's research back in 1941, had rekindled an interest in this drug and was testing it for Bayer while the company's labs were bombed and chemical supplies dried up. Rose didn't get very far and moved to typhus experiments, for which he would later be convicted at Nuremberg.

SCHILLING sat at a small desk answering questions from a string of prosecutors who changed throughout the day. By early evening, he grew weary and confused. His interrogators were shocked at what they heard. How could he so matter-of-factly talk about giving malaria to people? They, of course, had no way of knowing that Americans were employing very similar research techniques. So they pressed on, asking why he didn't immediately treat the prisoners' fevers.

Schilling tried to explain, in broken English, that many of his prisoners were in his ward simply to maintain strains—they served the same purpose as state hospital patients in the United States. Parasite counts were at their highest during an acute infection, so he waited until symptoms appeared. He treated the fevers only after he took blood and used it to infect more prisoners. "If I would not have done that, I would not have maintained my strains, you see. A great number of my patients were like this."

"Well, let me go back again, Doctor," one interrogator said. "Explain to me what you did when you inoculated a prisoner with malaria serum, and you waited until he got the chills or the fever, did you treat him—in other words, what did you do?"

"I transmitted the disease to him."

"What else did you do?"

"What else? Nothing. I waited, after a few days, until he had parasites and fever, and then I took his blood and inoculated it to a second man for maintaining the strain."

"You took the strain, and put it in the number two man—another man?"

"Yes." Schilling wondered why his questioners failed to understand him.

"What did you do to that second man?"

"The same thing, let him pass a few attacks, and take his blood, and inoculate number three, and treated number two."

In other words, his interrogators puzzled, "you continued the blood strain throughout the three?"

"In this kind of malaria I had the interest to continue the strain, and the effect of the treatment," Schilling answered.

The Americans looked at one another. They had heard and seen a lot in the past few days. This was a labor camp; unskilled prisoners built roads, worked in munitions plants, dug in gravel pits, made bricks, etc. Skilled prisoners were used as clerks, interpreters, accountants, cooks, doctors, nurses, and morticians. They survived on watery coffee for breakfast, watery soup for lunch, and watery soup with a piece of sausage or cheese for dinner. Female "skilled workers" were held in a barracks that functioned as a "bordello," where privileged German prisoners called "capos" paid the SS an entrance fee to rape them. Punishment for minor infractions—falling out of line, stealing bread, appearing dirty—included hanging by the wrists or twenty-five lashes with a cattail, or a severe beating of punches to the face and kicks to the groin. Political prisoners took their punishment in the bunker, where they were tortured and then put in a coffin-size cell called a "hatbox." Tall prisoners had to slouch to fit in a space too narrow to sit. There they stood for three days, sometimes longer. Many were removed dead. German shepherds, Great Danes, and boxers held in kennels next to the crematorium hunted down prisoners, sometimes for game. The SS starved the dogs and then set them free to see how long they took to tear a prisoner apart.

Other medical experiments were absolutely revolting. Dr. Sigmund Rascher wired about three hundred prisoners with rectal thermometers and heart monitors before submerging them in a tank of ice water to study them as they died. In a pressure chamber he watched men scream as they tore at their hair and beat their heads, trying to relieve pressure on their eardrums. He injected pus into the leg muscles of Catholic priests, then used an ineffective compound to try to stop the resulting sepsis. Later experts reviewed his work and found that no conclusion could be drawn from the torture of these men because of his poor study design and sloppy record keeping.

Stories of these atrocities hung in the air of the makeshift courtroom inside the old Landsberg Prison, where Hitler penned *Mein Kampf* twenty years earlier. Shocking stories went home with the Americans, who had nightmares and lost weight. The lead investigator dropped to 116 pounds. One prisoner painted Schilling with the same brush used for men like Rascher. But prisoners used in Schilling's experiments said he wasn't harsh. And records showed that while on Schilling's ward prisoners received an extra piece of bread daily, slept on linens, and escaped for long stretches the despicable sanitary conditions in the barracks. Still, Schilling's interrogators believed him to be as evil as the rest.

Dr. Franz Blaha, a prisoner who became a key witness in the Dachau convictions and the later war crimes tribunals at Nuremberg, said, "I can't make a statement about malaria cases . . . they would conduct them in a different block, and pertained to Dr. Schilling's department. Dr. Schilling was a civilian who worked here at the camp." Blaha said he conducted about forty autopsies on malaria patients and remembered that eight died of agranulocytosis following administration of pyramidon. About thirty died from neosalvarsan injections due to hemorrhagic encephalitis, jaundice, and liver necrosis. Many more died of secondary infections as a result of malaria. He said he talked to Schilling many times. "I think he is a bit senile," Blaha said. "I know, as a matter of fact, that, secretly, Dr. Schilling's patients obtained quinine from orderlies, and that many were saved by the fact that, without Dr. Schilling's knowing, quinine was administered to them."

This, of course, meant Schilling's data had been corrupted. The research he prized because he believed it could prove he had successfully inoculated prisoners was unreliable. Perhaps they secretly had been given quinine by the orderlies. There was no way of knowing.

ONE prisoner, John Ashton Alpar from London, said he didn't think the malaria experiments were dangerous. He said only fit men were chosen from every block, mostly Russians. They received malaria injections and then antitoxins. "Each of these men got a special nourishment. It was not such a dangerous experiment. The phlegmone [Rascher's pus injection] was the dangerous experiment."

DURING his initial interview, Schilling was asked whether prisoners were given an option.

"No, they were not asked."

Was he aware that prisoners were expected to obey absolutely?

"I thought that at any prisoners camp it would be a safe understanding, I think."

Could he say that no one died?

"Well, as a matter of fact, I can't say."

Did he know that some did die?

"Yes, some died, but not as a consequence of malaria. That is the great difference." He said if a man caught malaria the natural way, but then also contracted typhus and died from the typhus, malaria was just the secondary cause of death.

Schilling explained that he could cure his patients of malaria over the course of a few months using atabrine but that he had been trying to establish immunity by treating the fevers with this drug pyramidon, which wasn't active against the parasites but brought down fevers and worked as an anti-inflammatory. To him, this treated the symptoms of malaria but allowed the parasites to grow and, he hoped, create time for the host to build immunity against them.

"If we can combine Pyramidon or a similar drug, combined with a state of allowable infection, then we would proceed and come to the end of immunization!"

Was he trying to say that pyramidon was an answer to the treatment of malaria?

"Yes, certainly it is. It can, you see. I used Pyramidon for a few days, and then, when the effect of the Pyramidon was digressing, then I gave neosalvarsan, Atabrine, or quinine to treat them."

What if it didn't work?

"I think I had two or three cases where I used Pyramidon for ten days, after which they had their agranulocytosis [reduced white blood cells compromising their immune systems]. But I gave them the specific treatment for malaria, and then they were cured. As long as I could follow them they were cured."

DR. Adam Czercovitz had worked with Schilling as a prisoner assistant. He confirmed that Schilling used only the three strains of *vivax*, from Madagascar, Russia, and Crete. But he said Schilling's data was useless. "In my opinion his research upon malaria is valueless . . . he was able to produce this latent immunization only in nine out of two thousand cases. In my opinion he is somewhat of a monomaniac, in that he has a one-track mind and in spite of the fact that defeat stares him in the face, he continues on in an automatic fashion." He didn't believe Schilling was forced into the work. "He had made the statement to me that he was going to make the discovery so that he would put Robert Koch to shame."

SCHILLING was asked whether he was God-fearing.

"In God I believe," he said.

When he took his oath as a doctor, did he subscribe to the oath of Hippocrates?

"We didn't take any oath. . . . I can't remember."

Did he know what the oath was?

"No. This is the first I heard about it. It is a law, that is?"

By now it was late in the day and he had been answering questions for many hours. At seventy-four years old, still recovering from surgery, he started to falter.

"I have held the law code in my hands many, many years ago, but I cannot remember the details now. It is a little late." He was surprised to be put to such a test.

Wasn't it true that he was not to experiment or to operate upon a man without his consent, unless it was to save that man's life?

"This is well-known."

What was Dr. Conti's rank in the SS or in any German organization that he could order Schilling?

"He could order me because he was secretary of state in the Ministry of the Interior, and he could order me."

Wasn't it true that he gave to those healthy persons at Dachau malaria germs until it took hold, once or twice or even three times?

"It did happen."

In other words, he subjected a man to malaria until he got it?

"My intention was—vaccination—of treating the man in a preparatory way without producing fever. I had many cases where they had been inoculated. They had no fever, because I prepared them with inoculation treatment, and I tried many different treatments once, twice, three, and even four times."

Wasn't it true that if it did not take he would inoculate the man again and again until he did get the malaria germ?

"No, not so. Well, if I had inoculated him twice, and he did not take the disease, then my conclusion was that he had malaria before."

Did he at any time, while he was at Dachau, give malaria to a sick man?

"No."

Every man was healthy, to his knowledge?

"As much as I knew."

If a man were sick, could Schilling send him back?

"Yes, certainly. I would not inject a sick man in the interest of my experiments." The results would be skewed by the other infection.

Malaria was pretty trying upon a man's body, was it not?

"Yes, yes, very, very trying. I have had malaria myself. I know perfectly well how trying it is. I get shakings and what-not."

If he knew that it was against the law to perform experiments upon healthy men who did not consent, why did he perform his experiments?

"Because I tried to find a method which would save millions of lives. That was the reason."

Did he also believe that if he did not perform the experiments it would cost him his life?

"Cost my life? I don't think so. They wouldn't have killed me, but they would have certain[ly] put me through . . ."

"You are a very aged man, Doctor; how long do you think you would have lasted in that concentration camp?"

"I can't say."

Did he feel that Dachau was a great field in which to explore his malaria experiments?

"Yes."

And he wished to explore his work with these healthy men?

"Yes. These experiments—you don't do malaria experiments on other living beings because malaria is not transmittable to lower animals."

So he didn't care how many healthy bodies he worked on, as long he got to do his experiments?

"No."

In other words, he was going to explore this to the fullest?

"It was a serious and great enterprise, and I tell you sincerely that I knew I had to make these experiments on living and healthy men, and I did."

How much care did he give them? How much quinine and atabrine did he have for them?

"Oh, any amount. I was not hindered by lack of medicine. . . . I had it from Berlin."

So the men never had any ill effects after he was finished working on them?

"I have observed these men for two or three years; some of them are reexamined, asked when they were brought to another camp. I asked, later, is the man healthy, or has he relapsed, so I did what I could do to cure, to be sure a man is good."

"Doctor, this is a very ticklish question, one which you should weigh carefully before answering: It is wintertime, here in Dachau. You have infected a healthy man with malaria. That man, after a while, say thirty days, goes back to his block and he lies down among men who have typhus dysentery. That man gets the same food rations that the other men in the block get. Would you say that that man was going to live?"

"If he gets dysentery or typhus or so, his chances to live or die are smaller."

Would he say that a man's physical resistance, after being subjected to malaria experiments, was so lowered that typhus could easily infect him?

"When I let a man go out I was certain that his general condition was good. I didn't leave him out before I had a conviction that he was cured from malaria."

When he used pyramidon and reduced a man's white blood cells, and sent him back to his block, was his resistance compromised and was he left open to disease?

"Probably, you see I couldn't follow these experiments to the end."

SCHILLING was held at Dachau from May until his conviction in December. The judges took ninety minutes to find all forty defendants on the stand guilty as charged. Schilling and twenty-seven others were sentenced to death. He cried, then begged for his records so he could write up his research.

A year after the Americans entered the camp, he walked up a dozen steps to an elevated platform in a small courtyard of the Landsberg Prison. A breeze caught his wavy gray hair as he blinked repeatedly. His neatly trimmed goatee exaggerated a foreboding grimace, as he stood there amid witnesses and his executioners. He looked frail and sad. The executioners covered his head in a black hood, and then he was gone, just a faceless man about to die. Without pomp or ceremony, the noose was tightened around his neck and in an instant he fell through the trapdoor. Underneath the platform, a doctor declared him dead. His body was dragged into a pine coffin and hammered shut with a lid bearing a cross and a label that read "Schilling."

There is no better metaphor for what happened to Germany's place in malaria research and drug development. It was dead. And into this enormous vacuum walked the Americans.

CHAPTER 30
A Miracle of Sorts

In May 1945, a final test run of SN-7618 was needed and it had to be done in a large population of people who naturally acquired the disease—to mimic conditions encountered by troops coming home from war and to determine whether this drug could be used postwar to reduce infection rates in tropical places within the Allies' colonial empires. Here was an opportunity to finish testing a drug not just for use in returning troops but also for overseas industries. Chosen as a test site was the eighteen-thousand-acre Hacienda San Jacinto sugar plantation in Peru, owned by a British family named Lovett, about 250 miles from Lima.[1] For a hundred years, the Lovetts ran this compound like a "small feudal state," with more than five hundred permanent laborers working the sugar processing plant and commercial alcohol distillery. Malaria usually spiked after the winter rains.[2]

George Carden went to inspect the study, but because he wasn't a malaria expert Dr. R. B. Watson from the Rockefeller Foundation accompanied him.[3] They flew on Pan Am from Miami, over the snowcapped peaks of the Andes, to Peru's coast, where they landed amid hundreds of derricks drilling in the Talara oil fields. They refueled, then flew inland over rice plantations and up into thick fog to land in Lima. Then they drove the Pan American Highway, past adobe villages and moonlike landscapes that, in the sun, glistened with brilliant reds, purples, greens, and yellows from eroding mineral beds.

They turned off the highway onto a rutted road to enter the verdant valley owned by the Lovetts. Carden and Watson spent their days on horses, riding out to the three villages used in this big, expensive study, which the Malaria Project hired Rockefeller scientists to run.

Their second night there, word arrived that Japan had surrendered. An all-night party of heavy drinking and gift giving filled the square. The next day, hungover and tired, Watson and Carden analyzed two months of data collected by Rockefeller's investigators. But Carden found mathematical errors. Not accustomed to Rockefeller's usual handling of such things—which was to never embarrass the intelligence of country partners and always be cordial and helpful—Carden openly criticized the Peruvian epidemiologist, a Dr. Alberto Valderrama, who was sent from Lima to assist with the cultural and technical aspects of collecting data on local laborers, many of whom were Quechua Indians. The project would not have been doable without him, and his feelings were seriously hurt.

But the math wasn't the real problem. The much bigger problem was that the study started two months after the valley's malaria season began. Rockefeller's lead investigator, Osler Peterson, arrived as instructed, on May 5, but his car, drugs, and microscopes didn't get there until the end of the month. His study finally began on June 11 with five hundred people on SN-7618; another five hundred on atabrine (as the gold standard); and five hundred on aspirin (as a control).[4] He abandoned atabrine, because his small team couldn't get five hundred villagers *every day* to take a pill that tasted like poison. Many village residents hid in the cane fields or ducked behind barns, or stayed with family in other villages, and were gone for days.

So the team focused on delivering SN-7618, which was delivered just once a week. The children were lined up at schools and on the same day each week given a pill. For fieldworkers, hacienda bosses withheld pay until each took his pill. Every Saturday night men and older boys lined up for the week's $2 in wages, along with a pill and a draft made of alcohol, lime juice, and sugar.[5] This usually kicked off a night of heavy drinking in the hacienda-owned saloon. Seeing this made Watson wonder whether some of the toxic effects attributed to SN-7618, like blurred and double vision, were due to copious alcohol consumption, not the drug.[6]

Everyone took the same dose—0.5 gram—even small children, who were as young as six. They had severe nausea, cramps, headaches, vomiting, and difficulty seeing. Eighty percent of them got sick and half missed school.[7] So their dose was reduced. Women suffered similar symptoms, but not as severe. They and the men stayed on this dose—which Alf had already shown was twice the amount needed to suppress malaria in Stateville's inmates, who were almost twice the weight of these Quechua Indians.

Nowhere in the report written by Watson did he mention consent forms. The Quechua were told the drug would prevent malaria and that they had to take it. Some already had malaria that year and refused. Most, however, complied.

From early June to August, Peterson studied a thousand people living in a region with naturally occurring malaria, where adults and older children worked long, strenuous days.[8] It was perfect for a toxicity study, and would have been perfect for an efficacy study, too, but for the late start and loss of the atabrine group.

"It seems fairly obvious from the data that nothing much will be learned about the effectiveness of the drug as a suppressive," wrote Watson in his report.[9] The data on toxicity, however, was "quite valuable." Which made the study a big success. James Shannon wanted to know whether an exceedingly high dose, even in children, would create irreversible toxicity. The answer was no.[10]

Watson and Carden returned to Lima for meetings and dinners with Peruvian public health officials. A representative from the country health ministry, Dr. Villalobos, told Watson he was anxious to get started on a malaria control campaign. The war was over and these two new weapons—DDT and SN-7618—presented great hope. He asked whether they wouldn't be easier to use than a plan Rockefeller helped devise before the war, which focused on ditching and irrigation controls, in conjunction with hydroelectric projects to reduce mosquitoes—modeled after the Tennessee Valley Authority, in a plan they called "the TVA of Peru." Politics and logistics had created hurdles, and that work "turned out not too satisfactorily."[11] This Peruvain official saw DDT and SN-7618 as cheap alternatives to expensive land reforms.

Watson raised an eyebrow. His greatest fear had become a reality. Public health officials all over Lima now eyed these wartime technologies as miracle weapons against malaria. "Everyone tended to seek an easy way out of the malaria control problem," he lamented in the diary he kept for this trip. "[Their] use might prevent the development of permanent means for control."[12]

No one would appreciate just how right he was for another twenty-five years.

. . .

ON the flights home that August, Watson and Carden penned a rough draft of their report, to be submitted to the OSRD and the Rockefeller Foundation. Carden took the handwritten notes and had them typed up. But in the process he added statements that Watson "could not accept." Those statements claimed SN-7618 was indeed the miracle drug the project had worked so hard to find, and that it promised to be a great tool in controlling malaria in civilian populations in the tropics.

In a private note to Dr. E. L. Bishop, a scientific director for Rockefeller's International Health Division and a TVA authority, Watson wrote: "I cannot wholly subscribe to statements made in the last two sentences [of the first paragraph on page 8 of the report]; as a matter of fact, I had suggested they be deleted from the final draft." Those sentences said that if SN-7618 could be started before the malaria season, it alone could "go far toward the complete control of malaria."[13] Watson said nonexperts, like Carden, "almost invariably" think a drug is "an easy way out of malaria control problems." Even if SN-7618 were a perfect drug—which it was not—its use as a "suppressive medication in civil populations" would lead unwitting officials down the wrong path. This drug and DDT should be used to complement larger control programs, not replace them. On this point, Watson was emphatic.[14]

But Carden wasn't interested in having this larger debate. For him, the most important result of the trip was that he had what he needed to close out the Malaria Project: a drug he could call a miracle with a now famous name, chloroquine.[15]

MOST Malaria Project investigators, at the end of the war, returned to their Ivy League universities and government agencies to resume their prewar work. Carden went back to Columbia to become a renowned cancer researcher, Shannon became head of NIH, Coatney ran one of Shannon's institutes, Butler returned to Harvard to teach, Marshall went back to being the most skilled chemist at Johns Hopkins, and Robert Loeb became chairman of the Department of Medicine at Columbia.

Alf Alving was an exception. He had started something valuable at Stateville. Carden had his drug, SN-7618. But Alf was onto something that might be far better. He finally worked on an 8-aminoquinoline that had potential. Of three hundred plasmochin analogues made by project scientists, only twenty-one made it to human testing. Alf studied eighteen of them.[16] He understood these analogues better than anyone. The baseline drug, plasmochin, could be tolerated in small doses (15 milligrams per day) but required six times that dose (90 milligrams per day) to cure an infection. So he focused on finding an analogue six times more active than plasmochin.

Four of them proved particularly dangerous, causing potentially deadly blood disorders—including the dreaded agranulocytosis that Schilling encountered at Dachau and that autopsies later showed had been responsible for killing some of Schilling's prisoners. In Alf's inmates, the blood disorder was spotted quickly, the drug was stopped or the dosage dramatically reduced, and no one was harmed. The drug that killed Milam, SN-8233, was among these four. It caused agranulocytosis, although that's not what caused his heart attack. Alf shelved it, along with the other three: SN-11,191, SN-11,226, and SN-1452. All were just too toxic to fool around with at high doses.

The most common symptom was pain—in the stomach, lower intestines, back, neck, and shoulders. So Alf gave his volunteers painkillers.[17] He told them they could, at any time, stop taking the drug. None opted out. Instead they stayed and suffered through the other usual symptoms, like loss of appetite, nausea, vomiting, headache, weakness, skin eruptions, and diarrhea. When doses were too high, they developed cyanosis.[18]

Five of the eighteen analogues, however, looked good. The most exciting drug, SN-13,276 (pentaquine), appeared to cure fourteen out seventeen severe Chesson infections, and twenty-six of twenty-seven moderate infections—Leopold was among the cured. "Appeared" was the best Alf could claim, because of *vivax*'s quirks. But Chesson *tended* to relapse quickly until it burned out. Some of his potentially cured cases had been observed for a year without a relapse.[19] *This* was exciting!

But there were caveats. For this drug to work, it had to be taken with quinine, and it could be used only by white inmates, as these 8-aminoquinolines kept triggering severe, life-threatening blood disorders in black volunteers. Nor did SN-13,276 work well against intense infections. Alf demonstrated this by having eighty infectious mosquitoes bite his inmates, instead of the usual ten.[20]

While other project investigators packed for home, Alf continued to look for that perfect analogue.

. . .

LOWELL'S trip through American fir forests and then the Great Plains, on his way home from Klamath Falls, was mostly uneventful, but for a frustrating flat tire.[21] The townspeople had thrown him a party before he packed his family into their station wagon and headed for Ann Arbor. On the way home, Lowell stopped in Chicago. He wanted to talk to Alf about this amazing study he had done curing inmates with SN-13,276. Until then, no one was able to study how long a person remained immune to Chesson after an infection—because no one but Alf had cured it. Would Alf allow Lowell to study the men he had cured?

Lowell Coggeshall, by now, was pretty famous—for a public health expert. During the war, he decided who received federal support for malaria work, and brought huge grants back to the University of Michigan (Lowell's projects absorbed a fifth of the grant money during the early years). When he stopped in on Alf at Chicago, he held a Gorgas Medal for outstanding work returning marines and sailors back to duty, after they had been written off as invalids. He was often quoted in the newspapers, and he had just run a five-thousand-bed facility that put Klamath Falls on the map.

The University of Chicago wanted Lowell back. To top his tropical disease chair at Michigan, Chicago offered him a position as dean of their medical school, and Lowell accepted. He bought a house on Blackstone Avenue, in South Chicago, a few doors down from Alf, in a middle-class section of Hyde Park. Alf and Lowell's children played together and the two men became great friends and confidants, as Lowell's wife also had multiple sclerosis.

LOWELL'S Stateville study didn't pan out.

He took seventeen prisoners Alf had cured and reinfected them with the Chesson strain. These men included Charles Ickes, Alf's favorite inmate assistant. Ickes, as it turned out, became one of Lowell's star guinea pigs. When Alf had infected him in 1945 with Chesson, Ickes had a rough ride, with seven relapses and fifty days of high fevers and severe symptoms, before taking SN-13,276. When Lowell *reinfected* him with Chesson, Ickes showed "considerable" evidence of immunity, in that his fever never went above a hundred degrees Fahrenheit, and his worst symptom was a mild headache. Had everyone else shown similar immunity, Lowell's study would have been extremely exciting. But the other inmates' immune responses were inconsistent, and Lowell couldn't draw many conclusions, except that immunity to reinfection with Chesson, after being cured of it, did not last long—except in isolated cases, like Ickes's. This strain was like any other—tricky and persistent.[22] Decades later, scientists would learn that malarial parasites shed their identity as they morph through their different life-cycle stages, making each antigenically unique. The old-school vaccines—using the germ to confer immunity against future infection—would never work against this disease.

WHEN the Nuremberg trials unfolded overseas, lawyers for Gerhard Rose, Schilling's successor at the Koch Institute, did better for him than Schilling had done for himself. Defendants in the so-called doctors' trials at Nuremberg benefited from the experiences at Dachau, where every person charged was convicted. Many wondered whether this perfect score was, perhaps, a result of emotions, not due process. Those first encounters with the concentration camps were intense. Now, however, law and order had to prevail. Defendants were assigned talented German lawyers to represent them. The lawyers pushed back on generalizations about what constituted a violation of medical oaths, and established that there really were no set lines between right and wrong when it came to human experiments, but rather norms that were fuzzy and ill defined. For years, medical experts had discussed clinical results at international conferences without much concern for how the data were collected. Study design—were enough people used, was there a control, and was there a gold standard for comparison—stood paramount when judgments were made about the value of results. So while some Nazi-run human studies clearly crossed the line of decency, others teetered between ethical and unethical—in a medical community that hadn't fully defined what made the difference.

The argument so confounded U.S. prosecutors that they called in an ethicist to draw some lines. They thought they had their perfect witness: Werner Leibbrandt, a German doctor and medical historian who had been persecuted by the Nazis for "racial reasons."[23] But Leibbrandt's testimony backfired. One of the Germans' lawyers asked a generic question: Could prisoners, in general, give consent, to which he said, "No." Then he was asked whether research on willing prisoners was admissible and Leibbrandt answered, "No," again, because prisoners "were already in a forced situation."[24]

ALL this went on overseas, with little to no consequence for Lowell and Alf at Stateville, who were still infecting their prisoners. That is until the Germans' chief lawyer, Robert Servatius, brought out a *Life* magazine article, with pictures of Warden Ragen looking on as little jars of mosquitoes bit the bare legs and arms of prisoners sprawled on metal-framed hospital beds. One shot was a close-up of Nathan Leopold. Alf had arranged for the magazine article way back in June 1945 because he felt the publicity was warranted. The war was about to end and the National Research Council "lifted the curtain of secrecy" on the Malaria Project's prison work, which led *The New York Times* to write: "[S]everal hundred prisoners at three of the country's large penal institutions have volunteered to serve as living test-tubes for new potent drugs developed by our chemists against malaria."[25] The newspaper said controlled tests "on a scale never tried before" were doable because prisoners were told the risks, signed consent forms, and even lined up to participate, to do their part for the war. News stories were superficial, with little detail. But just knowing that government-sponsored scientists were infecting convicts was shocking enough. Reporters wanted to get inside the prisons and see murderers being bitten by malarial mosquitoes.

Life magazine asked for access to the Stateville experiments. Alf liked the idea; it would give the men due credit and satisfy Warden Ragen's interest in changing public opinion of his prison. Alf sought permission and got it. A team of reporters and photographers came for "bite day," and witnessed inmates joking around about their dates with "Ann," and the coming fevers. Alf made himself scarce; he didn't want to be in the pictures. But Warden Ragen did. So did Nathan Leopold. The story was glowing. Then the local radio station asked to air a program on Stateville's malaria ward. Again, Alf sought permission and got it. On January 3, 1946, Chicago's main radio station, WGN, broadcast a prime-time "exclusive" report. Governor Dwight Green, Warden Ragen, and Captain Branch Craige, an army doctor who did much of the work running Alf's studies, were interviewed, as were inmates, including Leopold.

Now, at Nuremberg, the Germans were building their defense on the premise that what went on there was no different than what went on in Stateville.

Servatius read the *Life* story aloud and used excerpts from the WGN radio broadcast. Then he asked whether the expert ethicist Leibbrandt thought the work was ethical. And he said, "No." Leibbrandt called the experiments "excesses" of "biological thinking" that allowed researchers to treat patients poorly; the "human relation no longer exists and a man becomes a mere object like a mail package."[26]

The Americans scrambled to control the damage by hiring another expert ethicist. This time they chose an American, Dr. Andrew C. Ivy, who had been sent to Nuremberg as a consultant for the American Medical Association. In 1944, he had actually approved the experiments at Stateville in his position as Illinois's top authority on medical ethics. He was a respected research physician and vice president of the University of Illinois.

To prepare to testify, Ivy returned to the United States and immediately formed a committee to examine the Stateville experiments. Then he flew back to Germany to rebut Leibbrandt's testimony, using a report from this committee, even though the committee never met.[27] Ivy even read findings of the committee: that prisoner experiments were ethical provided the reward of time off was not so much "as to exert undue influence." And prisoners guilty of "atrocious" crimes "should be told in advance . . . not to expect any drastic reduction in sentence."[28] This articulated exactly the setup he had approved for Stateville three years earlier. Ivy, testifying in Nuremberg, concluded that Alf Alving's project was ethical.

In the end, Gerhard Rose was found guilty of committing crimes against humanity in medical experiments involving typhus, not for his malaria work. And he was imprisoned, not hanged.

When Ivy returned to the United States and actually met with his committee, they drafted a report that concluded Alf's work at Stateville was not just "ethically acceptable" but "ideal," because the work conformed "with the highest standards of human experimentation." Prisoners gave consent without coercion and were cared for by an ethically competent and qualified researcher.[29] This, historian Jon Harkness concluded, "contributed to a widespread failure among U.S. medical scientists to grapple with the difficult ethical questions about their work that the Nuremberg Medical Trial might have raised."[30]

THAT Alf's name came up at Nuremberg in the same breath as Schilling's would haunt him to the day he died. He understood the charge of "biological thinking" but felt he was also a humanitarian. Lowell and Alf had that in common. They liked science *and* people. Some on the Malaria Project didn't care much for people, which made it easier for them to turn up their biological thinking and just work. But this wasn't Alf. He had pushed for the *Life* story and the WGN radio broadcast *because* he cared for his volunteers, and for that, he was implicated by name in history's most infamous court proceedings.

In retrospect, this seems deeply unfair. Alf stood at project meetings in

Washington to get for his inmates higher payments for participating in the studies, certificates of merit for them to use at parole hearings, special medical coverage for when they were paroled and still relapsing, and official recognition of the one man who died on his watch. He even gave jobs to two of his best inmate technicians—Charles Ickes and a speed typist—which got them out of prison, as parole was contingent upon having a job offer. The typist worked in Alf's secretary pool at the University of Chicago, while Ickes worked in his lab there (Ickes was later arrested for robbery and thrown back in jail in 1962, breaking Alf's heart).

Alf, in the beginning, even tried to name his prisoner assistants in his published reports. Draft papers he wrote included the names of all his assistants, including Leopold. But they were edited out by censors and replaced with a generic thanks to all prisoner volunteers.

A year after Alf was shamed at Nuremberg, President Harry Truman awarded him a Presidential Certificate of Merit for his work at Stateville.

ALF would stay at Stateville testing drugs on inmates, looking for the holy grail of a malaria cure, until his death in 1965.

In that time he made two important discoveries that make him famous to this day in research circles. The first was in 1949; he finally found a true cure for malaria. Newspapers caught wind of it and reported that the malaria scourge would soon end, thanks to the University of Chicago's prison-based studies. But Lowell Coggeshall, now Alf's dean, had to correct the record again. He said reports of malaria's demise were "premature" and "inaccurate." Lowell was trying to contain enthusiasm over an 8-aminoquinoline made in 1945, cataloged as SN-13,272, and finally vetted by Alf. Called primaquine, this drug cured nearly 100 percent of his inmates induced with the virulent Chesson strain.

But as with all malaria drugs, there was a catch. Alf kept seeing dangerous destruction of red blood cells when he used it on certain ethnic groups, especially darker-complexioned people. While whites were totally cured of relapsing malaria, with no side effects, blacks were almost killed by it.

The second important discovery he made while trying to figure out why his black inmate volunteers were so different in their response to the drug. The answer he stumbled onto would forever change the field of chemotherapy—and the underlying assumptions about the value of "magic bullet" pharmaceuticals.

Alf Alving, with two colleagues, Paul Carson and Larkin Flanagan, discovered what the *Chicago Tribune* called "the most common clinically significant enzyme deficiency of humans."[31] He had a hunch that the massive death of red blood cells was actually a residual effect of the drug, not the drug itself. And it looked like favism, the potentially deadly reaction to fava beans recognized as far back as Pythagoras, who ordered his students to never eat them. Alf, twenty-five hundred years later, figured out the problem with fava beans while studying primaquine.

He ran an ingenious investigation that involved centrifuging, washing, freezing, centrifuging again, then dialyzing blood to remove substrates and coenzymes. He found that blood taken from primaquine-sensitive inmates lacked sufficient levels of a special "ubiquitously expressed enzyme" with important "housekeeping" duties.[32] This special enzyme, called glucose-6-phosphate dehydrogenase—or G6PD—maintains the integrity of red cells when they are put under stress by outside elements, such as primaquine and, as it turned out, fava beans.

This enzyme deficiency, Alf inferred, was genetic.

Nature magazine reported that Alf's work helped launch a new field called pharmacogenetics, the study of genetic differences and how they affect a person's response to drugs. These links now inform how many classes of drugs are handled. For example, the efficacy of many cancer drugs depends on certain genetic factors in patients. And two different genetic distinctions (an allele of the CYP2D6 gene and serotonin reuptake inhibitors, or SSRIs) make it difficult to metabolize certain antidepressants, like Zoloft and Prozac, effectively triggering overdoses on small amounts.[33]

For malaria, Alf's dueling discoveries continue to fascinate and occupy malaria researchers. Primaquine, on the one hand, is truly miraculous, in that it kills the liver-stage parasites that cause relapses *and* the sexual-stage parasites that mosquitoes drink in. And it causes few, if any, side effects—in normal people. Primaquine, therefore, has all the ingredients of a magic bullet—with powers to end transmission of this disease.

But, as it turns out, people with the G6PD deficiency—who cannot take the drug—tend to come from, or have ancestors from, regions with malaria. So the people most in need of primaquine can't take the drug because they lack this special enzyme to protect their red cells from being destroyed by it. To Alf, this must have felt cruel, like the heavens were playing a trick. The one drug to beat relapsing malaria and end transmission was the one drug that couldn't be used where it was needed most.

Today, the G6PD deficiency is recognized as an inherited trait carried by women and passed on to sons, mostly in people living in or having ancestors from Africa, the Mediterranean, and Asia. Researchers believe that the deficiency was a random mutation that helped protect early man from the deadly symptoms of malaria. Like with the Duffy negativity and sickle cell trait, over thousands of years populations with this G6PD deficiency have evolutionarily been selected for. Today, populations no longer living with malaria have the deficiency because it made their ancestors stronger and more able to survive, procreate, and pass it on to the next generation. The deficiency becomes apparent only when stressed by drugs like primaquine, plus fava beans and a few other substances that include aspirin and mothballs.

Had Alf stopped with primaquine's quirky toxicity, it might have been shelved with all the other Malaria Project compounds labeled as great drugs but too toxic to use. Instead, Alf put a fine point on the dilemma. His determination to figure out *why* you couldn't use primaquine allowed today's scientists to focus on finding a way around this enzyme deficiency. If they figure it out, the world will have a very good drug, and maybe even a magic bullet.

IN 1968, Stateville's inmates put up a bronze plaque in the prison hospital in memory of their favorite "blusteringly" praise-avoidant malaria expert, Alf Sven Alving.[34] The hospital has since been torn down and the plaque long lost. But his name is well remembered. Forty-five years later, in 2013, the warden of Stateville, Michael Lemke, knew it well. His father had been a prisoner there and volunteered for Alf's malaria experiments. Lemke said Alf's time at Stateville was considered its heyday, when the cells were full, Ragen was a national icon, the roundhouse architecture and birdcage view into cells were seen as ingenious for keeping unscrupulous convicts in line, and prisoners were forced to remain engaged and useful—assuming these were best practices for curbing recidivism. Lemke said prisoners have rights now. None of what Ragen instituted to keep prisoners busy and out of trouble is allowable. This, of course, includes letting in scientists to experiment on them in exchange for time off. The ethics are clearly defined and, interestingly, resemble the standards articulated during the Nuremberg trials by Werner Leibbrandt—that prisoners are in "a forced situation." They lack the free will so necessary to making an informed, uncoerced decision around volunteering for medical experiments. It's simply forbidden. As are malaria experiments on mental health patients.

Today, regular people volunteer and receive as much as $6,000 or more to participate in NIH-sponsored malaria vaccine trials. They help establish safety and some level of efficacy. Then the vaccines are sent to Africa to be tested on large numbers of volunteers infected with naturally occurring malaria.

NATHAN Leopold tried many times to get Alf to help him win parole, but the doctor refused. All he got from Alf was a generic letter sent to all volunteers, dated June 25, 1945. It said Leopold had voluntarily submitted to malaria experiments. "These experiments cause considerable physical discomfort and are not without danger. . . . The medical officers assigned to this work by the Surgeon-General of the United States Army, join me in requesting that our appreciation of this man's loyalty and patriotism be incorporated into his official record."[35] Leopold also received the standard certificate of merit, acknowledging his contribution to the war effort.

But Leopold needed more; he saw the Malaria Project as "the finest thing" that happened to him in prison. It also presented an opportunity he "could not afford to miss."[36] All the publicity he received, up to the Malaria Project, had been negative—about how he had kidnapped, murdered, and stuffed into a drainpipe poor Bobby Franks. But now he had helped the Malaria Project, not just by recruiting inmates, but by making sure volunteers cooperated, which preserved the integrity of the results. He kept the cons in line. When Governor Green had announced that all volunteers would be given a chance at parole, Leopold, like the others, had lined up with anticipation. He had wanted his work on the Malaria Project to convey how sorry and repentant he was.[37]

But Leopold was denied parole because of the special circumstances of his crime. He sat out as two rounds of hearings held specifically for malaria volunteers released some two hundred inmates. He was allowed to submit a petition for executive clemency, and in 1948 his ninety-nine-year sentence was reduced to eighty-five years, which meant he'd be eligible for parole in 1953 instead of 1957. With that, he launched a letter-writing campaign to line up support, convincing two of Alf's colleagues to write on his behalf, John H. Edgcomb and Clay Huff. Both stated he had been reformed by the malaria work and was no longer a dangerous man. Leopold's lawyer, Elmer Gertz, included an April 1947 *Time* magazine article that said his client helped with the "birth of new knowledge" when "the war was on!" and malaria was "causing more casualties in the South Pacific than Japanese bullets." The article said: "[T]he malaria project is probably the best thing to ever happen in the Illinois State Penitentiary, and the beneficiary is society!"

Also included was a letter from J. B. Rice, a vice president of Winthrop. He wrote of Leopold: "I think it entirely possible that because of his profound intellectual attainments he may yet make some very worthwhile contribution to society, if given the opportunity."[38] Well-known journalist John Barlow Martin wrote that time had cured the immaturity and rebellion that had led Leopold to kill. His was a "situational" murder committed by a boy. As a man, he was no longer dangerous.[39]

Leopold also wrote an autobiography designed to convince the world that he had changed. Carl Sandburg, the famous poet, helped him get a contract with Doubleday and assured Leopold that the book would make him "many friends. Long after we, the living, are all vanished, your book will be telling your story."[40]

In 1957, as he finished his manuscript, he was granted another hearing. Again his lawyers petitioned Alf to support Leopold's release, and this time Alf relented and gave his verbal consent. By now Leopold had amassed hundreds of letters of support (that today fill twenty-four files at the Northwestern University archives). Included in his package was a job offer in Puerto Rico to work as an X-ray technician at the Brethren Service Commission Hospital, run by a Protestant organization out of Illinois.[41] This time he won parole, and in 1958, shortly after his release, he fled a crush of publicity to settle in a quiet village near San Juan. There he married and lived out his remaining years working, traveling, and birding—a member of society's rich but infamous. While he tried to forget what he had done, writers and filmmakers kept reminding him (including Alfred Hitchcock's movie *Rope*, starring James Stewart, and Meyer Levin's book *Compulsion*, which Leopold tried but failed to legally block, and which was made into a movie).

Leopold kept trying to use the Malaria Project to earn legitimacy. Dr. Geoffrey Jeffery worked for Coatney in Atlanta during the war and first met Leopold at Stateville in 1945, when he flew up to Chicago to bring Alf mosquitoes infected with the Chesson strain. Nearly fifteen years later, just after Leopold was released from prison, Jeffery bumped into him while in San Juan for an annual meeting of the American Society of Tropical Medicine and Hygiene. "[Leopold] asked to be an honorary member," said Jeffery, who was secretary of the society at the time. The members voted down Leopold's request, feeling his reasons were more political than scientific. "He had no degree or position," Jeffery said.[42] Plus Leopold was blatantly self-serving. Jeffery remembered that about him, and remembered that his personality had interfered with his usefulness on the Malaria Project. He wasn't accepted by the society.

Leopold died in 1971 from heart failure, the first signs of which started shortly after the 8-aminoquinoline was used to cure him of the Chesson strain. The combination of drug and disease had probably weakened his heart.

LOWELL Coggeshall had done his last malaria experiment at Stateville in 1946. In taking over as chair of the University of Chicago's medical school and dean of its entire Division of Biomedical Sciences, he finally stepped away from malaria for good, and focused more broadly on the roles of medical education and medicine in society. In 1949 he was elected to the National Academy of Sciences. He was recruited in 1956 to serve as a special assistant to the U.S. secretary of health, education, and welfare. In 1962 he chaired a seventeen-committee commission to investigate drug safety, following the thalidomide debacle in which thousands of pregnant women who were prescribed the drug for morning sickness gave birth to deformed babies.

In 1965, the year Alf Alving died, Lowell wrote what he is best known for today: the so-called "Coggeshall Report" (formally called "Planning for Medical Progress Through Education"). He had been asked to chair a group of "luminaries" to come up with an analysis of where medical schools should be in the betterment of society. The report, which he ended up writing alone, reflected deep concerns about an increasing focus on specialties. He called for medical educators to train doctors to be agents of society, "licensed by society to serve society before serving himself."[43] Historian Joel D. Howell said Lowell Coggeshall was "by all accounts, a marvelously likable person and a born executive" who wanted medical education "to devote increasing attention to the requirements of the nation" and train medical students in the art of health-care delivery services.[44] These ideas took on the American Medical Association—run by specialists—because they sought to create a counterbalance to the AMA's enormous lobbying influence by strengthening the American Association of Medical Colleges. To do this, Lowell proposed that the AAMC include the professions of nursing and public health, and shift medical education to a team approach—versus each practice for itself or focusing on a niche body part or type of disease.

This threatened to upset the status quo and was met by hyperbolic accusations that he was staging a takeover of medical education, and trying to plug it with political power. Howell documented how some likened Lowell's work to the Nazi takeover of education in prewar Germany, and trends developing in Communist China. One detractor said, "[T]he old alliance of Church and State, fortunately gone with the Middle Ages, could be replaced by Education and State."[45]

This, of course, was not what Lowell expected or proposed. He simply saw danger in a medical profession so focused on specialties. He was accused of trying to use federal support of medical education to socialize it. But, according to Howell, Lowell simply articulated concerns of the time. Lowell went out on a limb to say what needed to be said: that there existed a disconnect between technologically elite medical professionals and a responsibility to apply their skills in the service of mankind.

This was Lowell's last act before retiring to Alabama, where he died on November 11, 1987, at age eighty-six.

EPILOGUE

In August 2005, a twenty-four-year-old farmer named Ysufu fell onto a grass mat on the dirt floor of his mud hut in the Pombwe Delta of Tanzania, burning with malaria.

I met him about a week later. His lips were so severely swollen that he couldn't speak. Instead he nodded that, yes, he had taken a sulfa-based antimalarial called "SP" (sulfadoxine and pyrimethamine, also known as Fansidar). In the 1990s, SP was brought to Tanzania to replace chloroquine (SN-7618), which after decades of use had stopped working. SP, however, was far from ideal for Ysufu—he clearly suffered from a common allergy to sulfa. He didn't know it until he broke out in a fiery rash; then his skin fell off in sheets. This severe reaction, called Stevens-Johnson syndrome, is gruesome around the orifices, and made Ysufu's face look inside out.

I was with Joseph Njau, a Tanzanian public health researcher, and Rene Gerretts, a Dutch anthropologist, when we found Ysufu resting at a rustic clinic in the village of Mohoro. He had been there four days and needed to get to a regional hospital two hours away for any hope of surviving. Salimani Muhade, the clinic's director, claimed he had no ambulance to transport him. For forty thousand shillings ($40) he said he could hire a truck. Shifting his eyes he said, "Or maybe fifty thousand." An hour later it was seventy thousand. So we decided to drive Ysufu. And the price fell back to fifty thousand shillings, which Joseph paid.

Two men lifted Ysufu's bloodied body onto a mattress in the bed of a small Toyota pickup truck. They gave him a bedsheet that he pulled over his head and off he went, covered like a corpse. We watched the truck bounce over rutted clay roads until it disappeared in a cloud of dust; then we three drove off in the opposite direction, bouncing over more rutted roads. Dust stuck in my teeth as Rene and I accused Muhade of using poor Ysufu to extort cash. But Joseph just sat there, unfazed. I energetically asked why he wasn't mad. And he chuckled at me. As he wiped dust from his brow, he smiled and said, "*This* is Africa." Muhade probably got into health care for the right reasons. But imagine running a clinic with almost nothing—no bandages or medications, short on staff, empty promises from the government, and the throngs begging you to save them. This can turn a good man bad. Wise Joseph then used his friendliest smile to deliver another pearl. While he agreed that Muhade probably took a cut of the cash, he asked: "How can we know why?" What if it bought supplies for his empty shelves, or paid staff, would I still be angry? We Americans are quick to judge, Joseph said, still smiling. We should have faith in the good in people.

Joseph told me a year later that Ysufu was hospitalized for weeks, but survived. We had saved him for $50. He probably never again sought treatment for malaria. He certainly had no idea that the drug that almost killed him came out of the World War II program. SP's two parts were derived from sulfamerzine (Mary) and paludrine (SN-12,837). Its very existence is a testament to the project's success at rooting out different drug options. But his reaction exemplifies how flawed those options continue to be.

AFTER the war, chloroquine (SN-7618) was the primary option all over the world, used along with DDT in the World Health Organization's Global Malaria Eradication Programme, which was the agency's first major public health campaign after it was created in 1948. The existence of chloroquine and DDT made the effort possible; politicians claimed they could be used in targeted ways until malaria was wiped from the planet. It didn't happen, as Paul Russell and Dr. Watson (who had gone to Peru with George Carden) had predicted. They strenuously advocated for longer-term strategies that involved antipoverty programs along with hydroelectric projects, ditching, and drainage work, as was done in the U.S. South.[1] But the WHO had financial and political constraints that made the larger work impossible. So the agency marched forward and, at first, depended largely on DDT. Men with knapsack sprayers doused villages with it to kill mosquitoes. It also wiped out horrid flies that carried dysentery, and ubiquitous bedbugs, typhus-carrying lice, and fleas that left itchy, painful bites.[2] The flies came back first. Then mosquitoes developed resistance and exhibited what entomologists call "refractory behavior." That is, they avoided anything that smelled of DDT.

That's when the WHO turned to mass treatment using chloroquine. Throughout the 1950s and 1960s, it was the second-most-commonly-taken medicine in the world, after aspirin, "[A]nd it saved millions of lives."[3] Some scientists say it may have saved more lives than any synthetic drug ever made because of how broadly it was used in global treatment programs. Malaria cases in India, for example, went from millions to less than a thousand, and in Sri Lanka to just a few dozen. The same happened in countries all over the globe.

Eventually parasite resistance cropped up and began to spread.[4] By 1969, the WHO admitted defeat. The plan changed from a targeted, short-term push for eradication to one with long-term goals integrated into socioeconomic development, as Russell, Watson, and others had originally proposed. Countries with populations too poor to afford screens, bed nets, and proper disease surveillance and mitigation ended up worse off than ever. Malaria rates in Sri Lanka and India, for example, exceeded those before the campaign. Death rates were higher too because people had lost their acquired immunity.

The world needed more drugs, so they took out Wiselogle's catalog of compounds and began testing the leftovers from World War II, stored at the NIH.

MOST drugs used today come from this catalog of over 14,000 compounds made during the war, now in the public domain. The worst is a drug called Lariam, the one I took while backpacking with friends through Africa in 1993 (discussed in the Prologue). We each suffered our own personal hell of paranoia. After I endured hallucinations, vertigo, and wildly violent dreams, I tossed my pills into the sands of Monkey Bay, Malawi, not knowing I was following tradition. Nor did I know that Lariam's prototype was derived in 1945 and cataloged by Wiselogle as SN-10,275.[5] Alf used this 4-quinoline methanol at Stateville to cure five prisoners with the Chesson strain.[6] One of two hundred analogues made in this series, SN-10,275 was the most exciting, because it acted three times faster than quinine and remained in the body a long time.[7] "The information at hand indicates that SN-10,275 could be used as a suppressive, administered perhaps as infrequently as once monthly. In addition, the finding of curative activity in a completely new class of compounds in itself warrants further study of the group," read the minutes of the last meeting of the Board for Coordination of Malaria Studies, on June 4, 1946.[8]

The new compounds weren't explored because chloroquine was the better drug. SN-10,275 sat on a shelf until the 1960s, when chloroquine resistance first began to spread. That's when scientists at the Walter Reed Army Institute of Research reexamined it, and eventually developed a variation of it with funding from WHO in partnership with the drug company Hoffman–La Roche. They called it mefloquine and sold it under the name Lariam. Walter Reed did the initial legwork because the military always must have a good antimalarial for troops sent overseas.

For military use, however, Lariam ended up being a bad drug. Troops experienced "neuropsychiatric effects" that "negatively impact compliance with treatment," wrote recent researchers.[9] As if the ghosts of World War II were upon them, U.S. troops sent to Liberia, Somalia, Afghanistan, and elsewhere tossed their pills, then contracted malaria. Experts have since studied this drug and documented its effects. Dr. Remington Nevin listed them in testimony he gave on June 6, 2012, before the U.S. Senate Appropriations Committee during a hearing on the defense budget: "vivid nightmares, profound anxiety, aggression, delusional paranoia, dissociative psychosis, and severe memory loss." He called the symptoms "severe intoxication syndrome"—a type of brain injury—and said they were responsible for violent acts committed by U.S. troops overseas. So widespread is the syndrome, he said, that it should be included with post-traumatic stress and traumatic brain injury as "the third recognized signature injury of modern war."[10]

In 1994, after I returned from Africa, I had several dramatic episodes of dizziness and vertigo, as if my brain were flipping over inside my skull. Eighteen months after discarding my pills at Monkey Bay, I had a flashback while hiking Mount Cotopaxi in Ecuador. The world tipped to one side—my brain did another flip. I drove my ice pick into the glacier and held on, feeling like I was about to fall into the barren valley thousands of feet below. For days afterward, just as I drifted into sleep, I'd fall off a cliff or the wing of a plane, and awake with a violent jerk, scared out of my mind. My symptoms were mild compared to those of an American Fulbright scholar who took his Lariam while in India and developed a rare side effect that wiped his memory clean. The harrowing story is told by the victim, David Stuart MacLean, in his memoir *The Answer to the Riddle Is Me.*

An alternative for travelers is an antibiotic called doxycycline, by Pfizer. Lowell Coggeshall was the first researcher in the United States to test an antibiotic against malaria. G. Robert Coatney tested several of them in 1946 on inmates at a federal prison in Seagoville, Texas.[11] Nearly a half century later, in 1992, Walter Reed Army Institute of Research helped develop doxycycline for FDA approval as an alternative to Lariam, and to be used in areas with chloroquine resistance. This antibiotic works well but has its drawbacks. The biggest, by military standards, is that it has to be taken every day, which is hard to enforce. It also causes stomach problems, diarrhea, and phototoxicity, which discourage compliance—troops refuse to swallow the pills. What's more, the drug can cause discoloration to teeth if taken by young children or pregnant women.[12]

The drug I chose to take while in Tanzania for the CDC in 2005 was called Malarone, made by GlaxoSmithKline. This is a combination of atovaquone and proguanil. Proguanil's predecessor is paludrine, which was made by the British during World War II and cataloged as SN-12,837.[13] Alf ran the first big study of it with assistants Nathan Leopold and Charles Ickes. The FDA approved paludrine in 1948, but it wasn't used much because chloroquine was the better drug. In 2000, with chloroquine crippled by resistance, Lariam making people crazy, and SP triggering hideous Stevens-Johnson syndrome, researchers took paludrine off the shelf. They rotated a few molecules to make proguanil, added it to atovaquone (developed to treat pneumonia), and marveled at how the oddly coupled drugs showed "profound antimalarial synergy" with almost no side effects—and 98 percent efficacy.[14] The one drawback is that it's too expensive to use in the military or in large treatment programs in poor countries.

I knew none of this in 2005. I was just grateful to have it as an option. I took it for six weeks and had no side effects while being fully protected in areas with measured infectious bite rates that reached two thousand per year (five per day!). Malarone remains effective because it's too expensive to use broadly enough to trip the dynamics of resistance.

TODAY, to treat the masses, WHO uses a fairly new drug called artemisinin. In the 1960s, Chairman Mao ordered Chinese chemists to scour the countryside for an effective herb against malaria. After testing two hundred plants they found qing hao, also known as sweet wormwood, and scientifically cataloged as *Arte-*

misia annua. They claimed it cleared parasites faster than any drug ever made. But Western researchers ignored them. That is, until the 1990s, when they finally studied and accepted the value of this Chinese discovery. Another decade passed as officials worked through the hurdles of making it cheaply enough to use in WHO programs. That finally happened in 2006. Had Ysufu sought medicine a year later, he would have been given this nontoxic drug rather than SP.

Artemisinin-based combination therapies (or ACTs) remain the frontline drugs for treating the poor. But drug resistance that started in Cambodia is now spreading fast in the blood of migrating people, marching militias, and international travelers. The WHO forbids countries to use artemisinin as a monotherapy. It must be combined with other drugs, a strategy that tends to slow the spread of resistance. But the fact that resistance has already appeared means new drugs will be needed soon.

Primaquine—Alf's drug—continues to be among the most important medicines available because of its ability to radically cure relapsing *vivax* and block transmission to mosquitoes. It's the only 8-aminoquinoline approved by the FDA. But its dangerous blood-destroying side effect in G6PD-deficient people severely limits its use overseas. At present, scientists are trying to devise a cheap way to diagnose this enzyme deficiency so people with it living in the tropics can be identified, and everyone else can be safely treated. Colin Ohrt, an expert on this topic, said, "[G]ood tests exist for those in the Western world," but they're too expensive to use in mass treatment programs. Scientists are also trying to get around the problem by combining chloroquine with low doses of an analogue of primaquine called tafenoquine. Phase III studies are under way in Asia to see whether this reduces "hemolysis" (death of red blood cells due to G6PD deficiency) while also remaining powerful enough to destroy parasites in the blood and liver. The study is being watched closely because, if it works, FDA approval would follow, and this combination therapy could be used to drive down transmission of *vivax* in Asia and the Pacific, maybe even dramatically.

As for sontochin, Jean Schneider's drug—SN-6911—it recently made a comeback. Researchers in Portland, Oregon, were "surprised" to discover that it "retains significant activity" against strains of *falciparum* that are resistant to chloroquine and other drugs.[15] This ignited new interest in developing sontochin as a better antimalarial than chloroquine—which Lowell and the navy felt to be the case way back in 1945. After the United States put its weight behind chloroquine, the French marketed sontochin as Nivaquine. It wasn't available in the United States, but it was in Britain, and many there immediately following the war preferred it over chloroquine.[16] Then the WHO endorsed chloroquine and sontochin fell into obscurity. Now sontochin may finally have its turn at malaria.

THE Malaria Project's legacy lives on in global health strategies that are dominated by drug and vaccine development. Billions of dollars committed by the Bill and Melinda Gates Foundation and donor nations have placed a disproportionate emphasis on magic-bullet thinking. The Gates Foundation even used the word *eradication* as the goal for their efforts, which is admirable but seen by many malariologists as naive. Without job-generating investments and better overall health care for the rural poor, these high-tech products are severely handicapped and set up to fail. Everyone agrees that entrenched malaria exists only where populations are too poor to protect themselves from mosquitoes. Why not concentrate all this financing on programs that help lift people from poverty, knowing that that alone will eradicate malaria? This is a question that, in different forms, applies to assistance programs for many diseases circumscribed by poverty—for which billions more are spent. Dysentery, filariasis (and elephantiasis), dengue, schistosomiasis, leishmaniasis, cholera, even HIV/AIDS, each invokes its own version of this question. Each is preventable in countries that have the resources and systems in place to diagnose, isolate, and treat patients. A serious discussion of eradication, therefore, must be embedded in the larger matter of why we fail to reform policies that allow so few of us to have so much while so many are left with so little—not even screens to keep out mosquitoes.

But that's beyond the scope of public health practitioners. So they are left with little choice but to isolate malaria, call it a disease, and hope that funds continue to support the search for the elusive magic bullet.

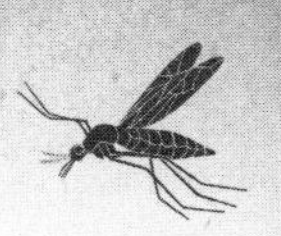

ACKNOWLEDGMENTS

Many good people helped me with this project and I owe them all a debt of gratitude. First and foremost I want to thank my husband and best friend, Tom Wang, to whom this book is dedicated. I couldn't have written it without his love and support. Many thanks also go to my agent, Jeff Kleinman, for talking me through the story line early on, and helping me out of the weeds; and to my editor, Brent Howard, for having faith in this book and in me as an author, and for his excellent edits and amazing patience.

Many additional people had a hand in this book, to whom I am indebted.

On the science, I want to thank Johns Hopkins University and Ann Finkbeiner for awarding me a graduate fellowship to explore in coursework the history of public health—which is largely about malaria. Special thanks go to Ann for her support and sage advice on moving forward with a book project of this size and scope. I also owe many thanks to the John S. and James L. Knight Foundation and Charlie Haddad for awarding me a journalism fellowship to study malaria at the CDC, and to Charlie for becoming my friend and guide on all things related to Atlanta. I'm also deeply grateful to the generous and dedicated experts I learned from while at CDC, especially Bill Collins (may he rest in peace), John Barnwell, Mark Eberhard, Peter Bloland, Patrick Kucher, and CDC's partners in Tanzania, Hassan Mshindafor, Gerry Killeen, Thomas Smith, and, especially, Joseph Njau and Rene Gerretts. I must also thank malaria drug and vaccine experts at Walter Reed Army Institute of Research and NIH, especially Colin Ohrt, Wolfgang Leitner, and Elke Bergmann-Leitner. Wolfgang and Elke took time from their busy lives to translate German papers. And they read early drafts, then offered edits on the science and political history of Austria that saved me from embarrassing mistakes (those that remain are entirely my doing).

Dr. Carl Alving and Ambassador Theodore Russell, sons of Alf and Paul, were invaluable in providing me with out-of-print and unpublished papers, diaries, photos, insights, personal stories, and encouragement. I cannot thank them enough. They helped me bring their fathers back to life with as much accuracy as possible, and served as valuable sounding boards for matters of interpretation. They invited me into their homes and served me many cups of coffee. Carl very generously invited me to dinner with today's top malaria researchers at NIH and the Walter Reed research institute, and took me on a tour of Walter Reed. Then we flew to Chicago and with his brother Donald Peter Alving toured Stateville Prison. For that tour, special thanks go to the then warden, Michael Lemke, his assistant, Kevin Senor, and the communications director for the Illinois Department of Corrections, Tom Shaer. Special thanks also go to the inmate volunteers at Stateville who went into the prison's damp, disheveled basement to look for records from the 1940s.

Thanks to Carol Govan and Richard Coggeshall, children of Lowell, for sharing stories and some pages from his unpublished autobiography. I hope they find in this story a careful and accurate handling of their father's great works.

I also want to thank Ira Chinoy for teaching me how to navigate the National Archives and for support early on in my research efforts. I'm also deeply grateful to the excellent archivists I had the pleasure of working with at Northwestern University, the University of Miami, the University of Chicago, and Columbia University. Special thanks go to Lisa Pearl at the Holocaust Museum, and, most of all, Richard Boylan at the National Archives and Records Administration at College Park. When I couldn't find what I was looking for, Richard used his decades of experience with the archives' World War II records to intuit that some "refiles" not listed in the reading room's reams of finding aids might have helpful records, and he was right.

A very special thank-you goes to the Stimson Center, especially Cheryl Ramp, Julie Fischer, and Ellen Laipson, for giving me an office in which to do the bulk of my research, an office setting filled with brilliant and interesting people, and two excellent interns, Helen Thompson and Jessica Vineberg. Many thanks go to Helen and Jessica for their hard work, and special thanks to Helen for coming back to me to help with photos and endnotes, even though she was no longer my intern.

On the history, I need to thank Doug Snoeyenbos for his muscular edit of my use of military terms. Thanks to Leo Slater for trekking up to the archives with me when he was still working on his book about the drug companies involved in this project, for sharing research materials with me, and for reading an early version of this book—without complaining too much. Thanks also to Joel Howell for very generously sharing with me his files on Lowell Coggeshall at the University of Michigan.

I also want to thank Amy Allina, Tori Holt, Ted Yoder, and Margie Finn for reading concept ideas and early drafts of chapters, and for pushing me along. Thank you also to Betsy Kramer for astute comments as a reader and help with time management, and to Jordan Schwartz for helpful edits and insights on closing out the story. And I thank the following people for a whole range of help and support: Joel Achenbach, Gawain Kripke, Lynn Erskine, Gene Roberts, Alison Piantedosi, Alexandra Gomez, KinZ Neal, Eric Roston, Joanne Krause, Laura Krause, Shannon O'Boye, Kevin Smith, Lara D'Agaro, Anne Hiller Clark, Larry W. Price, Thomas Culbert, Laura Champe Mitchell, Judith Hassen, and Elizabeth Schill.

And, finally, I want to thank my two daughters, Evie and Sidney, for agreeing to color and draw and play quietly in my office while I wrote and edited and struggled through endnotes. They livened up my days and made writing fun. They are the light in my heart.

BIBLIOGRAPHY

Books

Abraham, George. *The Belles of Shangri-La.* Springfield, MA: Vantage Press, 2000.

Atkinson, Rick. *An Army at Dawn: The War in North Africa, 1942–1943.* New York: Henry Holt and Company, 2002.

———. *The Day of Battle: The War in Sicily and Italy, 1943–1944.* New York: Henry Holt and Company, 2007.

Bagby, Wesley Marvin. *The Eagle-Dragon Alliance: America's Relations with China in World War II.* Newark, Delaware: University of Delaware Press, 1992.

Bender, Marylin and Selig Altschul. *The Chosen Instrument: Pan Am, Juan Trippe: The Rise and Fall of an American Entrepreneur.* New York: Simon and Schuster, 1982.

Bierman, John and Colin Smith, *The Battle of Alamein: Turning Point, World War II.* New York: Viking, 2002.

Chaves-Carballo, E. *The Tropical World of Samuel Taylor Darling: Parasites, Pathology and Philanthropy.* Eastborne, UK: Sussex Academic Press, 2007.

Coates, John Boyd, ed. United States Army Medical Department. *Preventive Medicine in World War II, Vol. VI: Communicable Diseases, Malaria.* Washington, DC: Office of the Surgeon General, 1963. Cited here as Coates, Vol. VI.

Coggeshall, Charles Pierce, and Thellwell Russell Coggeshall. *The Coggeshalls in America: Genealogy of the Descendants of John Coggeshall of Newport with a Brief Notice of Their English Antecedents.* Spartanburg, South Carolina: The Reprint Company, 1982.

Condon-Rall, Mary Ellen. *United States Army in World War II: Technical Services, The Medical Department: Medical Service in the War Against Japan.* Washington: Center of Military History, United States Army, 1998.

Conn, Stetson and Byron Fairchild. *United States Army in World War II, The Western Hemisphere, The Framework of Hemisphere Defense*, Chapter 10: "Air Defense Preparations in Latin America." Washington, DC: Center for Military History, United States Army, 1960.

Daley, Robert. *An American Saga: Juan Trippe and His Pan Am Empire.* New York: Random House, 1980.

de Kruif, Paul. *Microbe Hunters.* Boston: Houghton Mifflin Harcourt, 2002.

Dmitri, Ivan. *Flight to Everywhere.* New York: Whittlesey House, 1944.

Egeberg, Roger O. *The General: MacArthur and the Man He Called "Doc."* New York: Hippocrene Books, 1983.

Farley, John. *To Cast Out Disease: A History of the International Division of the Rockefeller Foundation (1913–1951).* New York: Oxford University Press, 2004.

Garland, Albert, and Howard McGaw Smyth. *United States Army in World War II: The Mediterranean Theater of Operations, Sicily and the Surrender of Italy.* Washington, DC: Center for Military History, United States Army, 1993.

Honigsbaum, Mark. *The Fever Trail: In Search of the Cure for Malaria.* New York: Farrar, Straus and Giroux, 2002.

Humphreys, Margaret. *Malaria: Poverty, Race, and Public Health in the United States.* Baltimore: Johns Hopkins University Press, 2001.

Jacobs, James B. *Stateville: The Penitentiary in Mass Society.* Chicago: The University of Chicago Press, 1977.

Jeffreys, Diarmuid. *Hell's Cartel: IG Farben and the Making of Hitler's War Machine.* New York: Metropolitan Books, 2008.

Kauffman, Sanford. B. *Pan Am Pioneer: A Manager's Memoir from Seaplane Clippers to Jumbo Jets.* Lubbock: Texas Tech University Press, 1995.

Kendrick, Douglas B. *Medical Department United States Army in World War II, Blood Program in World War II.* Washington, DC: Office of the Surgeon General, Department of the Army, 1964.

Kirk, Raymond E., ed. *Encyclopedia of Chemical Technology, Vol. 8.* New York: The Interscience Encyclopedia Inc., 1952.

Lane, Kerry L. *Guadalcanal Marine.* Jackson: University Press of Mississippi, 2004.

Leckie, Robert. *Helmet for My Pillow: From Parris Island to the Pacific.* New York: Bantam Books, 2010.

Leopold, Nathan. *Life Plus 99 Years.* Garden City, NY: Doubleday & Company, 1958.

Litsios, Socrates. *The Tomorrow of Malaria*. Wellington, New Zealand: Pacific Press, 1997.

Macintyre, Ben. *Operation Mincemeat: How a Dead Man and a Bizarre Plan Fooled the Nazis and Assured an Allied Victory*. New York: Broadway Books, 2011.

Miller, Edward S. *War Plan Orange: The U.S. Strategy to Defeat Japan, 1897–1945*. Annapolis: United States Naval Institute Press, 1991.

Nichols, David, ed. *Ernie's War: The Best of Ernie Pyle's World War II Dispatches*. New York: Random House, 1986.

Packard, Randall. *The Making of a Tropical Disease: A Short History of Malaria*. Baltimore: The Johns Hopkins University Press, 2007.

Pyle, Ernie. *Here Is Your War: Story of G.I. Joe*. Lincoln, NE: University of Nebraska Press, 2004.

Ray, Deborah. *Pan American Airways and the Trans-African Air Base Program of World War II*. New York University Press, UMI Dissertation Publishing, 1973.

Richardson, B.K. *A History of the Illinois Department of Public Health, 1927–1962*. State of Illinois, 1963.

Russell, Paul. *Man's Mastery of Malaria*. London: Oxford University Press, 1955.

Sandler, Stanley, *World War II in the Pacific: An Encyclopedia*. New York: Garland Publishing, Inc., 2001.

Schilling, Claus. *Die Methoden der experimentellen Chemotherapie*. Jena: Gustav Fischer, 1938.

Silverman, Milton. *Magic in a Bottle*. New York: The Macmillan Company, 1941.

Slater, Leo B. *War and Medicine: Biomedical Research on Malaria in the Twentieth Century*. New Brunswick: Rutgers University Press, 2009.

Slim, William (Field Marshal Viscount). *Defeat into Victory: Battling Japan in Burma and India, 1942–1945*. New York: Cooper Square Press, 2000.

Snowden, Frank M. *The Conquest of Malaria: Italy, 1900–1962*. New Haven: Yale University Press, 2006.

Snyder, Louis Leo. *Hitler's Third Reich: A Documentary History*. Chicago: Nelson-Hall, 1982.

Spielman, A., and M. D'Antonio, *Mosquito: The Story of Man's Deadliest Foe*. New York: Hyperion, 2001.

Tregaskis, Richard. *Guadalcanal Diary*. New York: The Modern Library, 2000.

Webb, James L. A., Jr. *Humanity's Burden: A Global History of Malaria*. New York: Cambridge University Press, 2009.

Whitman, John W. *Bataan: Our Last Ditch.* New York: Hippocrene Books, 1990.

Whitrow, Magda. *Julius Wagner-Jauregg (1857–1940).* London: Smith-Gordon, 1993.

Wiltse, Charles M. "Army Medical Service in Africa and the Middle East," *The United States Army in World War II: The Technical Services, The Medical Department, Medical Service in the Mediterranean and Minor Theaters.* Washington, DC: Office of the Chief of Military History, Department of the Army, 1965.

Archives

Leopold and Loeb Collection, Elmer Gertz Papers, Northwestern University Library, Archival and Manuscript Collection.

George and Katy Abraham Papers, 1915–2005, Cornell University Library, Division of Rare and Manuscript Collections.

William Mansfield Clark Papers, American Philosophical Society.

Lowell T. Coggeshall Papers and Leon O. Jacobson Papers, University of Chicago Library, Special Collections Research Center.

Reminiscences of Lowell T. Coggeshall: Oral History, 1964, Columbia University Libraries, Health Sciences Project.

Pan American Airways Archives, University of Miami Libraries, Special Collections.

Philip S. Hench Walter Reed Yellow Fever Collection, University of Virginia, Health Sciences Library, Historical Collections.

World War II Collections, U.S. National Archive and Records Administration, College Park.

The Wilbur A. Sawyer Papers, U.S. National Library of Medicine.

Journal Articles, Chapters, and Speeches

Barber, M. A. "The Present Status of Anopheles Gambiae in Brazil," *American Journal of Tropical Medicine and Hygiene,* Vols. 1–20, no. 2 (March 1940).

Beadle, Christine, and Hoffman, Stephen L. "History of Malaria in the United States Naval Forces at War: World War I through the Vietnam Conflict," *Clinical of Infectious Diseases*, Vol. 16, no. 2 (Feb. 1993).

Beare, Nicholas A.V., et al. "Perfusion Abnormalities in Children with Cerebral Malaria and Malarial Retinopathy," *Journal of Infectious Diseases* 199 (Jan. 15, 2009).

Birtle, Andrew James. *Sicily: The U.S. Army Campaigns of World War II* (Pamphlet, U.S. Center of Military History Publication, 2004).

Braslow, J. T. "Effect of therapeutic innovation on perception of the disease of the Doctor-Patient Relationship: A History of General Paralysis of the Insane and Malaria Fever Therapy, 1910–1950," *The American Journal of Psychiatry* 5 (May 1995).

Brown, Edward. "Why Wagner-Jauregg Won the Nobel Prize for Discovering Malaria Therapy for General Paresis of the Insane," *History of Psychiatry* 11 (2000).

Chernin, Eli. "The Malariatherapy of Neurosyphilis," *The Journal of Parasitology* 70, no. 5 (Oct. 1984).

Coatney, Robert G. "Pitfalls in a discovery: the chronicle of chloroquine," *American Journal of Tropical Medicine and Hygiene* 12 (March 1963).

———. "Reminiscences: My Forty-Year Romance with Malaria," *Transactions of the Nebraska Academy of Sciences and Affiliated Societies*, Paper 222 (1985).

Coggeshall, L. T. "The Cure of Plasmodium Knowlesi Malaria in Rhesus Monkeys with Sulfanilamide and Their Susceptibility to Reinfection," *American Journal of Tropical Medicine* 18, no. 331 (1938).

———. "Malaria as a World Menace," *Journal of the American Medical Association*, Vol. 122, no. 1 (May 1, 1943).

Coggeshall, L. T. (and personnel of the medical department). "Development of a Medical Service for Airline Operations in Africa," reprinted in *War Medicine*, Vol. 3 (May 1943).

Coggeshall, L. T., et al, *Science* 92, no. 2382 (Aug. 23, 1940).

Coggeshall, Lowell T. "The Occurrences of Malarial Antibodies in Human Serum Following Induced Infection with Plasmodium Knowlesi," *Journal of Experimental Medicine* 72, no. 21 (1940).

———. "Plasmodium lophurae, a New Species of Malaria Parasite Pathogenic for the Domestic Fowl," *American Journal of Hygiene*, Vol. 27 (1938).

———. "Relationship of Plankton to Anopheline Larvae," *American Journal of Epidemiology*, 6:4 (1926).

Darling, Samuel Taylor. "Factors in the transmission and prevention of malaria in the Panama Canal Zone," *Annals of Tropical Medicine and Parasitology*, Vol. IV, no. 2 (July 1910).

Das Gupta, B. M. "Transmission of P. inui to Man," *Proceedings of the National Institute of Science, India* 4 (June 1938).

de Kruif, Paul. "Enter Atabrine—Exit Malaria," *Reader's Digest* (Dec. 1942).

Diller, William F. "Notes on Filariasis in Liberia," *The Journal of Parasitology* 33, no. 4 (1947).

Doan, Charles A. "The First International Conference on Fever Therapy," *The Scientific Monthly* 44, no. 5 (May 1937).

Earle, David P. "Malaria and Its Ironies," *American Clinical Climatological Association*, Vol. 90 (1979).

Eaton, Monroe D., and Lowell T. Coggeshall. "Complement Fixation in Human Malaria with an Antigen Prepared from the Monkey Parasite Plasmodium Knowlesi," *The Journal of Experimental Medicine* 69 (1939).

———. "Production in Monkeys of Complement-Fixing Antibodies Without Active Immunity by Injection of Killed Plasmodium Knowlesi," *The Journal of Experimental Medicine* 70 (1939).

Eckart, Wolfgang W., and H. Vondra. "Malaria and World War II: German malaria experiments 1939–45," *Parasitologia*. Vol. 42 (2000).

Fedunkiw, Marianne. "Malaria Films: Motion Pictures as a Public Health Tool," *American Journal of Public Health*, Vol. 93, no. 7 (2003).

Gartlehner, G., and K. Stepper. "Julius Wagner-Jauregg: Pyrotherapy, Simultanmethode, and 'racial hygiene,'" *Journal of the Royal Society of Medicine*, 105: 357–59 (2012).

Gladwell, Malcolm. "The Mosquito Killer," *The New Yorker*, July 2, 2001. http://www.newyorker.com/archive/2001/07/02/010702fa_fact_gladwell.

Greenwood, David. "Historical Perspective: Conflicts of Interest: the genesis of synthetic antimalarial agents in peace and war," *Journal of Antimicrobial Chemotherapy* 36 (1995).

Hamilton, Walton. "The Strange Case of Sterling Products," *Harper's Magazine*, Vol. 186, no. 1112 (Jan. 1, 1943).

Howes, Gliatto, et al, "Neurosyphilis: A history and clinical review," *Psychiatric Annals* 31, no. 1 (March 2001).

Howes, Oliver D., et al. "Julius Wagner-Jaruegg, 1857–1940," *The American Journal of Psychiatry* 166, no. 4 (April 2009).

Huggins, Charles B. "The Business of Discovery in the Medical Sciences," *Journal of the American Medical Association*, 194, no. 11 (Dec. 13, 1965).

Hulverscheidt, Marion. "German Malariology experiments with humans, supported by the DFG until 1945," ed. In Wolfgang U. Eckart, *Man, Medicine and the State: The Human Body as an Object of Government Sponsored Medical Research in the 20th Century*. Stuttgart: Franz Steiner Verlag, 2006.

Hutchisson, J. M. "Sinclair Lewis, Paul de Kruif, and the Composition of 'Arrowsmith,'" *Studies in the Novel*, Vol. 24 (1992).

Joy, Robert T. "Malaria in American Troops in the South and Southwest Pacific in World War II," *Medical History* 43 (1999).

Killeen, Gerry, et al. "Eradication of Anopheles gambiae from Brazil: lessons for malaria control in Africa?" *The Lancet* 2, no. 10 (Oct. 2002).

Kitchen, Lynn W., David W. Vaughn, et al. "Role of U.S. Military Research Programs in the Development of U.S. Food and Drug Administration-Approved Antimalarial Drugs." *Clinical Infectious Diseases*, 43, no. 1 (July 1, 2006).

Koch, Robert. "Remarks on Trypanosome Diseases," *British Medical Journal* 2, no. 2291 (Nov. 26, 1904).

Lacroix, Renaud, et al. "Malaria Infection Increases Attractiveness of Humans to Mosquitoes," *PLoS*, 3, no. 9.

Maren, Thomas H. "Eli Kennerly Marshall, Jr.: 1889–1996." In *Biographical Memoirs*. Washington, DC: National Academy of Sciences, 1987.

Martin, J. Purdon. "Conquest of General Paralysis," *British Medical Journal* 3 (1972).

Marx, Otto M. "Book Reviews" (review of *Julius Wagner-Jauregg* by Magna Whitrow), *Isis* 85, no. 4 (Dec. 1994).

Miller, Louis H., et al. "The Pathogenic Basis of Malaria," *Nature* (Feb. 7, 2002).

Pou, S., R. Winter, et al. "Sontochin as a Guide to the Development of Drugs against Chloroquine-Resistant Malaria," *Antimicrobial Agents and Chemotherapy*, 56, no. 7 (April 16, 2012).

Rajala, Hope. "Joliet and Stateville: The Twin Prisons," *Quarterly Publication*, Will Country Historical Society (Summer 2000).

Rose, Gerhard. "Prophylactic Treatment of Malaria with Atabrine: Its dosage and alleged complications," *Deutsche Medizinische Wochenschrift*, 67, no. 48 (Nov. 1941).

Rosenman, Stanley. "Review Essay: Freud Contesting the Predatory System" (review of *Freud as an Expert Witness: The Discussion of War Neuroses Between Freud and Wagner-Jauregg*, by K. R. Eissler), *The American Journal of Psychoanalysis* 49, no. 2 (1989).

Russell, Paul. "Introduction," *Preventive Medicine in World War II: Communicable Diseases, Vol. VI*, ed. John Boyd Coates. Washington, DC: Department of the Army, 1963.

Sweeney, Anthony, W. "The Malaria Frontline: Pioneering malaria research by the Australian Army in World War II," *Medical Journal of Australia*, 166, no. 17 (March 1997).

Talmadge, David W. "William Hay Taliaferro," in *Biological Memoirs*. Washington, DC: National Academy of Sciences, 1983.

Warren, Don C. "The Egg and I for 50 Years," Proceedings of the Twenty-Sixth Annual National Poultry Breeders' Roundtable, Hotel Continental, Kansas City, Missouri (May 5–6, 1977).

Weindling, Paul. "From Medical War Crimes to Compensation: The Plight of the Victims of Human Experiments." In *Man, Medicine and the State: The Human Body as an Object of Government Sponsored Medical Research in the 20th Century*, ed. Wolfgang U. Eckart. Heidelberg, Germany: Franz Steiner Verlag, 2006.

Welch, S. W. "Malaria Control Work in Alabama," *Southern Medical Journal* 20: 6 (June 1927).

Whitrow, Magda. "Wagner-Jauregg and Fever Therapy," *Medical History* 34 (1990).

Newspapers, Magazines, News Journals, and Proceedings, Organized by Publication, Then Date

The British Medical Journal

"A Synthetic Remedy for Malaria," Vol. 2, no. 3434 (Oct. 30, 1926).

"Reports of Societies," Vol. 1, no. 3453 (March 12, 1927).

"Nobel Prizes," Vol. 2, no. 3489 (Nov. 19, 1927).

"The First International Malaria Congress," Vol. 2, no. 3386 (Nov. 21 1925).

"Reports of Societies: Synthetic Anti-Malaria Drugs," Vol. 1, no. 3706 (Jan. 16, 1932).

"A Malaria Treatment Centre: The Unit at Horton Medical Hospital," Vol. 1, no. 3985 (May 22, 1937).

"Treatment of Malaria," Vol. 2, no. 4845 (Nov. 14, 1953).

Chicago Daily Tribune

"Ex-Kaiser Joins Fascisti; Finds Cool Welcome," May 19, 1924.

"Bare Foiled Plot for an Outbreak at New Prison: National Guard Ready to Curb Convicts," March 17, 1931.

"Two Gun Killer is Sentenced to 99 Years in Cell: Called a 'Cold Blooded Bandit,'" Dec. 13, 1933.

"Kills Loeb; Prison Scandal," Jan. 29, 1936.

"Prisoner Power of Loeb Told: Fight to Death with Razor in Bath Described," Jan. 30, 1936.

"Jurors Acquit Razor Slayer of Young Loeb," June 5, 1936.

"Tighten Rules at Stateville; Favoritism Goes," March 7, 1937.

"Illinois Prison Employees Put on Merit Basis," Sept. 22, 1937.

"Warden Perfects Plan to Guard Roads in Prison Breaks," Jan. 4, 1938.

"4 Young Killers Sentenced; 3 Are Doomed to Chair: Policeman's Slayer Gets Life Prison Term," Jan. 21, 1940.

"Ex-Convict Gets Life in Filling Station Killing," Jan. 8, 1941.

"Bells Regulate Life in Prison at Stateville," Nov. 25, 1945.

"Front Views & Profiles," July 26, 1948.

"Notorious Chicago Badmen Get Time Cut for Guinea Pig Roles," Jan. 8, 1949.

"Malaria Hero at U.C. Seized—as a Robber," Feb. 7, 1962.

"Dr. Ernst Beutler: 1928–2008: Researcher Pioneered Treatments for Blood Diseases," Oct. 10, 2008.

Christian Science Monitor

"Guilt Admitted by Adolf Hilter," Feb. 27, 1924.

"Air Ferry Firm to Africa Formed by Pan American," Aug. 21, 1941.

"A 'Benevolent Neutral' Drawn Closer by President's Visit," Jan. 29, 1943.

The Manchester Guardian

"German Treason Trial: Hitler's Frothy Defence," Feb. 28, 1924.

"Election Prospects in Germany: 25 Parties, Many of Them Freaks," Dec. 5, 1924.

"Prisoners Released in Bavaria," Dec. 22, 1924.

The New York Times

"New Discoveries May Eliminate Quinine for Malaria," July 13, 1913.

"World Doctors Hail Ehrlich as Hero," Aug. 9, 1913.

"Germany Seeks Arrest of Four Dye Experts Who Accepted Jobs Here—Two Have Landed," Feb. 21, 1921.

"Chemist to Fight German Charges: Jordan, Now in Holland, Ready to Face Accusers in Cologne," March 18, 1921.

"Ask Jarres to Act Against Racialists: German Moderates Say Speeches at Weimar Convention Were Treasonable," Sept. 7, 1924.

"Hitler Tamed by Prison," Dec. 21, 1924.

"Austrian Gets Nobel Prize," Oct. 28, 1927.

"Austrian Psychiatrist Wins Nobel Prize for Medicine," Nov. 6, 1927.

"Anschluss Appeal Is Signed by Many," Nov. 10, 1928.

"Death at Joliet Heightens Unrest: Precautions Intensified as the Second Prisoner Shot in Mutiny Succumbs," March 17, 1931.

"German Dyes Trust Reports Big Gains: 5,000 More Men Hired in Last Quarter of 1932 to Meet Rise in Home and Export Demand," Jan. 28, 1933.

"German Dye Trust Lists Export Drop," Nov. 15, 1934.

"Convict Kills Loeb, Franks Boy Slayer: Chicago Murderer Serving Life for 'Perfect Crime' Slashed with Razor in Brawl," Jan. 29, 1936.

"Experts on Fever to Meet this Week: 80th Birthday of Prof. Jauregg Nobel Prize Winner, to Be Commemorated," March 28, 1937.

"Notes for Travelers," Nov. 30, 1941.

"Troops on Bataan Routed by Malaria; Final Defeat Laid to Disease That Felled Thousands as Quinine Supply Ran Out," April, 18, 1942.

"Quinine Loss Offset by Synthetic Drug: Atabrine Largely Voids Japan's Gain," Dec. 18, 1942.

"New Drugs to Combat Malaria Are Tested in Prisons for Army," March 5, 1945.

"Lolin Lopez Engaged: She Will Be the Bride of Maj. Harold M. Jesurun, Army," March 29, 1947.

Time

"Manteno Madness," Vol. 34, issue 17 (Oct. 23, 1939).

"Medicine: Screen Salesman," issue 56 (June 14, 1943).

"Mumu & Virility," Vol. 46, issue 27 (Dec. 31, 1945).

Miscellaneous

"Flachslander Denies Charges," *The (Baltimore) Sun*, Feb. 21, 1921.

"The Ecological Society of America, Program of the Meeting at Cincinnati, December 27–31, 1923," *Bulletin of the Ecological Society of America*, Vol. 5 (Jan. 1924).

"Nazi Plague Plan Fails: Italian Front," *Daily Mail*, Paris Edition (Sept. 27, 1944).

"New Attack of Illness of Manteno Hospital Believed Food Poison," *The Freeport Journal-Standard,* Jan. 4, 1940.

"Programs and Abstracts of the 13th Annual Meeting of the American Society of Parasitologists, Indianapolis, Indiana, Dec. 28–30," *The Journal of Parasitology*, Vol. 23, no. 6 (1937).

"Professor J. Von Wagner-Jauregg: Nobel Prize Winner, 1927," *The Scientific Monthly* 26, no. 2 (Feb. 1928)

"News Series," *Science*, 64, no. 1654 (Sept. 10, 1926).

"Prize Method Used Here," *The Science News-Letter* 12, no. 349 (Dec. 17, 1927).

"The Manteno State Hospital Tragedy in Illinois," *Social Service Review*, 13, no. 4 (Dec. 1939).

"Quinine Substitute Being Produced on Record Scale," *Wall Street Journal,* Sept. 25, 1942.

"Mosquito Killed in Hordes by Poison from Marine Plane: Quantico Camps Make War on Malaria Pest of Two Creeks," *The Washington Post*, July 2, 1926.

"Defeating the Curse: How science is tackling malaria," *BBC,* Sept. 2005: http://www.bbc.co.uk/sn/tvradio/programmes/horizon/malaria_prog_summary.shtml

Miscellaneous Books and Video Interviews

Building the Navy's Bases in World War II, Vol. 1. Washington, DC: Bureau of Yards and Docks, Navy Department, GPO, 1947.

Nobel Lectures, Physiology or Medicine 1922–1941. Amsterdam: Elsevier Publishing Company, 1965, online at: www.nobelprize.org/nobel_prizes/medicine/laureates/1927/wagner-jauregg-lecture.html

Franklin D. Roosevelt, Message to Congress Suggesting the Tennessee Valley Authority, April 10, 1933 (http://docs.fdrlibrary.marist.edu/odtvacon.html).

G. Robert Coatney, interviewed by Leon H. Schmidt, A National Medical Audiovisual Center production, National Institutes of Health, National Library of Medicine, 1979 (catalog record: 7901256A).

Lowell T. Coggeshall, interviewed by John Z. Bowers, Leaders in American Medicine, U.S. National Library of Medicine, 1975 (catalog record: 7601824A).

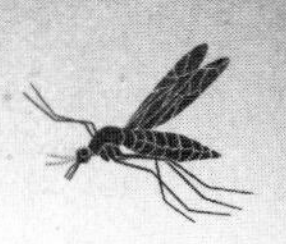

NOTES

The following is a key to understanding citations for archived materials. Records from the National Archives and Records Administration, College Park, came from more than a hundred boxes in a handful of record groups and are cited by record group and box number.

NACP—National Archives and Records Administration II, College Park.

NACP Record Groups, noted as RG:

RG 52—Records of the Bureau of Medicine and Surgery. Boxes cited are from stack 49A, unless otherwise noted.

RG 112—Records of the Surgeon General, U.S. Army, World War II, Administrative Records, Zone of the Interior, dec. 710. Boxes cited are from entry 31, stack 390, unless otherwise noted.

RG 169—Records of the Foreign Economic Administration, Board of Economic Warfare.

RG 227—Records of the Office of Scientific Research and Development, Committee on Medical Research. Boxes are from NC-138, entry 165, unless otherwise noted.

RG 492—Records of Special Staff, Theater Surgeon, MTO, General Correspondence. Boxes are from entry 290 (decimal 441), stack 54, unless otherwise noted.

BMR—Board for the Coordination of Malaria Studies, *Bulletin on Malaria Research: Comprising Minutes of Meetings of the Board and Its Panels and the Various Malaria Committees which Preceded the Board*, 2 volumes (Washington, DC, 1943–1946). These are available at several libraries and archives. The copy accessed for this book came from the National Archives, College Park, RG 227, NC-138, entry 167, Boxes 1–4. Cited here as BMR with page number from the meeting minutes or report number.

Introduction: A Brief History

1 *World Malaria Report: 2013* (Geneva: World Health Organization, 2013).
2 For a fuller history of malaria, read Paul Russell's *Man's Mastery of Malaria* (London: Oxford University Press, 1955).
3 Nicholas A. V. Beare, et al, "Perfusion Abnormalities in Children with Cerebral Malaria and Malarial Retinopathy," *Journal of Infectious Diseases* 199 (Jan. 15, 2009), 263–71.
4 Renaud Lacroix, et al, "Malaria Infection Increases Attractiveness of Humans to Mosquitoes," *PLoS* 3(9): (Aug. 9, 2005).
5 James L. A. Webb Jr., *Humanity's Burden: A Global History of Malaria* (New York: Cambridge University Press, 2009), 21; Richard Carter, "Speculations on the origins of Plasmodium vivax malaria," *Trends in Parasitology* 19, no. 5 (May 2003), 214–19.
6 Louis H. Miller, et al, "The Pathogenic Basis of Malaria," *Nature*, Feb. 7, 2002, pp. 673–79; and Webb, 26.
7 Webb, 28.
8 Miller, 675–76.
9 Webb, 73.
10 Ibid., 75.
11 Margaret Humphreys, *Malaria: Poverty, Race, and Public Health in the United States* (Baltimore: Johns Hopkins University Press, 2001), 21–22.

Chapter 1: Lowell T. Coggeshall

1 Unless otherwise noted, all details of Lowell Coggeshall's life were derived from the following sources: an unpublished autobiography in the Lowell T. Coggeshall Papers, Special Collections Research Center, the University of Chicago Library; "Reminiscences of Lowell T. Coggeshall," oral history, 1964, Columbia University Library, Archival Collections; a 1975 videoed interview with Coggeshall by John Z. Bowers, U.S. National Library of Medicine, Leaders in American Medicine series (catalog record: 7601824A); the personal files of University of Michigan professor Joel Howell, historian Leo Slater, and Carol Govan, which were shared with the author, with permission; and phone interviews with Carol Govan and Richard Coggeshall.
2 Leon O. Jacobson Papers, "Childhood," *Coggeshall Forebears*, p. 10 [Box 36, Folder 3], Special Collections Research Center, University of Chicago Library.
3 E. Chaves-Carballo, *The Tropical World of Samuel Taylor Darling: Parasites, Pathology and Philanthropy* (Eastborne, UK: Sussex Academic Press, 2007), 11.
4 For a full history of the Rockefeller Foundation's public health work see: John

Farley, *To Cast Out Disease: A History of the International Division of the Rockefeller Foundation (1913–1951)* (New York: Oxford University Press, 2004).

5 "New Discoveries May Eliminate Quinine for Malaria," *New York Times*, July 13, 1913.

6 "World Doctors Hail Ehrlich as Hero," *New York Times*, Aug. 9, 1913.

7 Paul de Kruif, *Microbe Hunters* (Boston: Houghton Mifflin Harcourt, 2002), 236–37.

8 Farley, 107–11.

9 Ibid., 6–7.

10 Ibid., 108–13.

11 Ibid. Also, see Frank M. Snowden, *The Conquest of Malaria: Italy, 1900–1962* (New Haven: Yale University Press, 2006).

12 Jacobson Papers, "Childhood," 13.

13 L. T. Coggeshall, "Malaria as a World Menace," *Journal of the American Medical Association* 122, no. 1 (May 1, 1943: 8–11) 9; Paul Russell, "Introduction," *Preventive Medicine in World War II: Communicable Diseases, Vol. VI*, ed. John Boyd Coates (Washington, DC: Department of the Army, 1963), 3–4.

14 Ibid.

15 Snowden, 128.

16 Don C. Warren, "The Egg and I for 50 Years," Proceedings of the Twenty-Sixth Annual National Poultry Breeders' Roundtable, Hotel Continental, Kansas City, Missouri, May 5–6, 1977, 59–67.

Chapter 2: Fever Therapy

1 G. Gartlehner and K. Stepper, "Julius Wagner-Jauregg: pyrotherapy, Simultanmethode, and 'racial hygiene,'" *Journal of the Royal Society of Medicine 2012;* 105: 357–59.

2 Magda Whitrow. *Julius Wagner-Jauregg (1857–1940)* (London: Smith-Gordon, 1993), 161.

3 Ibid., 31.

4 Ibid., 3.

5 Ibid.

6 Ibid., and "Prize Method Used Here," *The Science News-Letter* 12, no. 349 (1927).

7 Whitrow, 2.

8 Ibid., 47.

9 Ibid., 49.

10 Ibid., 4.

11 Ibid., 151–54.

12 Oliver D. Howes, et al, "Julius Wagner-Jaruegg, 1857–1940," *The American Journal of Psychiatry* 166, no. 4 (April 2009): 409.

13 Eli Chernin, "The Malariatherapy of Neurosyphilis," *The Journal of Parasitology*, 70, no. 5 (Oct. 1984): 611–17.

14 Whitrow, 157–58.

15 Ibid., 156.
16 Norman Tobias, "Making Malaria Work for the Doctor," *The Scientific Monthly* 41, no. 3 (Sept 1935).
17 Whitrow, 157.
18 "Professor J. Von Wagner-Jauregg: Nobel Prize Winner, 1927," *The Scientific Monthly* 26, no. 2 (Feb. 1928), 192.
19 Stanley Rosenman, "Review Essay: Freud Contesting the Predatory System" (review of *Freud as an Expert Witness: The Discussion of War Neuroses Between Freud and Wagner-Jauregg*," by K. R. Eissler), *The American Journal of Psychoanalysis* 49, no. 2 (1989), 169–80.
20 Ibid., 172.
21 Whitrow, 198–99.
22 Rosenman, 173.
23 Whitrow, 160.
24 Ibid.
25 Ibid.
26 Gliatto Howes, et al, "Neurosyphilis: A history and clinical review," *Psychiatric Annals* 31, no. 1 (March 2001), 153–61.
27 Ibid.
28 Ibid.
29 Ibid.
30 J. T. Braslow, "Effect of therapeutic innovation on perception of the disease of the Doctor-Patient Relationship: A History of General Paralysis of the Insane and Malaria Fever Therapy, 1910–1950," *The American Journal of Psychiatry* 5 (May 1995), 660.
31 Whitrow, 162.

Chapter 3: Making of a Malaria Warrior

1 Also called Eagle Lake.
2 "The Ecological Society of America, Program of the Meeting at Cincinnati, December 27–31, 1923," *Bulletin of the Ecological Society of America,* Vol. 5 (January 1924).
3 Samuel Taylor Darling, "Factors in the transmission and prevention of malaria in the Panama Canal Zone," *Annals of Tropical Medicine and Parasitology*, Vol. IV, no. 2 (July 1910), 179–223; and Chavel-Carballo.
4 The Wilbur A. Sawyer Papers, Correspondence, 1911–1995, U.S. National Library of Medicine, LWBBKX, Letter from Wilbur A. Sawyer to his wife, Margaret Sawyer, Sept. 3, 1924.
5 Charles Pierce Coggeshall and Thellwell Russell Coggeshall, *The Coggeshalls in America: Genealogy of the Descendants of John Coggeshall of Newport with a Brief Notice of Their English Antecedents* (Spartanburg, South Carolina: The Reprint Company, 1982).
6 "Eyes of a poet," E. Chaves-Carballo, 15.
7 Sawyer letter.

8 From a photo caption of Coggeshall, the "limnologist," from the personal collection of retired Ambassador Theodore (Ted) Russell, son of Paul Russell, with permission.
9 S. W. Welch, "Malaria Control Work in Alabama," *Southern Medical Journal* 20, no. 6 (June 1927), 482; University of Virginia, Philip S. Hench Walter Reed Yellow Fever Collection, letter from T. H. D. Griffitts to Henry Rose Carter, March 12, 1913.
10 Lowell T. Coggeshall, "Relationship of Plankton to Anopheline Larvae," *American Journal of Epidemiology* 6, no. 4 (1926), 556–69.
11 Interview with Ted Russell.
12 Ibid.
13 Rockefeller's International Health Commission was renamed in 1916 the International Health Board, and renamed again in 1927 the International Health Division.
14 Chaves-Carballo, 130; Diary of Paul Russell, courtesy of and with permission from Ted Russell, hereafter referred to as Russell diary.
15 Ibid.
16 Ibid.
17 Chaves-Carballo, 139.

Chapter 4: From Insects to Medicine

1 "Mosquitoes Killed in Hordes by Poison from Marine Plane: Quantico Camps Make War on Malaria Pest of Two Creeks," *The Washington Post* (July 2, 1926), 1.
2 "Poison From Plan Fatal to All Kinds of Mosquito Larvae," *The Washington Post* (July 6, 1926), 8.
3 *The Washington Post* (July 3, 1926), 6.

Chapter 5: A Nobel Prize

1 Braslow, 660.
2 Ibid., 660 and 662–63.
3 Ibid., 663.
4 "Reports of Societies: Synthetic Anti-Malaria Drugs," *The British Medical Journal* 1, no. 3706 (Jan. 16, 1932), 101.
5 Ibid.
6 J. Purdon Martin, "Conquest of General Parlaysis," *British Medical Journal* 3 (1972), 159–60.
7 *The Science News-Letter* (Dec. 17, 1927).
8 Ibid.
9 Ibid.
10 Ibid.
11 Braslow, 661.
12 Until 1918 his name was Julius Wagner Ritter von Jauregg ("Knight of Jauregg"), which had to be changed after the war to Julius Wagner-Jauregg.

13 Whitrow, 163 and 166–67.
14 *Nobel Lectures, Physiology or Medicine, 1922–1941* (Amsterdam: Elsevier Publishing Company, 1965), online at: www.nobelprize.org/nobel_prizes/medicine/laureates/1927/wagner-jauregg-lecture.html
15 Whitrow, 163.
16 Edward Brown, "Why Wagner-Jauregg Won the Nobel Prize for Discovering Malaria Therapy for General Paresis of the Insane," *History of Psychiatry* 11 (2000), 371–82.
17 Magda Whitrow, "Wagner-Jauregg and Fever Therapy," *Medical History* 34 (1990), 310.
18 "Nobel Prizes," *British Medical Journal* 2, no. 3489, Nov. 19, 1927.
19 Whitrow (1993), 169.
20 *Nobel Lectures.*
21 Whitrow (1993), 190–92.
22 "Anschluss appeal is signed by many," *New York Times,* Nov. 10, 1928, 6.
23 Otto M. Marx, "Book Reviews," *Isis*, review of *Julius Wagner-Jauregg* by Magna Whitrow, 85, no. 4 (Dec. 1994), 719.
24 Ibid.
25 "Austrian Gets Nobel Prize," *New York Times*, Oct. 28, 1927, 25.
26 "Austrian Psychiatrist Wins Nobel Prize for Medicine," *New York Times,* Nov. 6, 1927, 7.
27 "A Malaria Treatment Centre: The Unit at Horton Medical Hospital," *The British Medical Journal* 1, no. 3985 (May 22, 1937), 1081.
28 Ibid.

Chapter 6: Divided Loyalties

1 Charles B. Huggins, "The Business of Discovery in the Medical Sciences," *Journal of the American Medical Association* 194, no. 11 (Dec. 13, 1965).
2 Humphreys, *Malaria: Poverty, Race, and Public Health,* 94.

Chapter 7: Germany and Magic Bullets

1 Diarmuid Jeffreys, *Hell's Cartel: IG Farben and the Making of Hitler's War Machine* (New York: Metropolitan Books, 2008) 114.
2 "Ex-Kaiser Joins Fascisti; Finds Cool Welcome," *Chicago Daily Tribune*, May 19, 1924, 12.
3 "Guilt Admitted by Adolf Hitler," *Christian Science Monitor*, Feb 27, 1924, 1.
4 "German Treason Trial: Hitler's Frothy Defence," *The Manchester Guardian*, Feb. 28, 1924, 10.
5 Louis Leo Snyder, *Hitler's Third Reich: A Documentary History* (Chicago: Nelson-Hall, 1982), 41.
6 "Prisoners Released in Bavaria," *The Manchester Guardian*, Dec. 22, 1924, 8.
7 "Hitler Tamed by Prison," *New York Times*, Dec. 21, 1924, 16.
8 "Election Prospects in Germany: 25 Parties, Many of Them Freaks," *The Manchester Guardian*, Dec. 5, 1924, 13.

9 Ibid.

10 "Ask Jarres to Act Against Racialists: German Moderates Say Speeches at Weimar Convention Were Treasonable," *New York Times*, Sept. 7, 1924, S6.

11 "Election Prospects in Germany: 25 Parties, Many of Them Freaks," *The Manchester Guardian,* Dec. 5, 1924.

12 NACP, RG 169, Box 3, Bureau of Economic Warfare, "confidential" internal document generated by Justice Department staff: "The IG-Sterling Partnership: Sterling, I.G. and the Nazi Government," Sept. 22, 1943, 33–36. Note: RG 169 holds a deep history on the I.G.-Sterling relationship, much of it in boxes 3 and 22.

13 Jeffreys, 106.

14 "The IG-Sterling Partnership," 37.

15 "Flachslander Denies Charges," *The (Baltimore) Sun*, Feb. 21, 1921, 1; "Germany Seeks Arrest of Four Dye Experts Who Accepted Jobs Here—Two Have Landed," *New York Times*, Feb. 21, 1921, 12; "Chemist to Fight German Charges: Jordan, Now in Holland, Ready to Face Accusers in Cologne," *New York Times,* March 18, 1921, 30

16 *The British Medical Journal* (Jan. 16, 1932), 100.

17 Ibid.

18 Leo B. Slater, *War and Medicine* (New Brunswick, NJ: Rutgers University Press, 2009), 60.

19 "Reports of Societies," *The British Medical Journal* 1, no. 3453 (March 12, 1927), 466.

20 "A Synthetic Remedy for Malaria," *The British Medical Journal* 2, no. 3434 (Oct. 30, 1926), 798–99.

21 Ibid. Investigators included Prof. Muhlens of the Hamburg Tropical Disease Institute; Prof. Memmi and Dr. W. Schulemann, director of the pharmaceutical laboratory of the dye industries at Elberfeld, running tests in Italy; and Dr. Roehl at Bayer.

22 *British Medical Journal* (Jan. 16, 1932), 101.

23 Ibid.

24 Ibid., 100.

25 Jeffreys, photo caption.

26 "German Dyes Trust Reports Big Gains: 5,000 More Men Hired in Last Quarter of 1932 to Meet Rise in Home and Export Demand," *New York Times*, Jan. 28, 1933, 6.

27 "German Dye Trust Lists Export Drop," *New York Times*, Nov. 15, 1934, 9.

28 Jeffreys, 162–74.

29 Slater, 71.

30 Ibid.

31 Wolfgang W. Eckart, H. Vondra, "Malaria and World War II: German malaria experiments 1939–45," *Parassitologia* Vol. 42 (2000), 53–58; Marion Hulverscheidt, "German Malariology experiments with humans, supported by the DFG until 1945," ed. Wolfgang U. Eckart, *Man, Medicine and the State: The Human Body as an Object of Government Sponsored Medical Research in the 20th Century* (Stuttgart: Franz Steiner Verlag, 2006), 221–35.

32 Snowden, p 147.
33 "The First International Malaria Congress," *The British Medical Journal* 2, no. 3386 (Nov. 21, 1925), 970–71.
34 Snowden, 145.

Chapter 8: Eradication

1 Interviews with Lowell Coggeshall's daughter Carol Govan, November 2009.
2 Franklin D. Roosevelt, Message to Congress Suggesting the Tennessee Valley Authority, April 10, 1933 (http://docs.fdrlibrary.marist.edu/odtvacon.html).
3 David P. Earle, "Malaria and Its Ironies," *American Clinical Climatological Association,* Vol. 90 (1979), 1–26, p. 23.
4 For a fuller discussion of the difficulties of knowing how successful TVA was in controlling malaria, see Margaret Humphreys, *Malaria: Poverty, Race and Public Health in the United States*, Chapter 5, "A Ditch in Time Saves Quinine" (Baltimore: The Johns Hopkins University Press).
5 For details, see: Lowell T. Coggeshall, "The Occurrences of Malarial Antibodies in Human Serum Following Induced Infection with Plasmodium Knowlesi," *Journal of Experimental Medicine*, 72, no. 21 (1940); Coggeshall, "The Cure of Plasmodium Knowlesi Malaria in Rhesus Monkeys with Sulfanilamide and Their Susceptibility to Reinfection," *American Journal of Tropical Medicine* 18, no. 331 (1938); Monroe D. Eaton and Lowell T. Coggeshall, "Production in Monkeys of Complement-Fixing Antibodies Without Active Immunity by Injection of Killed Plasmodium Knowlesi," *Journal of Experimental Medicine* 70 (1939), 141–46; Eaton and Coggeshall, "Complement Fixation in Human Malaria with an Antigen Prepared from the Monkey Parasite Plasmodium Knowlesi," *Journal of Experimental Medicine* 69 (1939), 379–98.

Chapter 9: Claus Schilling

1 Charles A. Doan, "The First International Conference on Fever Therapy," *The Scientific Monthly* 44, no. 5 (May 1937), 483–86.
2 "Experts on Fever to Meet this Week: 80th Birthday of Prof. Jauregg, Nobel Prize Winner, to be Commemorated," *New York Times*, March 28, 1937, 34.
3 Ibid.
4 Hulverscheidt in *Man, Medicine and the State*, 225.
5 Paul Weindling, "From Medical War Crimes to Compensation: The Plight of the Victims of Human Experiments," *Man, Medicine and the State: The Human Body as an Object of Government Sponsored Medical Research in the 20th Century*, ed. Wolfgang U. Eckart (Heidelberg, Germany: Franz Steiner Verlag, 2006), 246.
6 Hulverscheidt, 223.
7 Weindling, 246.
8 de Kruif, 236–37 and 330–33.
9 Robert Koch, "Remarks on Trypanosome Diseases," *British Medical Journal* 2, no. 2291 (Nov. 26, 1904), 1445–49.

10 "News Series," *Science*, 64, no. 1654 (Sep. 10, 1926), 245–47.
11 Eckart, Vondra, 56.
12 Ibid., 53–55.
13 Hulverscheidt, *Man, Medicine and the State*, 224.
14 Claus Schilling, *Die Methoden der experimentellen Chemotherapie* (Jena, Germany: Gustav Fischer Verlag, 1938), kindly translated for the author by Elke Bergmann-Leitner.
15 Hulverscheidt, 226.
16 Nov. 16, 2013, e-mail exchange with Wolfgang Leitner, malaria expert with NIH.
17 Hulverscheidt, 227.
18 Eckart, Vondra, 56.
19 Ibid.
20 In person interview, Nov. 10, 2013.
21 Hulverscheidt, 228.
22 Ibid.

Chapter 10: A New Plan

1 W. Mansfield Clark Papers, "A Survey of Antimalarial Drugs, 1941–1945, History of the Organization (Rough Tentative Draft)," American Philosophical Society.
2 NACP, RG 52 Records of the Bureau of Medicine and Surgery, Entry 49A, Box 6, letter from Mark F. Boyd to W. A. Sawyer, the Rockefeller Foundation, September 28, 1940; and letter to Boyd from Coggeshall, August 5, 1940.
3 L. T. Coggeshall, et al, *Science* 92, no. 2382 (Aug. 23, 1940), 176–78.
4 Ibid.
5 Milton Silverman, *Magic in a Bottle* (New York: The Macmillan Company, 1941), 285–97.
6 NACP, RG 227, Entry 165, Box 23, letter to Coggeshall from A. N. Richards, Committee on Medical Research, May 22, 1942; letter to Richards from Coggeshall, May 18, 1942.
7 B. M. Das Gupta, "Transmission of P. inui to Man," *Proceedings of the National Institute of Science, India* 4 (June 1938), 241–44.
8 David W. Talmadge, "William Hay Taliaferro," *Biological Memoirs* (Washington, DC: National Academy of Sciences, 1983), 386.
9 NACP, RG 227, Entry 165, Box 23, report by Coggeshall and W. H. Taliaferro, "The Need for Plasmodium Gallinaceum in Experiment Investigations on the Chemotherapy of Malaria."
10 Lowell T. Coggeshall, "Plasmodium lophurae, a New Species of Malaria Parasite Pathogenic for the Domestic Fowl," *American Journal of Hygiene*, Vol. 27 (1938), 615–18.
11 Ibid.
12 Hulverscheidt in *Man, Medicine and the State*, 233.
13 Files of University of Michigan professor Joel Howell.
14 NACP, RG 52, 49A, Box 7, memorandum for Admiral McIntire, Malaria Conference, Sept. 11, 1941.

15 NACP, RG 52, Box 6, Memorandum for Admiral McIntire, Subject: Committee on Tropical Diseases, meeting October 13, 1941, at 2101 Constitution Ave.
16 NACP, RG 112, Entry 165, Box 23, letter from A. N. Richards to Coggeshall, October 16, 1941.
17 Ibid. location, letter to Coggeshall from E. Cowles Andrus, assistant to Richards.

Chapter 11: Africa

1 Stetson Conn and Byron Fairchild, *United States Army in World War II, the Western Hemisphere, the Framework of Hemisphere Defense* (Washington, DC: Center for Military History, United States Army, 1960), 238–52.
2 Robert Daley, *An American Saga: Juan Trippe and His Pan Am Empire* (New York: Random House, 1980), 304.
3 John Bierman and Colin Smith, *The Battle of Alamein: Turning Point, World War II.* (New York: Viking, 2002), 70.
4 University of Miami Libraries, Special Collections: Pan American Airways, Inc. Cited hereafter as "PAA." Accession I, Box 22, folder 4.
5 Marylin Bender and Selig Altschul, *The Chosen Instrument: Pan Am, Juan Trippe: The Rise and Fall of an American Entrepreneur* (New York: Simon and Schuster, 1982), p. 350.
6 "Air Ferry Firm to Africa Formed by Pan American," *Christian Science Monitor*, Aug 21, 1941, 10.
7 PAA, Box 23, Folder 10, Typescript.
8 "Notes for Travelers," *New York Times,* Nov. 30, 1941, p. 5.
9 Deborah Ray, "Pan American Airways and the Trans-African Air Base Program of World War II" (New York: New York University Press, UMI Dissertation Publishing, 1973), 112.
10 PAA, Box 23, Folder 10, Typescript.
11 Daley, 335.
12 PAA, Box 74, Folder 21.
13 Ibid.
14 Ibid.
15 Ibid.
16 S. B. Kauffman, *Pan Am Pioneer: A Manager's Memoir from Seaplane Clippers to Jumbo Jets* (Lubbock: Texas Tech University Press, 1995), 83.

Chapter 12: Bataan

1 Russell diary; John Boyd Coates, ed., United States Army Medical Department, *Preventive Medicine in World War II, Vol. VI: Communicable Diseases, Malaria* (Washington, DC: Office of the Surgeon General, 1963), 498. Cited hereafter as "Coates, Vol. VI."
2 John W. Whitman, *Bataan: Our Last Ditch* (New York: Hippocrene Books, 1990), 11–13; and Edward S. Miller, *War Plan Orange: The U.S. Strategy to Defeat Japan, 1897–1945* (Annapolis: United States Naval Institute Press, 1991).

3 Coates, Vol. VI, 499; also see Robert T. Joy, "Malaria in American Troops in the South and Southwest Pacific in World War II," *Medical History* 43 (1999), 192–207.
4 Coates, 501–4.
5 Ibid., 503.
6 Ibid., 509.
7 Ibid., 505.
8 Whitman, 446.
9 Ibid., 398.
10 Ibid., 399.
11 Coates, 507.
12 Ibid., 507–8.
13 "Troops on Bataan Routed by Malaria; Final Defeat Laid to Disease That Felled Thousands as Quinine Supply Ran out," *New York Times,* April, 18, 1942, 5.
14 "Bataan Was No Picnic," *New York Times,* April 23, 1942.

Chapter 13: War on Mosquitoes

1 PAA, Box 22, Folder 4.
2 Ivan Dmitri (a pseudonym), *Flight to Everywhere* (New York: Whittlesey House McGraw-Hill Book Company, 1944).
3 PAA, Box 22, Folder 4.
4 Pan Africa, 52.
5 Ibid., 49–51.
6 PAA, Box 23, Folder 10, Typescript.
7 L. T. Coggeshall (and personnel of the medical department), "Development of a Medical Service for Airline Operations in Africa," reprinted in *War Medicine,* Vol. 3 (May 1943), 489–97.
8 Ibid.
9 PAA, Box 23, File 9, Typescript.
10 Ibid.
11 Ibid.
12 Gerry Killeen, et al, "Eradication of Anopheles gambiae from Brazil: lessons for malaria control in Africa?" *The Lancet* 2, no. 10 (Oct 2002), 618–27.
13 A. Spielman and M. D'Antonio, *Mosquito: The Story of Man's Deadliest Foe* (New York: Hyperion, 2001), 134.
14 M. A. Barber, "The Present Status of Anopheles Gambiae in Brazil," *American Journal of Tropical Medicine and Hygiene,* Vols. 1–20, no. 2 (March 1940), 249.
15 Malcolm Gladwell, "The Mosquito Killer," *The New Yorker,* July 2, 2001, http://www.newyorker.com/archive/2001/07/02/010702fa_fact_gladwell
16 PAA, Box 23, Folder 9.
17 Pan Africa, 57.
18 Ibid, 59–63.
19 PAA, Box 23, Folder 9.

Chapter 14: Guadalcanal

1 U.S. Navy Medical Department, Administrative History, 1941–45, Vol. 1, Narrative History, Chapters I–VIII, BuMed, Navy Department, 1946, unpublished, held at BuMed; Commanding General, 1st Marine Division, Fleet Marine Force, "Final Report on Guadalcanal Operation," Phase I, Annex H, Medical Experience. Historical Division, United States Marine Corps. Cited hereafter as the "Final Report."
2 Ibid.
3 Kerry Lane, *Guadalcanal Marine* (Jackson: University Press of Mississippi, 2004), 3.
4 Richard Tregaskis, *Guadalcanal Diary* (New York: The Modern Library, 2000), 50–51.
5 Final Report, 7–8.
6 Ibid.
7 Ibid., 11.
8 Ibid.
9 Ibid.
10 Ibid.
11 Coates, Vol. VI, 413.
12 Ibid., 426.

Chapter 15: A Malaria Manhattan Project

1 Earle, 12.
2 Thomas H. Maren, "Eli Kennerly Marshall, Jr.: 1889–1996," *Biographical Memoirs* (Washington, DC: National Academy of Sciences, 1987), 336.
3 For records pertaining to the different research institutions involved in bird malarias, see NACP, RG 227, Box 94.
4 NACP, RG 112, Box 1160, Minutes of the Subcommittee of Tropical Diseases, June 1942.
5 NACP, RG 227, Box 94, "Reports of Investigations of Malaria Research Projects under OSRD Contracts, January–April 1943," for A. N. Richards by Henry E. Meleney, Special Adviser. WOC, CMR, OSRD.
6 NACP, RG 52, 49B, Box 3, Minutes of the Fifth Meeting of the Malaria Conference, June 3, 1942.
7 Ibid.
8 Ratcliff, J. D., "Brains," *Collier's*, Jan. 17, 1942.
9 Maren, 335.
10 Earle, 18.
11 NACP, RG 227, Box 53, letter from Shannon to Andrus, Oct. 26, 1942; and letter to A. N. Richards from H. Beckett Lang, assistant commissioner, New York Commission of Hygiene, Nov. 4, 1942.
12 NACP, RG 227, Box 94, memorandum to A. N. Richards from Henry E. Meleney, Jan. 1943.

Chapter 16: Better to Be Yellow

1 NACP, RG 112, Box 1179, memorandum for the Surgeon General from D. C. McDonald, Colonel, G.S.C, Military Intelligence Division G-2, Washington, July 15, 1942.
2 Mark Honigsbaum, *The Fever Trail: In Search of the Cure for Malaria* (New York: Farrar, Straus and Giroux, 2002), 221.
3 RG 52, Box 5, Minutes of the Meetings of the Subcommittee on Tropical Diseases of the National Research Council, June 3–4, 1942, Colonel A. F. Fischer presentation to the subcommittee, June 4, 4.
4 NACP, RG227, Box 21, Carden File, cinchona, Memorandum to Dr. Andrus from George Carden, Nov. 6, 1942.
5 NACP RG 112, Box 1179, memorandum for the chief, Finance and Supply Service, Subject: Quinine, from James Simmons, colonel, Medical Corps, July 2, 1942.
6 NACP, 49A, Box 6, Joint Meeting of Subcommittee on Tropical Diseases and Conference Group on Malaria Research, National Research Council, Dec. 30, 1941, minutes, 7.
7 NACP, RG 112, Box 1177, memo to chief health officer, Balboa Heights, Canal Zone, from R. D. Harden, superintendent, Gorgas Hospital, Ancon, Canal Zone, May 12, 1943, giving the history of a year trying to use Winthrop-made atabrine without success; same box, letter to the surgeon general, USA, from M. C. Strayer, brigadier general, U.S. Army, Chief Health Office, the Panama Canal Zone, May 14, 1943, in which the general complained: "It has come to our attention that certain lots of atabrine, furnished by the Winthrop Chemical Company, Inc., do not dissolve properly, and that certain cases of malaria under atabrine treatment did not do well. These people promptly recovered when we issued quinine." He sent bottles back to have them studied by Malaria Project investigators.
8 Final Report, 16.
9 Coates, Vol. VI, 417.
10 Final Report, 24–26.
11 Ibid., 27–28.
12 Ibid., 26.
13 Ibid., 30–31.
14 Robert Leckie, *Helmet for My Pillow: From Parris Island to the Pacific* (New York: Bantam Books Trade Paperbacks, 2010), 92.
15 Final Report, 29–30.
16 Ibid.
17 Leckie, 153.
18 Stanley Sandler, *World War II in the Pacific: An Encyclopedia* (New York: Garland Publishing, Inc., 2001), 617.
19 Leckie, 252.
20 Final Report, 1.

Chapter 17: The Other Side

1 NACP, RG 112, Office of the Surgeon General (Army) Refiles from the U.S. Center for Military History; 014 Public Health through 1121.6, Supply Costs, Box 2. Cited hereafter as "RG 112, Refiles."
2 David Nichols, ed, *Ernie's War: The Best of Ernie Pyle's World War II Dispatches* (New York: Random House, 1986), 69.
3 NACP, RG 492, Box 2572, letter to Lieutenant Colonel Earle Sandlee, MC, Office of Chief Surgeon, ETO, from Colonel F. C. Tyng, MC War Department, Aug. 1, 1942.
4 Ibid. (NACP location and box), cable, "emergency" request to fly Atabrine to APO 512, from Colonel J. F. Corby, deputy surgeon, Allied Force HQ, Office of the Force Surgeon, Sept. 20, 1942.
5 NACP, RG 112, Refiles, letter from Colonel Francis C. Tyng, MC, Office of Surgeon General, U.S. Army, Washington, to Hawley, Oct. 18, 1942.
6 Ibid., letter to Tyng from Hawley, Nov. 7, 1942.
7 Ibid.
8 Rick Atkinson, *An Army at Dawn: The War in North Africa, 1942–1943* (New York: Henry Holt and Company, 2002), 163–64.
9 All details about this study came from the following: Memo to the Surgeon General, U.S. Army, Preventive Medicine Division, Washington, DC, from F. A. Blesse, brigadier general, surgeon, headquarters, North African Theater of Operations, APO 534 (July 8, 1943) and attached report, "Report on Sontoquine," by Jean Schneider, translated (May 31, 1943), NACP, RG 492, Records of the Special Staff, theater surgeon, general correspondence (dec. 441), Box 2523. Cited hereafter as "Schneider report."
10 *Ernie's War*, 102–3.
11 Schneider report.
12 Letter to Tyng from Hawley, Nov. 7, 1942.
13 Ibid.
14 PAA, Box 22, Folder 4, transcript of speech, Major General Harold L. George, Oct. 1, 1942.
15 NACP, RG 319, Records of the Army Staff, Intelligence, Incoming and Outgoing Messages, Accra, 1942–1945, Entry 57, Box 1, "secret" cable to Milid, from Accra (Major Thomas L. Dawson), April 6, 1942.
16 Ibid., NACP location, radiogram from Accra via ACFC, Intelligence Branch, MID, Jan. 29, 1942.
17 *Ernie's War*, 67.
18 NACP, RG 492, Box 2592, memorandum to Surgeon General, U.S. Army, Washington from Headquarters II Corps, Office of the Surgeon, Subject: Annual Report, January 10, 1943.
19 Douglas B. Kendrick, *Medical Department United States Army in World War II, Blood Program in World War II*, (Washington, DC: Office of the Surgeon General, Department of the Army, 1964), 143–44.
20 PAA, Box 206, Folder 22.

Chapter 18: Taking Tunis

1 Ernie Pyle, *Here Is Your War: Story of G.I. Joe* (Lincoln: University of Nebraska Press, 2004), 73.
2 Coates, Vol. VI., 256.
3 NACP, RG 492, Box 2592, memorandum from Richard Arnest, colonel, Medical Corps, Headquarters, First Floor, Maison Darmon Annex, 4 Rue de Revolution, Oran.
4 Atkinson, 507.
5 NACP, RG 492, Box 2594, "Toxic reaction to Atabrine: case study," Annual Report 1943, Medical Station NATOUSA.
6 NACP, RG 112, Box 1177, "Preliminary Notes Sent by Brig. General H. L. Morgan, reference his observations in N.A.," excerpts, no date, stamped "confidential."
7 NACP, RG 227, Box 102, September 1943, Reports of American-made Atabrine causing toxic side effects for *War Medicine*; RG 112, Box 1177, memorandum to the surgeon general, U.S. Army, Subject: Prophylaxis Atabrine Therapy, from H. R. Melton, Colonel, Medical Corps, Port Surgeon, New York Port of Embarkation, June 4, 1943; ibid, location, "Report of Voyage" aboard the Station Hospital USAT *Shawnee*, sent to the Port Surgeon, New York Port of Embarkation, from John B. Barrett, Captain, Medical Corps, Transport Surgeon, discusses ship-wide reactions, beginning May 7, 1943.
8 NACP, RG 52, Box 5, copy of article by Gerhard Rose, "Prophylactic Treatment of Malaria with Atabrin: Its dosage and alleged complications," *Deutsche Medizinische Wochenschrift*, Vol. 67, no. 48, 1306–8, Nov. 1941. This was translated and routed to key Nation Research Council investigators; date of translation unspecified.
9 NACP, RG 492, Box 2647, "Intelligence Reports, excerpts of captured memorandum, Instructions for Malaria Protection," signed by Kesselring, no date.
10 Schneider report.
11 Earle, 18.

Chapter 19: Mosquito Brigades

1 Coates, Vol. VI, 112.
2 Coates, Vol. VI: tables from p. 83 (military installations) and p. 97 (MCWA acreage around the military installations).
3 Russell diary.
4 Ibid.
5 Coates, Vol. VI., 8.
6 Interview with Ted Russell, who recalled family conversations after the war about the role played by Allan Dulles's operation in Switzerland in securing samples of DDT for American researchers.
7 For original details on the history of this research station, see RG 112, Box 1142, memorandum for General Simmons from A. L. Ahnfeldt, MC, Subject: Early history of insecticide and insect repellent investigations, April 8, 1942.

8 NACP, RG 492, Box 2636, Report on DDT in Morocco prisoners, Oct. 29, 1943, to Lieutenant Colonel Stone, Medical Section, NATOUSA, from Fred L. Soper, Representative, Rockefeller Health Commission in North Africa, Civil Affairs Section. Subj: Demonstration of Louse Powder at Prison Camps in Oran and Casablanca. The report said prisoners were anxious to be powdered; Dobbins dusters were used in prisoners' tents, and applied to their clothes and all extra clothing and bedding.
9 NACP, RG 112, Box 1147 contains records of how these units were set up, where they trained, and how they were deployed.
10 Coates, Vol. VI, 25.
11 Denton Crocker. Senior Sanitary Technician, Malaria Survey Unit 31. Library of Congress, American Folklife Center, Oct. 26, 2011. Memoirs: "My War on Mosquitoes, 1942–1945." Cited hereafter as LOC, Crocker.
12 Ibid.
13 Ibid.
14 Ibid.
15 Ibid.
16 Ibid.
17 Ibid.
18 Coates, Vol. VI, 400.
19 Ibid., 401.
20 Ibid., 400.
21 Ibid., 406 and 408.
22 NACP, RG 112, Box 1153: Memo for "The Surgeon General; Subject: Mosquito repellents and sprays for the Pacific theaters" (March 22, 1943), from Karl R. Lundeberg, lieutenant colonel, Medical Corps; telephone conversation (transcript) Lundeberg and Major Case (March 17, 1943, 12:20 p.m.); telephone conversation (transcript) Lundeberg and Paul Russell (March 18, 1943, 12:45 p.m.).
23 NACP, RG112, Box 1153, Memo for Surgeon General, March 22, 1943.
24 NACP, RG112, Box 1144, "Incidence of certain insect borne diseases in the U.S. Navy (for Naval & Marine Corps personnel)," May 6, 1946; and "Diagnosed Malaria and Undiagnosed Fevers, admissions rates/1000/Year: Total Army (Overseas)."
25 Ibid., NACP location, memorandum for Medical Consultants Division, to Colonel Dieuaide from Harold F. Dorn, major, MC, director Medical Statistics Division, July 7, 1945, with tables of overseas malaria rates by month and year for U.S. Army and Navy.
26 NACP, RG 112, Box 1189, memorandum to commanding generals from Norman T. Kirk, the Surgeon General, Nov. 13, 1943. He describes the problem: The men were warned that malaria would kill them. But most got it and lived. Kirk worried that malaria "may appeal" to soldiers "as a means of avoiding combat and an easy ticket home."
27 Russell's diary. Note: A slightly different version of this quote in the Walter Reed Archives, which has been broadly cited, does not contain the reference to New Zealand. I cite from this diary on the assumption that "New Zealand" was deleted by censors from the original quote.

Chapter 20: The Jump

1 One documented exception occurred in mid-1945, when a small party of U.S. Army pilots and nurses flew over the mountains to gawk at the naked, fuzzy-haired tribes scattering under the roar of the engine, until the plane crashed. Rescue of the three survivors took weeks, and involved paratroopers and a glider—a story well told by journalist Michael Zuckoff in his 2012 book *Lost in Shangri-La.*
2 Roger O. Egeberg, *The General: MacArthur and the Man He Called "Doc"* (New York: Hippocrene Books, 1983), 10.
3 Ibid., 11.
4 Ibid.
5 Ibid., 12.
6 Ibid.
7 Ibid., 12.
8 Russell diary.
9 Mary Ellen Condon-Rall, *United States Army in World War II: Technical Services, The Medical Department: Medical Service in the War Against Japan* (Washington, DC: Center of Military History, United States Army, 1998), 59.
10 Christine Beadle and Stephen L. Hoffman, "History of Malaria in the United States Naval Forces at War: World War I through the Vietnam Conflict," *Clinical of Infectious Diseases,* Vol. 16, no. 2 (Feb. 1993), 323.
11 NACP, RG 112, Box 1153, memorandum for General Somervell from Norman T. Kirk, major general, U.S. Army, the Surgeon General, July 5, 1943.
12 NACP, RG 112, Box 1162, "secret" cable to MacArthur, Eisenhower, Stilwell and Harmon from G. C. Marshall, July 13, 1943.
13 *Ernie's War*, 103.
14 Ibid., 116.
15 Ibid.
16 Rick Atkinson, *The Day of Battle: The War in Sicily and Italy, 1943–1944* (New York: Henry Holt and Company, 2007), 146.
17 NACP, RG 492, Records of the Special Staff, dec. 710, Box 2544, "Malaria in the Sicilian Campaign 9 July–10 September 1943," 1–2; and Coates, Vol. VI, 262–63.
18 Ben Mcintyre, *Operation Mincemeat: How a Dead Man and a Bizarre Plan Fooled the Nazis and Assured an Allied Victory* (New York: Broadway Books, 2011).
19 Andrew James Birtle, *Sicily: The U.S. Army Campaigns of World War II* (Pamphlet, U.S. Center of Military History Publication, 2004), 14.
20 Ibid., 20.
21 Atkinson, *Day of Battle,* 133–35.
22 Albert Garland and Howard McGaw Smyth, *United States Army in World War II: The Mediterranean Theater of Operations, Sicily and the Surrender of Italy* (Washington, DC: Center of Military History, United States Army, 1993), 242.
23 Birtle, 21.
24 Atkinson, *Day of Battle,* 143.
25 Ibid., 147.
26 Ibid., 148.
27 Ibid.
28 Ibid., 169.

Chapter 21: Malaria and the Madmen

1 NACP, RG 112, Box 1153, memorandum to commanding generals, War Department, from Surgeon General Norman Kirk, Nov. 10, 1943. Kirk was opposed to the blackout because he felt the public should know what was happening overseas, with respect to malaria, and that they were intelligent enough to handle the truth. To withhold the information would only create suspicion in the mind of the public.
2 Paul de Kruif, "Enter Atabrine—Exit Malaria," *Reader's Digest*, Dec. 1942.
3 Ibid.
4 J. M. Hutchisson, "Sinclair Lewis, Paul de Kruif, and the Composition of 'Arrowsmith,'" *Studies in the Novel*, 1992.
5 de Kruif, *Reader's Digest.*
6 "Quinine Substitute Being Produced on Record Scale," *Wall Street Journal*, Sept 25, 1942, 8.
7 de Kruif, *Reader's Digest.*
8 "Quinine Loss Offset by Synthetic Drug: Atabrine Largely Voids Japan's Gain," *New York Times,* Dec. 18, 1942.
9 NACP, RG 112, Box 1177, Minutes of the Committee on Drugs and Medical Supplies, National Resource Council meeting, January 27, 1943, 5.
10 Lowell T. Coggeshall, "Malaria as a World Menace," *Journal of the American Medical Association*, Vol. 122, no. 1, (May 1, 1943), 8–11.
11 Ibid.
12 Walton Hamilton, "The Strange Case of Sterling Products," *Harper's Magazine*, Vol. 186, No. 1112 (Jan. 1, 1943), 123.
13 "Personal and Otherwise," *Harper's Magazine,* Vol. 186, no. 1114 (March 1, 1943).
14 Ibid.
15 Clark Papers, National Research Council, Report of the Committee on the Toxicity of Commercial Atabrine, submitted Oct. 3, 1942, to A. N. Richards, chair of CMR, from W. Mansfield Clark, chair of the chemical division, NRC, 12.
16 Ibid.
17 Ibid.
18 Ibid., 16.
19 Ibid., 32.
20 NACP, RG 112, Box 1144, memorandum to General Magee from C. C. Brigadier General Hillman, July 13, 1942 (Merck built a new plant in Elkton, Virginia, to manufacture Atabrine as part of this deal).
21 Alfred E. Sherndal, "Chemistry and Development of Atabrine and Plasmochin," *The Chemical Age,* Vol. 21, no. 14 (July 25, 1943), 1154–58; NACP, RG 112, Box 1177, copy of a company report: "Quinacrine Dihydrochloride—Occupational Disorders and Their Control," by R. M. Watrous, plant physician, Abbott Laboratories, North Chicago, Illinois, no date, stamped receipt Dec. 22, 1944.
22 BMR, Report #464.
23 NACP, RG 112, Box 1164, "secret" memorandum, "Measures to Reduce Person-

nel Wastage caused by Avoidable Tropical Diseases," Australian Army Staff, Australian House, London, W.C.2, report for September 1934–February 1944.

24 NACP, RG 112, Box 1177, memo for chief, military personnel, SGO, attention Colonel Hudnall, Subject: Orders, from Kirk T. Mosley, major, Medical Corps, May 17, 1943.

25 RG 52, 49B, Box 3, Minutes of the Fifth Meeting of the Malaria Conference, June 3, 1942.

Chapter 22: The Convicts

1 Robert G. Coatney, "Reminiscences: My Forty-Year Romance with Malaria," *Transactions of the Nebraska Academy of Sciences and Affiliated Societies*, Paper 222 (1985), 6.

2 G. Robert Coatney, PhD, ScD, interviewed by Leon H. Schmidt, a National Medical Audiovisual Center production, National Institutes of Health, National Library of Medicine, 1979, http://collections.nlm.nih.gov/catalog/nlm:nlmuid-7901256A-vid.

3 Coatney (1985), 7.

4 Coatney, NLM interview.

5 "Programs and Abstracts of the 13th Annual Meeting of the American Society of Parasitologists, Indianapolis, Indiana, Dec. 28–30," *The Journal of Parasitology*, Vol. 23, no. 6 (1937), 556.

6 Coatney (1985), 7.

7 Coatney, NLM interview.

8 Coatney (1985). 7

9 Ibid., and Coatney, NLM interview.

10 Coatney (1985), 7.

11 Coatney, NLM Interview.

12 Ibid.

13 NACP, RG 112, Box 1177, Malaria Report #111, National Research Council, Division of Medical Sciences acting for the Committee on Medical Research of the OSRD. Plan for Testing Antimalarial Drugs at the U.S. Penitentiary, Atlanta, Georgia, G. Robert Coatney, W. Clark Cooper, and W. H. Sebrell, National Institutes of Health, June 19, 1944.

Chapter 23: Supply and Demand

1 Coates, Vol. VI., 513.

2 Ibid., 514.

3 NACP, RG 112, Box 1164, letter to the *Yank, the Army Weekly*, 205 East 42nd Street, New York, from Harold M. Jesurun, major, Medical Corps, assistant malariologist, Headquarters USAFFE, August 20, 1943; *Yank*, Vol. I, no. 48, May 21, 1943.

4 "Lolin Lopez Engaged: She Will Be the Bride of Maj. Harold M. Jesurun, Army," *New York Times*, March 29, 1947.

5 NACP, RG 112, Box 1164, letter from Harold M. Jesurun, major, MC, assistant malariologist, HQ, USAFFE, to *Yank, the Army Weekly*, 205 East 42nd Street, New York, Aug. 20, 1943.
6 NACP, RG 112, Box 1157, Reports of the Medical Department Activities, Forward Echelon, APO 879, CBI for the year 1943, April 3, 1944, extracts.
7 For records and reports on the malaria situation in the CBI, see NACP, RG 112, Box 1155.
8 Wesley Marvin Bagby, *The Eagle-Dragon Alliance: America's Relations with China in World War II* (Newark, Delaware: University of Delaware Press, 1992), 27.
9 Ibid., 32.
10 Ibid.
11 Ibid., 29.
12 Ibid., 31.
13 William Slim (Field Marshal Viscount), *Defeat Into Victory: Battling Japan in Burma and India, 1942–1945* (Cooper Square Press, 2000), 109.
14 Ibid., 106–10.
15 Ibid., 112.
16 Coates, 357.
17 Ibid., 393–94.
18 Ibid., 391.
19 Ibid.
20 NACP, RG 112, Box 1157, Col. Earle Rice, CBI Theater malariologist, April 3, 1944, extracts, "secret."
21 Ibid.
22 NACP, RG 112, Box 1158, Louis Williams to Office of Surgeon General Norman T. Kirk, AGWAR, November 17, 1943.
23 NACP, RG 112, Box 1187, letter to Brigadier General P. J. Carroll from Paul Russell, August 9, 1943.
24 NACP, RG 227, Box 23, letter to A. N. Richards, Committee on Medical Research, from L. T. Coggeshall, University of Michigan, October 30, 1943. This box contains much correspondence to, from, and about Coggeshall, including letters that indicate that the army, especially Paul Russell, tried several times to recruit Coggeshall but that the National Research Council fought to keep him on the project as a civilian.
25 Photos and notes from Abraham's time in Liberia are in the George and Katy Abraham Papers, 1915–2005, Cornell University Library, Division of Rare and Manuscript Collections.
26 George Abraham, *The Belles of Shangri-La* (Springfield, MA: Vantage Press, 2000).
27 NACP, RG 112, Entry 302, Box 218, Interviews with Officers Visiting SGO Installations, 1943–1945, "Report of the Medical Department Activities in Liberia," by John B. West, major, MC, post malariologist, Roberts Field, April 14, 1944, 6.
28 Abraham, 3.

29 NACP, RG 112, 390, 18, 9, Box 105, Unit History, 25th Station Hospital, Liberia, by John E. Weigel, major, MC, commanding, 1943–1944.
30 "A 'Benevolent Neutral' Drawn Closer by President's Visit," *Christian Science Monitor*, Jan. 29, 1943, 11.
31 Abraham; Russell diary.
32 Abraham, 65.
33 William F. Diller, "Notes on Filariasis in Liberia," *The Journal of Parasitology* 33, no. 4 (1947), 363. At the time of publication, William F. Diller was faculty of the University of Pennsylvania, but wrote it when he was a major in the AUS Sanitary Corps attached to the 14th Malaria Survey Unit.
34 Diller, 363.
35 NACP, RG 112, Box 1153, Reports of malaria control, Liberia, October 5, 1942, 11, and July 1, 1943, 19, 23.
36 NACP, RG 112, Box 1153, "Essential Technical Medical Data Report for the Month of November, 1944, USAFIME," File 319.1, Sheet No. 5 of 9, labeled "Secret."
37 Abraham, 85. For classified records on "tolerated villages" for prostitution, see NACP, RG 112, Box 1144, Report on Malaria Rates, Liberia, August 1945.
38 Abraham, 90.
39 Ibid., 127–39.
40 Diller, 364.
41 Charles M. Wiltse, "Army Medical Service in Africa and the Middle East," *The United States Army in World War II: The Technical Services, The Medical Department, Medical Service in the Mediterranean and Minor Theaters* (Washington, DC: Office of the Chief of Military History, Department of the Army, 1965), 89.
42 Phone interview with Joyce Sherman (Joyce Abramson married Herb Sherman after the war, taking his name), November 2013.
43 Wiltse, 89.
44 Joyce Sherman phone interview.
45 For reports and records of malaria control efforts, including descriptions of conditions in West Africa, see NACP, RG 112, 390, Box 1152, and RG 112, Entry 31 (MTO), Box 14.

Chapter 24: Whoville

1 NACP, RG 112, Box 1160, Report enclosed in memo from Colonel Percy J. Carroll to General Douglas MacArthur, date unreadable.
2 Ibid. location, memo initialed "PJC" (Colonel Percy J. Carroll) to General Douglas MacArthur, July 8, 1943.
3 Ibid.
4 For records containing copies of these posters, booklets, and bumper stickers, see NACP, RG 112, Box 1142.
5 "Medicine: Screen Salesman," *Time*, July 14, 1943.
6 Marianne Fedunkiw, "Malaria Films: Motion Pictures as a Public Health Tool," *American Journal of Public Health*, Vol. 93, no. 7 (2003), 1047, http://www.ncbi.nlm.nih.gov/pmc/articles/PMC1447902.

7 NACP, RG 112, Box 1160, memo to officers, USASOS, all grades, Re: Training in Anti-Malaria Measures, by command of Major General Marshall.
8 NACP, RG 112, Box 1169, page from "Malaria Control," pamphlet, June 1943.
9 Ibid. location, folder 11, reports of the 6th Malaria Survey Unit.

Chapter 25: The Breakthrough

1 Atkinson, *Day of Battle*, 214.
2 Ibid., 235.
3 Snowden, 188.
4 Clark Papers, Series V, National Research Council Antimalarial Drug Program, Malaria Report No. 72, Jan. 20, 1944, Table IV, Comparative Antimalarial Activity of a Series of Agents Based on the oral dosage required to produce a given therapeutic effect in trophozoite-induced *P. vivax* infections.
5 NACP, RG 112, Box 1180, "confidential" memorandum to the secretary of war from Roger G. Prentiss Jr., lieutenant colonel, Medical Corps, director Research and Development Division, and Larry B. McAfee, brigadier general, Medical Department, acting Surgeon General.
6 NACP, RG 112, "Secret" agreement signed by seventeen medical officers, Ledo, Assam, April 29, 1943; for discussion of intentionally exposing troops in New Guinea, see NACP, RG 112, Box 1192, "Draft Testing of Anti-Malarial Drug, March 28, 1943," and "Protocol for Test of Sulfamerazine as a Causal Prophylactic Against Malaria," Feb. 22, 1943; RG 112, Box 1190, "Protocol for Test of Sulfamerazine Prophylactic," headquarters, Panama Canal Department, Office of the Surgeon, from Wesley C. Cox, colonel, acting surgeon, April 2, 1943.
7 NACP, RG 112, Box 1158, report attached to letter to Colonel R. P. Williams, MC, surgeon, APO 879, from O. R. McCoy, major, Medical Corps, Tropical Disease Section (August 12, 1943).
8 BMR, 130–40.
9 NACP, RG 112, Box 1158, classified letter to Colonel R. P. Williams, theater surgeon, APO 879, from Paul Russell (August 16, 1943).
10 NACP, RG 112, Box 1189, letter to Colonel George W. Rice, APO 500, from Kirk T. Mosley, major, Office of the Surgeon General, USA, September 1943, with attached worksheet from Heidelberger: "Simple procedure for concentration of malaria parasites for shipment."
11 Ibid.
12 Clark Papers, Series V, National Research Council, Antimalarial Drug Program, Malaria Report No. 72, Jan. 20, 1944. Comments upon Preliminary Evaluations of New Antimalarial Drugs, W. Mansfield Clark.
13 NACP, RG 52, 49B, Box 3, Minutes of the Subcommittee on the Coordination of Malaria Studies, September 3, 1943, Bulletin of Malaria Research, 116–23.
14 BMR, 134; NACP, RG 112, Box 1177, Memo to the Surgeon General, U.S. Army, Washington, from C. C. Hillman, brigadier general, U.S. Army, commanding (July 8, 1945).
15 NACP, RG 227, General Records, malaria 43–44, Box 53, Malaria Report #254, stamped "Secret," 24.

16 Coates, Vol. VI, 394.
17 BMR, 116–23.
18 NACP, RG 492, Box 2523, Memo from Brigadier General F. A. Blesse, surgeon, HG, North African Theater of Operations, Office of the Surgeon, APO 534, to Surgeon General, U.S. Army, Preventive Medicine Division, Washington, DC, Subject: Sontoquine—A New Antimalarial Drug, July 8, 1943.
19 NACP, RG 492, Box 2544, Memo to the surgeon, NATOUSA, subject: Typhus Fever and Malaria Survey of Tunisia, from Theodore E. Woodward, captain, Medical Corps, pursuant to orders dated May 13, 1943.
20 Schneider report.
21 BMR, 127.
22 NACP, RG 227, Box 53, Report on Malaria, China bark, Sinine, extract.
23 NACP, RG 112, Box 1144, Science Service news release, Dec. 31, 1945.
24 BMR, 138.
25 NACP, RG 227, Box 53, Memo to Vannevar Bush from A. N. Richards, "Outline of Malaria Research Program," Feb 10, 1944.
26 Robert G. Coatney, "Pitfalls in a discovery: the chronicle of chloroquine," *American Journal of Tropical Medicine and Hygiene* 12 (March 1963), 125.
27 Ibid.
28 BMR, 155.
29 Ibid., 165.

Chapter 26: Stateville

1 BMR, 163.
2 RG 112, Box 1178, Memo to: The Commanding General, Sixth Service Command; Subject: "Improvement of anti-malarial treatment," by Francis R. Dieuaide, lieutenant colonel, Medical Corps (November 18, 1943).
3 BMR, 164 and 187.
4 NACP, RG 227, Box 21, Carden file, letter to Clifton T. Perkins from George Carden, Dec. 6, 1943.
5 Ibid. location, report by W. B. Castle, special supervisor to the Committee on Medical Research, "Fatigue and Medical Nutrition: Survey of Contracts Chiefly in Clinical Investigations."
6 Bureau of Labor Statistics, CPI Inflation Calculator, 1944 in 2013 dollars.
7 RG 227, Box 20, letter from Cowles Andrus, MD, assistant to the chairman, to Allan Butler, Jan. 21, 1944.
8 University of Chicago Library, Special Collections Research Center: Leon O. Jacobson Papers, Box 32, File 11.
9 "Manteno Madness," *Time* 34, issue 17, Oct. 23, 1939.
10 For a full list of all projects and their OEMcmr numbers, plus funding amounts, see NACP, RG 227, Box 21, memorandum to division chiefs, Committee on Medical Research, from Chester S. Keefer, medical administrator, CMR, August 24, 1945.
11 "New Attack of Illness of Manteno Hospital Believed Food Poison," *The Freeport Journal-Standard*, Jan 4, 1940.

12 B. K. Richardson, *A History of the Illinois Department of Public Health, 1927–1962* (State of Illinois, April 1963), 34.
13 *Time* (Oct. 23, 1939).
14 "The Manteno State Hospital Tragedy in Illinois," *Social Service Review,* Vol. 13, no. 4 (Dec. 1939): 694–95.
15 Richardson, 35.
16 NACP, RG 227, Box 10, Alving file, letter from E. Cowles Andrus, assistant to the chair of the Committee on Medical Research, to Fisher, April 29, 1944.
17 NACP, RG 227, Box 21, Board of Coordination of Malaria Studies, Malaria Report #724, "Chemotherapy of the Human Malarias: Final Report," by James A. Shannon, David P. Earle, OSRD Contract OEMcmr-112, 10. Shannon's COs came from CPS Unit 115. Box 21 also has a full list of names of COs used in malaria experiments.
18 NACP, RG 227, Box 21, letter from Alf Alving to Dr. Chester S. Keefer, CMR, OSRD, Jan 28, 1946. See letter, "None of the conscientious objectors assigned to this project . . . have served as experiment subjects. None are, therefore, entitled to Certificates of Merit. All of the men on our project have worked as nurses' aides or technicians. Very truly yours, Alf S. Alving, MD, Assoc Prof Medicine," Jan. 28, 1946.
19 Interviews with Carl Alving, 2013.
20 NACP, RG 227, Box 10, letter to Robert Loeb, National Research Council, from Alf S. Alving, responsible investigator, OSRD Project Symbol No. M-4060, June 3, 1944.
21 George Wright, "Tighten Rules at Stateville; Favoritism Goes," *Chicago Daily Tribune*, March 7, 1937, 12; Jacobs, 25.
22 Hope Rajala, "Joliet and Stateville: The Twin Prisons," *Quarterly Publication,* Will Country Historical Society (Summer 2000).
23 Ibid; and "Prisoner Power of Loeb Told: Fight to Death with Razor in Bath Described," *Chicago Daily Tribune*, Jan. 30, 1936, A1.
24 "Kills Loeb; Prison Scandal," *Chicago Daily Tribune*, Jan 29, 1936, A1.
25 Ibid.; "Prisoner Power of Loeb Told: Fight to Death with Razor in Bath Described," *Chicago Daily Tribune* (Jan. 30, 1936); and "Jurors Acquit Razor Slayer of Young Loeb," *Chicago Daily Tribune*, June 5, 1936, A1.
26 "Convict Kills Loeb, Franks Boy Slayer: Chicago Murderer Serving Life for 'Perfect Crime' Slashed with Razor in Brawl," *New York Times,* Jan. 29, 1936, A1.
27 James B. Jacobs, *Stateville: The Penitentiary in Mass Society* (Chicago: The University of Chicago Press, 1977), 29.
28 *Chicago Daily Tribune* (March 7, 1937).
29 Ibid.
30 "Illinois Prison Employees Put on Merit Basis," *Chicago Daily Tribune*, Sept. 22, 1937, 3.
31 "Warden Perfects Plan to Guard Roads in Prison Breaks," *Chicago Daily Tribune*, Jan. 4, 1938.
32 "Tighten Rules at Stateville; Favoritism Goes," *Chicago Daily Tribune* (March 7, 1937). Additional sources: "Death at Joliet Heightens Unrest: Precautions Inten-

sified as the Second Prisoner Shot in Mutiny Succumbs Chicago," *New York Times*, Mar. 17, 1931, 27; and "Bare Foiled Plot for an Outbreak at New Prison: National Guard Ready to Curb Convicts," *Chicago Daily Tribune*, March 17, 1931, 9.

33 Nathan E. Leopold, *Life Plus 99 Years* (Garden City, New York: Doubleday & Company, 1958), 305.

34 BMR, Malaria Report #255, Oct. 13, 1944, "Drugs Under Clinical Trial in Chicago," Alf S. Alving, University of Chicago: Work done under contract No. OEMcmr-450.

35 Interview with Carl Alving.

36 Ibid.

37 "Malaria Hero at U.C. Seized—as a Robber," *Chicago Daily Tribune*, Feb. 7, 1962, 8.

38 "4 Young Killers Sentenced; 3 Are Doomed to Chair: Policeman's Slayer Gets Life Prison Term, *Chicago Daily Tribune*, Jan. 21, 1940, 12.

39 "Ex-Convict Gets Life in Filling Station Killing," *Chicago Daily Tribune*, Jan. 8, 1941, 7.

40 "Two Gun Killer is Sentenced to 99 Years in Cell: Called a 'Cold Blooded Bandit,'" *Chicago Daily Tribune*, Dec. 13, 1933, 2.

41 "Notorious Chicago Badmen Get Time Cut for Guinea Pig Roles," *Chicago Daily Tribune*, Jan. 8, 1949, 6.

42 NACP, RG 112, Box 1179, draft, Alving, A., et al, "The Clinical Trial of Eighteen Analogues of Pamaquine (Plasmochin) in Vivax Malaria (Chesson Strain)," eventually printed in the *Journal of Clinical Investigations*, Vol. XXVII, May 1948, no. 3, 34–35.

43 Leopold, 320.

44 NACP, RG 227, Box 10, Letter from Alf Alving to A. N. Richards, July 17, 1945.

45 George Wright, "Bells Regulate Life in Prison at Stateville," *Chicago Daily Tribune*, Nov. 25, 1945.

Chapter 27: Taking Command

1 "Nazi Plague Plan Fails: Italian Front," *Daily Mail*, Paris Edition, Sept. 27, 1944 (copy of clip found in NACP, RG 112, Box 1149).

2 Russell diary.

3 Ibid.

4 Snowden, 192.

5 Ibid., 194.

6 Coates, Vol. VI, 285.

7 Russell diary.

8 Russell letter, courtesy of Ted Russell.

9 Russell diary.

10 Darwin H. Stapleton, "Internationalism and Nationalism: The Rockefeller Foundation, public health and Malaria in Italy, 1923–1951," *Horizontes, Braganca Paulista*, Vol. 22, no. 2, Jul/Dec, 2004, 223.

11 Russell diary.
12 Snowden, 196.
13 Stapleton, 223; and Giancarlo Majori, "Short History of Malaria and Its Eradication in Italy," *Mediterranean Journal of Hematology & Infectious Diseases*, Vol. 4, no. 1, 2012; and Snowden.
14 Stapleton, 223.
15 NACP, RG112, Box 1157, letter to Major O. R. McCoy, Malaria Section, Surgeon General's Office, Washington, from Major Maurice Seltzer, base malariologist, June 22, 1944.
16 LOC, Crocker.
17 NACP, RG 112, Box 1157, letter to Major Maurice Seltzer, MC, Office of the Surgeon General, APO 689, July 6, 144, from O. R. McCoy, Major, MC, director Tropical Disease Division, explaining why New Guinea malaria rates were so low.
18 Ibid.
19 NACP, RG 112, Box 1156, Secret Document 1889, March 9, 1943.
20 NACP, RG 112, Box 1156, extract, headquarters Base Section 3, Office of the Surgeon, Subject: Malaria Control Report for 1943, Jan 10, 1944.
21 BMR, p. 942—lichen planus cases from New Guinea, the Solomons, and Assam-Burma.

Chapter 28: Klamath Falls

1 Anthony W. Sweeney, "The Malaria Frontline: Pioneering Malaria Research by the Australian Army in World War II," *Medical Journal of Australia,* Vol. 166 (March 19, 1997), 316–319.
2 Phone interview with Richard Coggeshall, Friday, Nov. 11, 2009.
3 Richard P. Matthews, "Taking Care of Their Own: The Marine Barracks at Klamath Falls, Oregon, 1944–1946," *Oregon Historical Quarterly* 93 (1992), 323, 359.
4 Robert H. Myers, "Debunking the Filariasis Myth," *Marine Corps Gazette*, Nov. 1944, 28, 11.
5 *Building the Navy's Bases in World War II, Vol. 1*, Bureau of Yards and Docks, Navy Department, GPO, 1947, 369–70.
6 Ibid.
7 Matthews, 347.
8 L. T. Coggeshall, "Malaria and Filariasis in the Returning Servicemen, the Ninth Charles Franking Craig Lecture," delivered in St. Louis, Mo., Nov. 13–16, 1944, printed in the *American Journal of Tropical Medicine* Vol. 25 (Jan. 1, 1945), 177–84.
9 Richard Coggeshall interview.
10 Matthew, 357.
11 Myers.
12 Dorothy L. Ebbert, "Klamath Commandos," *Journal of the Shaw Historical Library*, Vol. 7 (1993), 1.
13 Ibid.

14 Matthews, 360.
15 Ibid.
16 Ibid., 359.
17 Ibid.
18 BMR, 815.
19 Coggeshall (Jan. 1, 1945).
20 Ibid., 183.
21 "Mumu & Virility," *Time*, Vol. 46, Issue 27 (Dec. 31, 1945), 73.
22 Coggeshall (Jan. 1, 1945), 179.
23 BMR, 633.
24 NACP, RG 227, Box 102, memorandum to Captain Robert A. Lavendar, adviser on patent matters, from Lieutenant L. D. Diddle, Jan. 30, 1945.
25 Ibid.
26 NACP, RG 227, Box 21, Carden file, V. Bush to A. N. Richards, March 19, 1945.
27 Clark Papers, APS History 47–54, 54.
28 NACP, RG 227, Box 21, memorandum to A. N. Richards from George Carden, June 23, 1945, Subject: Patents, reporting on meeting with Carleton Palmer and George Merck, and morning conference with V. Bush (received a cool reception from Mr. Lescholier of Winthrop).
29 Ibid.
30 Slater, 146.
31 Mary Ellen Codon-Rall, "The Army's war against malaria: Collaboration in drug research during World War II," *Armed Forces and Society*, Vol. 21, no. 1 (Fall 1994), 129. Codon-Rall wrote that SN-7618 came out on top because it cured *falciparum* in one or two days, eliminated an attack of *vivax* in one or two days, was well tolerated at twice the therapeutic dose, could be taken once a week, didn't stain the skin, and caused no gastrointestinal problems. But "scientists could make similar claims for SN-6911 and SN-8137 (oxychloroquine)," and the navy liked SN-6911 better and "adopted it for routine use." The army adopted SN-7618.
32 Clark Papers, Series V, National Research Council, Antimalarial Drug Program, special report prepared by George Carden, no date.
33 Leopold, 315.
34 Ibid.
35 Ibid., 314.
36 Ibid., 316.
37 Ibid., 319.
38 Ibid., 321.
39 BMR, 1359.
40 Ibid., 760 and 966.
41 Ibid., 536.
42 Ibid.
43 Ibid., 769.
44 Ibid., 538.
45 Ibid., 529.

46 Ibid., 947.
47 Ibid.
48 Ibid., 968.
49 Ibid., 1132.

Chapter 29: Victory of Sorts

1 Unless otherwise noted, the one source on this chapter is a 343-page transcript digitized for open use by the Harvard Law School. See *Nuremberg Trials Project: A Digital Document Collection*, "Report of the Atrocities Committed at the Dachau Concentration Camp Volume II," a non-trial evidence files document from which a trial was derived, May 1945.
2 Bill Collins in 2005 had a photo of Claus Schilling on his wall, along with other malaria experts, including Coatney, Shannon, Laveran, Ross, R. B. Watson, Grassi, Louis Williams, among others. Bill, in an interview in 2005, said he thought Schilling probably didn't deserve to die for his malaria work.

Chapter 30: A Miracle of Sorts

1 Diary of Dr. R. B. Watson (Watson diary), Aug. 5–18, 1945, Rockefeller Foundation, Collection 248: Tennessee, Record Group 2, 1945, Series 248, Box 300, Folder 2038; NACP, RG 112, Box 1178, "Report to the Committee on Medical Research, Office of Scientific Research and Development and the Board for the Coordination of Malaria Studies on Current Status of a Field Study with SN 7618 in Peru," by George Carden and Robert Briggs Watson, August 20, 1945 (Watson and Carden report).
2 Watson diary.
3 Letter to Eugene Bishop, TVA, from A. J. Warren, July 13, 1945, Collection: Rockefeller Foundation, Collection, Health Commission, Malaria Drug Study, Peru, Record Group 1.1, Series 700, Box 9, Folder 58.
4 Letter from Peterson to Shannon, July 6, 1945, Rockefeller Foundation, Collection: Health Commission, Malaria Drug Study, Peru, Record Group 1.1, Series 700, Box 9, Folder 58.
5 Watson diary.
6 Ibid.
7 Rockefeller Foundation, Peterson's report—also in Peterson's report: Dr. Carden said other drugs were worthy of testing but that the project was winding down and he was trying to unload the position of the secretary "so that he will not have to preside over its demise."
8 Peterson's report, "Preliminary Report of the Study of SN-7618," same location, 1.
9 Watson diary.
10 Letter from Shannon to Andrew Warren, Rockefeller Foundation, Aug. 5, 1945, Collection RF Health Commission, Malaria Drug Study, Peru, Record Group 1.1, series 700, Box 9, folder 58. This includes discussions of people dying, but the records give no indication that SN-7618 was involved.
11 Watson diary.

12 Ibid.

13 Report by Watson and Carden.

14 "Current Status of a Field Study with SN-7618 in Peru," report by Dr. Watson, received Sept 25, 1945, Rockefeller Foundation, Aug. 5, 1945, Collection RF Health Commission Malaria Drug Study, Peru, Record Group 1.1, series 700, Box 9, folder 58.

15 For records on how the project was closed out, including details on pounds of excess drugs and where they went (mostly to the NIH), see NACP, RG 112, Box 1143; for records on additional tests in military personnel using SN-7618, details regarding SN-7618 versus SN-6911, and continued testing of the 8-aminoquinolines on black soldiers, including a large study of 750 men at Fort Knox, see RG 112, Box 1178.

16 NACP, RG 112, Box 1179, draft A. Alving, et al, "The Clinical Trial of Eighteen analogues of Pamaquine (Plasmochin) in Vivax Malaria (Chesson Strain)," reprinted from the *Journal of Clinical Investigation*, Vol. XXVII, no. 3, 34–35, May 1948, received for publication Feb. 14, 1947.

17 Ibid., 41

18 Ibid., 41–42.

19 Ibid., 41.

20 A. Alving, et al, "Pentaquine (SN-13,726): A Therapeutic Agent Effective in Reducing the Relapse Rate in Vivax Malaria," draft, from the Malaria Research Unit, Department of Medicine, University of Chicago.

21 Interview, Carol Govan.

22 NACP, RG 112, Box 1182, Ernest Yount and L. T. Coggeshall, draft report: "Duration of Immunity in Humans with Malaria Following Eradication of Infection." This box also includes letters pertaining to the draft, dated November 1947.

23 Jon M. Harkness, "Nuremberg and the Use of Wartime Experiments on U.S. Prisoners: The Green Committee," *Journal of the American Medical Association*, Vol. 276, no. 20 (Nov. 27, 1996), 1672–75.

24 Transcripts of Nuremberg testimony relating to Stateville accessed on nine microfiches at the Holocaust Museum Research Collections, Washington, DC. See finding aid: *The Nuremberg Medical Trial 1946/47: Transcripts, Material of the Prosecution and Defense, Related Documents, Guide to Microfiche Edition*, 517, Stateville (Illinois).

25 William Laurence, "New Drugs to Combat Malaria Are Tested in Prisons for Army," *New York Times*, March 5, 1945.

26 Ibid.

27 Ibid., 1674.

28 Ibid.

29 Ibid., 1675.

30 Ibid.

31 "Dr. Ernst Beutler: 1928–2008: Researcher Pioneered Treatments for Blood Diseases," *Chicago Tribune,* Oct. 10, 2008.

32 R. E. Howes and J. K. Baird, et al, "G6PD Deficiency: Global Distribution, Genetic Variants and Primaquine Therapy," *Advances in Parasitology,* Vol. 81 (2013), 133–201, 136.

33 Cassandra Willyard, "Blues Clues," *Nature Medicine*, Vol. 13, No. 11 (Nov. 2007), 1272–73.
34 Marci Winn, "Front Views & Profiles," *Chicago Daily Tribune*, July 26, 1948, 2.
35 Northwestern University, Elmer Gertz Special Collection (Elmer Gertz Collection), Box 30, file 30, 2, to Illinois State Board of Parole from Alf S. Alving, June 25, 1945.
36 Leopold, 328 and 332.
37 Ibid., 330.
38 Letter from J. B. Rice, MD, vice president of Winthrop, director of medical research, to Governor William Stratton, April 15, 1957, Elmer Gertz Collection, Box 5, folder 5, 2.
39 John Bartlow Martin, statement to the parole board, Feb. 5, 1958. Northwestern University, Gertz Special Collection, Series No. 85, Box 33, file 33, 13.
40 Letter from Carl Sandburg to Leopold, Elmer Gertz Collection, Box 31, folder 31, 56.
41 Hannon, 101.
42 Interview with Geoffrey Jeffery in Atlanta, July 10, 2007.
43 Joel T. Howell, "Lowell T. Coggeshall and American Medical Education: 1901–1987." *Academic Medicine*, Vol. 67 (1992), 711–18, 713.
44 Ibid.
45 Ibid., 715.

Epilogue

1 Socrates Litsios, *The Tomorrow of Malaria* (Wellington, New Zealand: Pacific Press, 1997), 77–79.
2 Ibid., 83. Also see *The Making of a Tropical Disease* by Randall Packard.
3 "Defeating the Curse: How science is tackling malaria," BBC (Sept. 2005), http://www.bbc.co.uk/sn/tvradio/programmes/horizon/malaria_prog_summary.shtml
4 Ibid., 86.
5 Lynn W. Kitchen, David W. Vaughn, et al, "Role of U.S. Military Research Programs in the Development of U.S. Food and Drug Administration-Approved Antimalarial Drugs," *Clinical Infectious Diseases*, Vol. 43, no. 1 (July 1, 2006), 67–71.
6 BMR, 1424.
7 *Encyclopedia of Chemical Technology*, ed. Raymond E. Kirk, Vol. 8 (New York: The Interscience Encyclopedia, Inc., 1952), 674.
8 BMR, 1458.
9 Kitchen, 70.
10 Department of Defense Appropriations for Fiscal Year 2013, U.S. Senate Committee on Appropriations, Wednesday, June 6, 2012 (Washington, DC: Government Printing Office), 103–5.
11 Coatney, "Reminiscences," 8.
12 Kitchen, 70.
13 Anthony W. Sweeney, "The Malaria Frontline: Pioneering malaria research by

the Australian Army in World War II," *Medical Journal of Australia,* Vol. 166, no. 17 (March 1997), 319.

14 Kitchen, 71.

15 S. Pou, R. Winter, et al, "Sontochin as a Guide to the Development of Drugs against Chloroquine-Resistant Malaria," *Anitmicrobial Agents and Chemotherapy*, Vol. 56, no. 7 (April 16, 2012), 3475.

16 N. Dugdale, "Treatment of Malaria," *British Medical Journal*, Vol. 2, no. 4845 (Nov. 14, 1953), 1102.

INDEX